TECHNOLOGY
AND THE
ENVIRONMENT

JEFFREY K. STINE AND
WILLIAM MCGUCKEN

SERIES EDITORS

MIXING
THE
WATERS

JEFFREY K. STINE

MIXING THE WATERS

ENVIRONMENT, POLITICS, AND THE BUILDING OF THE TENNESSEE-TOMBIGBEE WATERWAY

THE UNIVERSITY OF AKRON PRESS
AKRON, OHIO

Copyright © 1993
The University of Akron
Akron, OH 44305-1703
All rights reserved.

Manufactured in the United States of America
First Edition 1993
99 98 97 96 95 94 93   5 4 3 2 1

LIBRARY OF CONGRESS CATALOGING-IN-PUBLICATION DATA

Stine, Jeffrey K., 1953–
  Mixing the waters : environment, politics, and the building of the Tennessee-Tombigbee Waterway / Jeffrey K. Stine.—1st ed.
    p.   cm. — (The University of Akron Press series on technology and the environment)
  Includes bibliographical references and index.
  ISBN 0-9622628-5-4 — ISBN 0-9622628-6-2 (pbk.)
  1. Tennessee-Tombigbee Waterway (Ala. and Miss.)—History.
2. Water resources development—Government policy—United States—History—20th century. 3. United States. Army. Corps of Engineers—History. 4. Environmentalism—United States—History—20th century.   I. Title. II Series.
TC625.T43S75   1993
333.91′5′09730904—dc20   93-23563

The paper used in this publication meets the minimum requirements of American National Standard for Information Services—Permanence of Paper for Printed Library Materials, ANSI Z39.48-1984.

For MCL

Contents

Series Preface ... ix
Acknowledgments ... xi

Introduction ... 1
I. Marshalling Support ... 13
II. Design ... 33
III. Construction ... 65
IV. The Environmental Debate ... 83
V. In Court—Act I ... 109
VI. The Environmentalist-Railroad Coalition ... 130
VII. Economic Accountability ... 149
VIII. Social Justice: "A Cruel Hoax" ... 176
IX. In Court—Act II ... 198
X. The Last Congressional Challenge ... 219
Conclusion ... 246

Notes ... 257
Bibliography ... 309
Index ... 325

Series Preface

This book series springs from public awareness of and concern about the effects of technology on the environment. Its purpose is to publish the most informative and provocative work emerging from research and reflection, work that will place these issues in an historical context, define the current nature of the debates, and anticipate the direction of future arguments about the complex relationships between technology and the environment.

The scope of the series is broad, as befits its subject. No single academic discipline embraces all of the knowledge needed to explore the manifold ways in which technology and the environment work with and against each other. Volumes in the series will examine the subject from multiple perspectives based in the natural sciences, the social sciences, and the humanities.

These studies are meant to stimulate, clarify, and influence the debates taking place in the classroom, on the floors of legislatures, and at international conferences. Addressed not only to scholars and policymakers, but also to a wider audience, the books in this series speak to a public that seeks to understand how its world will be changed, for ill and for good, by the impact of technology on the environment.

Acknowledgments

This book is based on data I originally collected for a study sponsored by the U.S. Army Corps of Engineers Office of History. The interpretations of that data, however, represent analysis undertaken while I have been at the Smithsonian Institution. The conclusions are my own and do not necessarily represent the views of either the Smithsonian or the Corps of Engineers.

I am grateful to the many people who have helped me during the course of this project. In gathering the data, I was assisted by three principal researchers: Phyliss Waldman, Frances B. Marecki, and Timothy R. Mahaney. Frederick A. B. Dalzell, Arnita A. Jones, Jo Anne McCormick, Melissa S. Parker, Margaret Rung, Susan M. Stacy, and Robert A. Zebroski shouldered more targeted research assignments. Emory L. Kemp and Bruce D. Smith served as technical advisors to the Corps study, and I have benefitted from their comments on that work. Christine O'Sullivan transcribed the oral history interview tapes, most of which are cited in this history. I would also like to thank the staff of History Associates Inc., who managed the Corps of Engineers contract and facilitated the archiving of all the supporting documentation compiled for this book, including photocopied manuscripts from nearly seventy archival and records repositories, microfilmed documents from the Tenn-Tom litigation, and a multitude of articles, news stories, and reports. This material is open to the public at the Corps's Office of History archives in Fort Belvoir, Virginia.

Like all historians, I have gained from the knowledge and expertise of librarians. I thank the staff at the following institutions for being especially helpful: Watertown (MA) Free Public Library, Massachusetts Institute of Technology Libraries, U.S. Library of Congress, National Museum of American History Branch Library, U.S. Army Corps of Engineers Headquarters Library, and the Technical Library of the U.S. Army Corps of Engineers, Mobile District.

Daniel Cornford, Eugene S. Ferguson, Elton Glaser, Martin V. Melosi, Daniel Nelson, John Opie, Martin Reuss, Theodore M. Schad, Frank N. Schubert, and Phyliss Waldman read versions of the entire manuscript, and their many valuable suggestions have improved it and enriched my own understanding of the subject. Lonnie Bunch, Pete Daniel, Barton C. Hacker, Robert C. Post, and Carroll Pursell also offered insightful advice on individual chapters or sections of the manuscript. Marybeth Mersky of The University of Akron Press guided the book through all stages of production with great efficiency and good humor.

Colleagues are not our only sources of help. Many gave generously of their time to describe their role in building, defending, or opposing the waterway, and one of the most enjoyable aspects of the research was the opportunity it provided to meet an engaging array of people who live and work in the Tenn-Tom region. The bibliography includes the names of those individuals formally interviewed, but I have also conducted informal and off-the-record discussions with scores of others. I am particularly grateful to Sarah J. (Sissy) Scott of the Corps's Mobile District and to Martin Reuss of the Corps's Office of History for their unselfish assistance in facilitating my access to records and individuals.

I have dedicated this book to my wife, Marcel C. LaFollette, who has brought immeasurable joy to my life, and whose own intellectual accomplishments and integrity have served as constant inspirations for me. More than anyone else, she has shared in the pleasures and burdens of this project, whether that be to clarify my muddled thinking or to laugh with me at the sight of a lone, 14-foot bass boat riding the rising waters of one of the Tenn-Tom's huge, 110-foot by 600-foot locks.

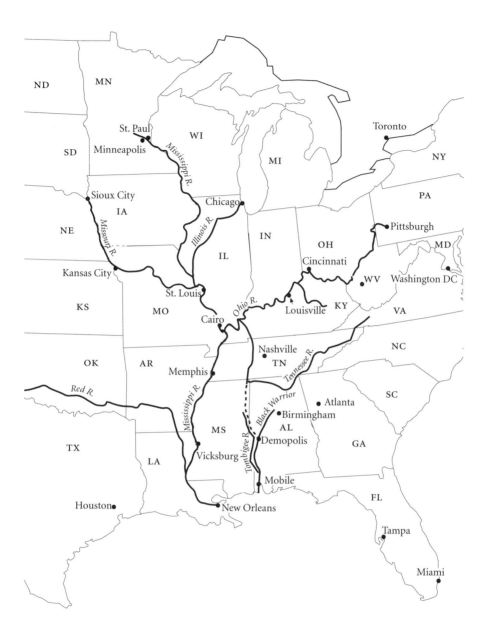

The Tennessee-Tombigbee Waterway, indicated by the dotted line through Mississippi and Alabama, was promoted as a navigational shortcut between the Gulf of Mexico and the Tennessee and Ohio rivers.

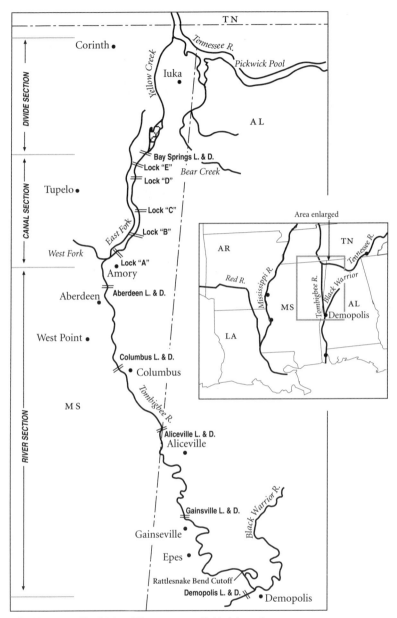

The Tennessee-Tombigbee Waterway was divided into three segments. The *river section* remained largely within the banks of the Tombigee; the *canal section* was an artificial channel paralleling the East Fork of the Tombigbee River; and the *divide section* consisted of the Bay Springs Lock and Dam and a massive, 27-mile-long cut through the hills separating the Tennessee and Tombigbee watersheds.

MIXING THE WATERS

Introduction

In January 1985, the towboat *Eddie Waxler* completed the first commercial passage through America's largest and most controversial inland navigation project, the Tennessee-Tombigbee Waterway—the "Tenn-Tom."[1] The 234-mile-long waterway represented a remarkable engineering achievement: ten locks, five dams, extensive river widening, deepening, and straightening, and a 27-mile-long cut through hilly terrain brought glory to the engineers involved. But of equal importance was the political effort to make that happen.

For over one hundred years, dreamers *and* schemers had proposed ways to connect the Tennessee River to the Tombigbee River, and hence to the Gulf of Mexico. Uncountable political battles had been waged over the most viable of these plans. Ironically, the last political skirmishes occurred just as the politics of public works in the United States was undergoing fundamental change. As a result, the issues of regional rivalry and even of cost diminished in importance: the focal point of opposition to the Tenn-Tom became its environmental impact. Moreover, the size of the project attracted such attention that the debate helped to accelerate a political trend, as environmental concerns further complicated the policy debates over other large-scale public works projects. Somehow, the Tenn-Tom survived the change of climate. The story of how that happened is the story of a symbol—possibly a monument—to the end of one era and the beginning of another.

Although first dreamed of in the eighteenth century, the waterway did not really receive serious attention until after the Civil War.[2] By then, engineering interventions to "improve" the rivers of the United States were gaining increasing political support within the federal government, and the U.S. Army Corps of Engineers emerged as the lead federal agency to carry out this mission, concentrating at first on navigational projects in the Ohio and Mississippi rivers.

To build this particular waterway, however, the Corps would have to embark on canal-building on a monumental scale. Nature brought the Tennessee River temptingly close to the Gulf of Mexico before the river veered northward to join the Ohio, which eventually added its waters to the Mississippi on its run to the Gulf. If engineers could only link the Tennessee with one of the many south-flowing rivers in Mississippi or Alabama, then boats traveling the Tennessee or the Ohio could move directly into the Gulf of Mexico at Mobile, Alabama.

The hilly, southernmost expanse of the Appalachian chain stood solidly in the way of these plans. Any connecting link would have to overcome differences in elevation that ranged from 340 to 400 feet. Engineers would have to dig hundreds of miles of canal and construct a series of complicated navigational locks. And nature is not pushed around easily, at least not on that scale.

By the late nineteenth century, local commercial interests in the Southeast had successfully lobbied for Corps of Engineers studies, but the enormous costs and the dubious economic returns forestalled the national support needed to secure federal involvement. Too many other public works projects (almost all on a smaller scale and promising greater benefits to more people in more influential parts of the country) competed for the limited federal resources. Even in the 1930s, when grand engineering projects generated political support as a part of the New Deal, the Tenn-Tom failed to gain Congressional approval.

The political climate changed after World War II. A host of public works projects—the Tenn-Tom included—won Congressional authorization in legislation intended to rekindle the New Deal spirit. Public works could always pull the nation back from economic depression. Yet, even with this renewed enthusiasm for big engineering schemes, the imbalance between the Tenn-Tom's potential cost and its estimated economic return discouraged the Congress from appropriating funds to start construction. Time after time, for nearly 25 years after receiving authorization, the waterway's supporters would push for approval; time after time, they had insufficient political muscle to win the appropriations battle. Gradually—aided by the seniority system in Congress, which placed Southern Congressional members at key points of power—the situation changed. By 1970, the South had sufficient political clout to get the waterway started, even though the benefit/cost ratio remained questionable.

New social and political forces were gaining strength by that time, however, and these unanticipated opponents used new techniques to fight the project. The post-World War II battles over western water projects had energized the American conservation movement and brought in new activists. By the 1960s the environmental movement had emerged as a force for political reform, and by 1970 it constituted a major threat to the plans of the old-style Southern politicians. The enactment of the National Environmental Policy Act (NEPA) in 1970 reflected the American public's profound reassessment of the relationship between nature and technology, and of the potential damage to intangible assets. The same dam that brought water and new economic life to one region could destroy the natural life in another. Balancing *these* costs and benefits became as important as standard economic formulas in deciding whether projects should be built. For the Tenn-Tom, its values as a public works project were antithetical to the values of the emerging environmental movement. The history of the waterway's design and construction is thus not only a history of what engineers can do, but also of how environmental politics came to influence what they may do. As such, the story has four primary themes: the engineering challenges, the role of regional commercial and political support, the development of grass-roots and national strategies for the environmental movement, and the controversy over waterway economics.

The Object of Debate

In a book that describes court fights, media campaigns, and behind-the-scenes political deals, it can be easy to lose sight of positive aspects of the same project. In this case, the design and construction of the Tenn-Tom represented a significant achievement of civil engineering, unrelated to whether it should or should not have been built on economic or environmental grounds. The twists and turns of the Tennessee River had attracted the attention of many people through the years. From its headwaters in east Tennessee, the river flows southwest through the state, then dips into and crosses the northern section of Alabama on a due-west course. At a point near the common borders of Alabama, Mississippi, and Tennessee, the river angles north, reenters the state of Tennessee, crosses Kentucky, where it feeds the Ohio River near Paducah. From there, the waters flow to merge with the Mississippi River at Cairo, Illinois, then head southward on their journey to the Gulf of Mexico.

It was at the northern turn of the Tennessee that visionaries saw the prospects of cutting an artificial channel southward to join the Tombigbee River, a tantalizing 39 miles away. Because the Tombigbee emptied into the Gulf of Mexico at Mobile, Alabama, the path to an international port would be shortened by several hundreds of miles, depending on a craft's point of departure.

As eventually constructed, the Tennessee-Tombigbee Waterway provided a 234-mile-long navigational channel running through Mississippi and Alabama, following a north-south course nearly parallel to the state border.[3] Its southern terminus, where the *Eddie Waxler* began its inaugural journey, is at Demopolis, Alabama; from that point, the waterway extends upstream on the Tombigbee River, along the East Fork of the Tombigbee, into Mackeys Creek, through a deep cut on the highland divide, into Yellow Creek, and ending at the Pickwick pool on the Tennessee River. The project was divided into three sections: the river section (149 miles, including four locks and dams); the canal section (46 miles, including five locks); and the divide section (39 miles, including Bay Springs Lock and Dam with its 84-foot lift, and the 27-mile-long cut). Appended to the Tennessee River in the north, the Tenn-Tom's southern extreme joined the existing Black Warrior-Tombigbee Waterway, which winded its way 217 miles to the port city of Mobile.

The Tenn-Tom is designed to accommodate two-way navigation by tows with a maximum overall dimension of 105 by 585 feet—or eight 35 by 195-foot barges plus a towboat. Each lock chamber, therefore, measures 110 by 600 feet. The channel depths are 9 feet in the river section and 12 feet in the canal and divide sections, and the bottom widths are 300 feet, except in the actual divide cut, where it measures 280 feet.

The Tenn-Tom provides a shorter, more direct water course between the Gulf of Mexico and the Tennessee, Ohio, and upper Mississippi River valleys. As such, its principal economic justification was the lowering of transportation costs for bulk commodities between the Gulf and mid-America, although benefits were also claimed for local recreation, fish and wildlife enhancement, and area redevelopment. Commercial shippers, depending on whether they pushed with or against the current, could save several days of travel time over the Mississippi River route. The justification was that these savings in time and fuel would be passed on to all American consumers in the form of lower prices.[4] As the Tennessee-Tombigbee Waterway Development Authority stated in 1980 during the heat of controversy over the project: "The Tennessee-Tombigbee Waterway is the single most important project since the St. Lawrence Seaway. It is a strong link in the national transportation system. We must build it for the U.S. to survive the energy crisis."[5]

The Tenn-Tom had powerful competitors in the transportation system, however. Tracing a parallel path only 150 miles to the west was the Mississippi River, easily the nation's most important watercourse. Two of its tributaries, the Tennessee River and the Ohio River, could accommodate tows of up to 15 barges; the Mississippi could handle tows of up to 25 barges below St. Louis, Missouri, and up to 45 barges below Cairo, Illinois.[6] This capacity to accommodate large tows—which on the Mississippi would average 8 to 9 miles per hour, thanks in part to its absence of navigation locks—was a distinct advantage over the Tenn-Tom, which could handle tows no larger than 8 barges, given the limits of its ten locks. The lock system limited Tenn-Tom average speeds to only 4 miles per hour. Moreover, traffic moving south on the waterway terminated into the older Black Warrior-Tombigbee Waterway, which was large enough only to accommodate 6-barge tows. In addition to the existing navigation systems, the Tenn-Tom faced competition from an extensive network of freight railroads; the Louisville and Nashville Railroad, for example, followed a north-south route nearly adjacent to that of the proposed waterway.

Regional Support

The prospect of spending money—especially other peoples' money—is a temptation hard for some people to resist. When seductively packaged in the form of federal public works, a project that might be considered of marginal value may be perceived as highly justifiable. Such was the case with the Tenn-Tom, which found strong support among regional boosters, chambers of commerce, investors, bankers, and commercial entrepreneurs in the Southeast. Local support sprang primarily from the promised economic benefits: the waterway's backers portrayed it as a panacea for regional unemployment and underemployment, a project that would promote job and business opportunities during construction, and permanent industrial development and enhanced commerce thereafter. Once the Tenn-Tom became equated with the hope for prosperity, the politicians who championed construction stood to gain substantial patronage benefits for their actions.[7]

The emphasis on industrial development grew even stronger after World War II, as the South's political and business leaders intensified their efforts to use the federal government as a stimulant for economic expansion. They sought to improve the South's image—to make it appear modern, developed, and attractive—and to lure new investors and industry to the region by creating a welcoming climate of low wages, passive state government, and sound infrastructure. Federal public works projects played a central role in this strategy, and as Southern Congressional members used the seniority system to move up the committee ranks and capture chairmanships, they became increasingly successful in bringing more federal dollars to their home states than those states sent to Washington in taxes. The Army Corps of Engineers, with its large water and flood control projects, became a major conduit for such federal expenditures. Indeed, the Corps had far more projects in the South than in any other part of the country. The Tenn-Tom's promotion was thus an expression of this broader effort to industrialize the region.[8]

A further argument for building the Tenn-Tom and a factor that also later affected the contracting process was the civil rights movement. Equality at the voting booth or lunch counter was only part of the new role for African-Americans in the South; economic opportunity—through job training programs and affirmative action requirements for federal contractors—represented another important aspect of change. As one of the biggest projects to be built in one of the poorest regions of the country, the Tenn-Tom provided a natural target for civil right advocates. Support of the project—or more precisely, lack of opposition—for organizations like the National Association for the Advancement of Colored People and the Congressional Black Caucus was another part of the formula for the proponents.

Environmental Opposition

Had the Tenn-Tom gone forward during the 1930s, 1940s, and 1950s when it was conceived, authorized, and planned, it might well have become one of the Corps's flagship efforts. The primary story to be told might have been the engineering accomplishment alone. But during that 25-year gap between authorization and first appropriation, public attitudes toward many things changed dramatically, including the social benefits and hidden costs of massive engineering works. It was sheer coincidence that the environmental movement first celebrated Earth Day in the same year (1970) that Congress appropriated the initial construction funds for the Tenn-Tom. This coincidence had profound implications for the waterway's public reception because it became one of scores of federal water projects caught up in the political debates between people who viewed engineering projects in terms of economic development and resource exploitation, and people who viewed them as dismissive of aesthetic and ecological concerns.

Although Southerners (especially those in rural areas) tended to hold a close affinity for the land, environmentalism as a political movement had not taken as deep a root in the South as it had elsewhere. Groups such as the Committee for Leaving the Environment of America Natural (CLEAN), the principal regional organization opposing the waterway, relied on contacts with national environmental organizations to make their voices heard in Washington.[9] Environmental politics in the American South assumed increasing importance during the 1970s and 1980s, in part because of the nationwide attention generated by the prolonged battle over the Tennessee-Tombigbee Waterway. The ecological and aesthetic impacts of the Tenn-Tom eclipsed those of any other project at the time, and the waterway quite naturally served as a lightning rod within the growing storm of environmental activism.

The conflict of values was heightened by greater citizen scrutiny of policy making at all levels of government. When the Corps was planning the Tenn-Tom during the late 1950s and 1960s, most of the activity was hidden from public view. That situation changed dramatically with the passage of the National Environmental Policy Act, which gave environmentalists the legal tools for obtaining information. As the burgeoning movement's leaders realized that change would come only after long, hard fights in the courts and legislatures, at both state and federal levels, they sought partners in the effort. By the mid-1970s, many U.S. environmental groups had formed coalitions with various economic stake-holders (such as the railroads that competed for freight business with the barge industry), mission-oriented state and federal agencies, and fiscal conservatives. These coalitions quickly learned how to coordinate their use of the news media, the courts, the agency review process, and other resources for lobbying and public education.

Economic Controversy

The railroads were particularly concerned about the damage to their business posed by a direct, government-subsidized transportation competitor like the Tenn-Tom. Their concern was matched by that of fiscal conservatives in the Congress who already opposed the waterway as an uneconomical waste of public funds, a political boondoggle in the classic sense.

The raw numbers appeared to support such criticisms. During construction, the cost of the project skyrocketed from an initial estimate of

$323 million to a final cost of nearly $2 billion. Because of the powerful combination of economic and environmental opponents to the project, however, Corps officials concealed known cost increases from Congress so that they could first quickly recalculate (and therefore increase) the estimates of benefits. During that process, the Corps enlarged the project beyond what was specifically authorized by Congress (to a width of 300 feet from 170), thereby raising additional questions about the constitutionality of its actions. Moreover, the Corps proposed linking the new 300-foot channel with an existing 200-foot-wide channel, causing some observers to note that even more construction south of the Tenn-Tom would be required later to enable the project to meet its stated goals.

Toward the end of construction, even as work continued at an accelerated pace, social aspects of the waterway came under fire. The Tenn-Tom's construction fell short of bringing the benefits initially promised to local citizens. The Corps thus drew allegations that it, among other things, lied to Congress and the American people, broke its promises to minority groups, exceeded its authority in enlarging the project, failed to abide by environmental legislation, and cooked its benefit/cost calculations.

A Window into Change

How this engineering anachronism came into being during the Age of the Environment is a story of intense political struggle. By the time the *Eddie Waxler* pushed its barges through the first lock, the waterway had survived: two major lawsuits and nine years of litigation; close appropriations votes in Congress which would have stopped the project *after* it was nearly half completed; reevaluation by the Carter Administration as part of a water projects "hit list"; probing assessments by the General Accounting Office, Army Audit Agency, and Congressional Research Service; prolonged scrutiny by the civil rights movement for its failure to provide equitable minority participation; and highly critical coverage by the national news media.

The history of the Tenn-Tom provides a window into the changes occurring in the relationship between environmental organizations and the federal government. It illustrates the maturation of the environmental movement and its growing facility in coalition-building, use of the courts, lobbying, and public education, just as it sheds light on the workings of the Corps of Engineers *and* the politics of the U.S. Congress. It also demonstrates the importance of seemingly intangible societal values on such tangible things as public works projects.

For years now, environmental historians and historians of technology have turned their attention to expanding the borders of U.S. history, attempting to bring to the fore new topics, people, and agents of change, fleshing them out and giving new meaning to the American past. Although naturally intertwined, the history of engineering and the history of the environmental movement too often escaped such sustained analysis. We focused on one or the other, but not on their reciprocal influence. As a result, the histories seemed like advocacy, not analysis.

In this book, I try to regain some of that balance, showing the negative and positive sides of all the various players, for study of the politics and engineering design of the Tennessee-Tombigbee Waterway reveals much about how the interplay between technology and the environment was assessed, misunderstood, and reassessed in the United States during the transitional decade of the 1970s. As increasing numbers of people came to question the benefits and costs of large-scale water resources development projects, those social attitudes became embodied in new sets of environmental laws and regulations. New laws, however, usually have to be implemented by existing bureaucracies. Old-line, mission-oriented agencies like the Army Corps of Engineers often found the philosophy of "environmental regulation" alien to their internal cultures. What ultimately made the difference—pushing even resistant bureaucracies to action—was the sustained involvement of citizens and citizen groups in the processes of federal decision making and program implementation. On the Tenn-Tom, as with hundreds of other federal and state projects across the land, activists literally inserted themselves—not in front of bulldozers, but in the courtrooms of federal judges and in the hearing rooms of Congress—and successfully forced government servants to reflect the changing concerns and desires of society, whether those be considerations of environmental values or of social justice issues.

Engineering projects like the Tenn-Tom demand that society confront the tensions in its sometimes conflicting, and usually evolving, attitudes toward the appropriate connection between technology and the environment. While often hidden and unrecognized, the process of revaluation has become increasingly central to modern society. This book recounts one significant episode along the route to change, a mixture of powerful forces—natural, social, political—whose turbulent story has not yet settled to an end.

CHAPTER I **Marshalling Support**

IN THE EARLY NINETEENTH CENTURY, canal construction spurred a transportation revolution in the United States, helping to solidify and unite the young republic.[1] Prior to this time, the French colonizers of the American Southeast also considered how the region's rivers could be exploited; their most grandiose fantasy was to link the Tennessee and Tombigbee rivers by canal, thus providing a shortcut from mid-America to the Gulf of Mexico. Later, during the early and mid-nineteenth century, residents of Tennessee, Alabama, and Mississippi periodically proposed building a waterway to join the two rivers.[2] Despite these early promotional efforts, the U.S. federal government did not act until 1874, when Congress directed the Secretary of War to survey the route from the Tombigbee River, up Big Bear Creek, and into the Tennessee River. In 1875, the U.S. Army Corps of Engineers completed a feasibility study for a channel 4 feet deep, 40 feet wide at the surface, and 28 feet wide at the bottom, and containing 43 locks. Prohibitive construction costs and limited commercial use (navigation could not be maintained during the wet weather months) led the Corps to recommend against the project.[3]

Congress accepted the Corps's negative recommendation, and proposals to build the project did not resurface at the federal level again until

1913 when the House of Representatives Committee on Rivers and Harbors authorized another feasibility study. This time the Corps evaluated the merits of a canal 6 feet deep with 65 low-lift locks. Again its engineers found the waterway economically untenable, and Congress shelved the proposal.[4] Over two decades later, when the Corps conducted a series of nationwide river system surveys, the agency's reports (known as "308" studies because of the House document that listed the initial survey requests) included an examination of joining the Tombigbee and Tennessee rivers with a 9-foot-deep canal and 20 locks. Even after considering a broadened range of benefits—such as navigation, flood control, and hydroelectric power generation—the Corps judged the project too costly.[5]

The New Deal provided a sympathetic political context for an expanded federal program of public works projects, which were seen as a means of mitigating the economic impact of the Great Depression by providing emergency employment and bolstering the national infrastructure. The House Committee on Rivers and Harbors directed the Corps to reexamine its previous reports on the Tenn-Tom in 1934, and in 1936 the committee authorized a new survey. The Corps completed its assessment in 1938 and, for the first time, found the project's benefits to outweigh—if only slightly—its costs.[6] This positive evaluation came more from a lowering of costs than it did from an increase in benefits. When the newly-created Tennessee Valley Authority (TVA) finished the Pickwick Dam across the Tennessee River in 1938, the resulting lake inadvertently lessened the greatest engineering challenge of building the Tenn-Tom: cutting through the hilly divide separating the Tennessee and Tombigbee rivers. Yellow Creek fed the Tennessee from its headwaters in the Tombigbee divide, and its waters were backed up 11 miles by the new TVA dam. The increased elevation of the Tennessee River and its tributaries would allow the Tenn-Tom's designers to reduce the depth of the divide cut by some 50 feet and would eliminate the need for locks on the Tennessee side of the cut. These engineering changes, combined with a revolution in earth-moving machinery, lowered the projected construction costs and led Corps officials to claim the waterway economically feasible.[7]

Although positive, the Corps's endorsement of Tenn-Tom was lukewarm compared to its reports on other projects and to the water resources development proposals advanced by the TVA and the Bureau of Reclamation. Even if the Tenn-Tom were to prove modestly successful, its enormous cost would mean forgoing many other water projects, and that prospect lacked political appeal. Moreover, federal planners in the 1930s favored multi-purpose river development. The Tenn-Tom was solely a navigational project and provided no flood control or hydroelectric power generation. Nevertheless, the Corps's favorable assessment of the waterway represented a milestone for the project's advocates. In 1939, Southern members of Congress once again brought the Tenn-Tom before the House Committee on Rivers and Harbors for consideration. Although national defense and recreation were added to the benefits package, Congress was still not persuaded and rejected the project.[8]

Federal navigation projects, no matter how economically and technically feasible they may appear to disinterested observers, have natural critics, and the most vocal of these prior to the environmental movement of the late 1960s and 1970s were the railroads. Recognizing the potential economic threat from the expanded waterway and highway development programs of the New Deal, the nation's freight railroads established the Association of American Railroads (AAR) in 1934 to serve as their principal trade association. From its headquarters in Washington, D.C., the AAR monitored federal expenditures on competing modes of transportation and lobbied for the implementation of highway and waterway user fees.[9] When the House Committee on Rivers and Harbors held hearings on the Tenn-Tom in 1939, for example, railroad representatives testified against the project.[10] Again in 1941 and 1943, Tenn-Tom proponents brought the waterway before the committee in unsuccessful attempts to gain project authorization, and again the AAR provided expert witnesses to testify against the project.[11]

External events also intervened. World War II necessitated the curtailment of large-scale domestic public works. Yet long before the conflict ended, concerns about the postwar economy drew the attention of elected officials and federal planners who feared the recurrence of a severe depression, which they believed could only be prevented through substantial government intervention. Federal public works projects were consequently considered important elements in avoiding an economic slump and as necessary measures in rebuilding and expanding the nation's engineering networks. Waterways had the special appeal of having been heavily used during the war, which attested to their contribution to the national defense. Acting within this spirit, the House Committee on Rivers and Harbors requested a reexamination of the Tenn-Tom's benefits and costs in January 1945. The Chief of Engineers submitted a report to Congress in February 1946. With the lock chamber dimensions increased to the Corps's new standard of 110 by 600 feet, the agency judged the project technically feasible and beneficial enough to warrant construction.[12] In the flush of postwar redevelopment, the Tenn-Tom gained Congressional authorization in the Rivers and Harbors Act of 1946—a legislative measure that authorized dozens of new water resources projects.[13] Because authorization did not ensure the appropriations required for construction, however, the Tenn-Tom had passed only the first of the two major Congressional hurdles.

The Role of the Economists

If authorization did not make construction funds immediately available for the Tenn-Tom, it did open the Congressional purse strings for advance planning and design, and the Corps took prompt advantage of the opportunity. Preconstruction planning for the waterway totaled $58,000 in 1946, $600,000 in 1947, $200,000 in 1948, and $120,000 in 1949–1950.[14] As planning progressed during this period, the cost estimate rose dramatically. By 1950, the Corps placed the cost at $169 million, a 45 percent increase over its 1946 projection.[15]

This steep increase prompted a formal investigation of the project by the House Appropriations Committee. Congressional staff members examined Corps files, interviewed potential users of the waterway, and visited the proposed project area. The four-person team of investigators submitted their report in January 1951. Four months later, the Subcommittee on Deficiencies and Army Civil Functions held a special hearing on the Tenn-Tom and found the Corps's claim of cost-savings to the Mississippi barge traffic to be grossly overstated. The special investigatory staff concluded that for every federal dollar spent on the waterway, the nation would reap only 27 cents in benefits. This type of benefit/cost ratio was politically unacceptable and, as a result, Congress revoked the funds available to the project and terminated further planning studies. The Corps thus tabled the Tenn-Tom, reclassifying the waterway's status from "active" to "deferred for restudy"—a bureaucratic category reserved for water projects suspended somewhere between authorization and deauthorization. The waterway remained on the books, but all advance planning and design were suspended and the project became ineligible for Congressionally-appropriated construction funds.[16]

And so, the Tenn-Tom lay dormant for five years, until the Southern coalitions within Congress began to flex their growing political muscle. In 1956, Mississippi Democrat Thomas G. Abernethy led the House Congressional delegations from Mississippi and Alabama in an effort to revive the project. They argued that changed economic conditions in the Southeast warranted a reevaluation of the waterway. Although they met initial resistance from many of their colleagues outside the South, they eventually succeeded in gaining a Congressional sanction for the restudy in the Public Works Appropriations Act, which President Dwight Eisenhower signed in July 1956.[17] The legislation included $160,000 for the Tenn-Tom economic study. The Corps's Mobile District completed the reevaluation in 1960, although it did not release it to the public until the following year. By raising the height of the locks, Corps designers reduced the number of locks from 18 to 10, thus helping improve the project's benefits by decreasing the time needed to navigate the system. The revised benefit/cost ratio of 1.08 to 1 led the agency to recommend a prompt reclassification of the project to "active" status.[18]

Upon the public release of the Corps's economic reanalysis in 1961, the railroads renewed their opposition to the waterway. The Association of American Railroads commissioned the engineering/transportation consulting firm Doswell Gullatt and Associates to evaluate the Tenn-Tom report. Doswell Gullatt himself knew the Corps well, for he had just retired after 27 years in the agency. In his published report, Gullatt severely criticized the Corps's economic evaluation. Echoing the 1951 findings of the House Subcommittee on Appropriations, he accused the army engineers of failing to include "the cost for improvement of the Tombigbee

Waterway below Demopolis," which was necessary to make possible the volume of traffic projected for the Tenn-Tom. This type of "piecemeal construction," he argued, hid many of the actual costs of the project.[19] He called the Corps's claim that 8,366,000 tons of commercial traffic would move through the Tenn-Tom during the first year of operation "wholly contrary to the history of existing inland waterways, in fact it is sufficiently absurd to warrant branding as completely ridiculous."[20] Gullatt found "no economic justification" to support construction of the proposed waterway. According to his calculations, the benefit/cost ratio was 0.24 to 1. "The project should be abandoned without further consideration by the Corps of Engineers and the Congress," he concluded, "and the restudy report . . . should be condemned."[21]

The Corps's Chief of Engineers, however, found the agency's economic report persuasive, and in April 1962 he reclassified the Tenn-Tom from "deferred for restudy" to "active," making it eligible once again for planning and construction appropriations.[22] Rather than allocate funds to resume planning the waterway, Congress called for yet another economic reevaluation. The Corps issued its findings in 1966 as part of the general design memorandum. This document expanded both the economic projections and the physical project itself.[23] The report analyzed two project dimensions, one with a 200-foot-wide channel (the width authorized by Congress in 1946) and another with a 300-foot-wide channel. The Corps assigned a benefit/cost ratio of 1.1 to 1 to the 200-foot project, while the 300-foot project received a ratio of 1.24 to 1.[24] Agency officials concluded that the 300-foot channel was "the most practicable plan of improvement" and that "the project is economically justified when all tangible, intangible, and secondary benefits are taken into consideration."[25]

The 1960 *Restudy* and the 1966 *Reevaluation* included several significant differences. The base year traffic estimate for the 1960 analysis incorporated data gathered in 1958—a year of recession—while the 1966 analysis used data gathered in the prosperous year of 1965, which helped increase traffic use projections. Non-transportation related benefits, such as recreation and area redevelopment, also increased substantially in the 1966 study. The most significant change, however, was the increase in channel width to 300 feet, which allowed for substantially higher traffic projections.

This is not to say that support for the project was unanimous within the Corps. Prior to reaching Congress, the agency's report had to follow the chain of command through the Secretary of the Army's Office of Civil Functions. James J. Tozzi, a senior staff member within the Army Secretariat, evaluated the document in early 1967. After examining construction cost estimates, traffic movements, area redevelopment benefits, and the economic assumptions underlying the Corps's benefit/cost analysis, Tozzi concluded that the Tenn-Tom "would not be justified on a cost basis," that even "a minuscule change in the traffic projections would result in a B/C ratio less than 1," and that some of the construction costs appeared to be "grossly understated." This all added up to one thing, he said: "that there is a reasonable doubt that the Tenn-Tom project is justified under the current evaluation criteria."[26]

The Army Secretariat submitted the Tenn-Tom economic restudy to the Bureau of the Budget for review. Donald G. Waldon—who began his professional career in the early 1960s as a civil engineer with the Corps's Mobile District—headed the Bureau's "Corps Unit" within the Natural Resources Programs Division. Waldon stated that, prior to receipt of the Tenn-Tom report, the Army Secretariat advised the Bureau that it "would be under considerable pressure by Congress" to release the report in time for the hearings scheduled for April 1967. This tight deadline prohibited the Bureau from raising a number of serious questions because the Corps would not have enough time to respond prior to the hearings. Consequently, as Waldon reported, "*it was agreed by all,* including the Deputy Director, to clear the report without raising the questions. Most of the questions were for qualification purposes but some could have had a significant impact on the feasibility of the project." Waldon then listed nineteen questions "for the record." Among them were:

> Is it within the discretion of the Sec. of Army to authorize modification of a project that has increased in cost from $126 million at the time of authorization to the presently estimated $312 million?
> How was the portion of costs allocated to Tenn-Tom for the proposed duplicate locks at the existing Demopolis and Jackson projects arrived at?
> If a portion of the first costs for improvement at Demopolis and Jackson

is assigned to the project, shouldn't some of the O&M [operations and maintenance] costs for these improvements also be allocated to Tenn-Tom? . . .

What is the estimated total costs of the duplicate locks at Demopolis and Jackson? . . .

Does the capacity of the waterway have any limiting effect on the benefits?

Is the waterway capacity based on ideal conditions or those that will actually exist?[27]

Secretary of the Army Stanley R. Resor also expressed reservations when he submitted the Corps's economic reevaluation to the Congress in March 1967. Resor explained that this represented the fourth economic evaluation of the waterway conducted by the Corps since World War II. "In each previous instance," he observed, "the proposed investment was found only marginally justified; and that is also the case with the present study." Although the agency estimated that a waterway with a 300-foot channel would have an overall benefit/cost ratio of 1.24, Resor worried about the meager 1.01 benefit/cost ratio for the navigational features of the project, which constituted the principal purpose of the waterway. Area redevelopment benefits helped raise the benefit/cost ratio to 1.14, but Resor cautioned: "While the inclusion of redevelopment benefits is not inconsistent with sound economic analysis, I have reservations about the wisdom of investing in a project which depends so heavily on counting local wage payments in calculating the benefit-cost ratio." In addition, he warned that "if the costs of the project prove to be even slightly underestimated, or the projected waterway traffic slightly overestimated, the present barely favorable benefit-cost ratio will be lost." Resor summarized his position: "All in all, the conclusion is inescapable that the Tennessee-Tombigbee navigation project continues to lack that margin of economic safety which typically marks federal investments in water resource development." The Bureau of the Budget, he said, shared that assessment.[28]

Although tepid and accompanied by numerous qualifications, Resor's approval of the Tenn-Tom economic study proved sufficient to revive the flagging project. In a move that came back to haunt the Corps in federal court, Resor also sent the Chief of Engineers a memorandum in March approving the expansion of the waterway channel from 170 to 300 feet, a change that most surely required Congressional sanction.[29]

The Role of the Railroads

Resor's actions took place amid the heated objections of the railroads. By the 1960s, this opposition was spearheaded by a special waterway analysis staff within the Association of American Railroad's Economics and Finance Department. The AAR's waterway analysis section coordinated the railroad industry's opposition to new navigational projects by monitoring developments within the Bureau of the Budget and the Congress, by testifying at Congressional hearings, and by organizing its own committees—made up of representatives of interested railroads—to oppose specific projects. The Tenn-Tom merited its own committee, the AAR Tennessee-Tombigbee Waterway Project Committee, whose most immediate task was to demonstrate the project's economic deficiencies. As a principal AAR member, the Louisville and Nashville (L&N) Railroad was particularly concerned about the Tenn-Tom's potential threat, for the proposed waterway would parallel the L&N's main north-south line, which ran from Nashville, Tennessee, through Birmingham, Alabama, and on to the Gulf port of Mobile. The L&N therefore agreed to provide the chairman of the Tennessee-Tombigbee Waterway Project Committee.[30] The twelve-member body consisted of representatives from the L&N Railroad, the Gulf, Mobile and Ohio Railroad Company, the St. Louis-San Francisco Railway Company, the Illinois Central Railroad Company, and the AAR.[31]

The committee released a formal response to the Corps's economic restudy in late 1966. Labeled a "preliminary review," the report charged that: navigation as the Tenn-Tom's principal purpose was not justified; the substantial changes made to the project's design required further Congressional authorization; benefits listed for transportation, recreation, and redevelopment were overstated; and the interest to be paid by the government during construction was grossly understated.[32] James G. Tangerose, who was AAR's Director of Waterway Analysis, drafted the document. He brought to the directorship valuable knowledge of the Corps's planning process; he had served as an economist with the agency's Board of Engineers for Rivers and Harbors between 1947–1951 and 1953–1965 and with the Corps's Facilities Engineer Support Activity from 1965–1966.[33]

As Tangerose and members of the AAR Tennessee-Tombigbee Waterway Project Committee continued to monitor developments associated with the Tenn-Tom, they sharpened their economic criticisms.[34] In addition to testifying at Congressional appropriations hearings, they lobbied Congress and the Bureau of the Budget to discontinue appropriations for preconstruction planning.[35]

When Congress assured the Corps of funds to complete preconstruction planning in 1968, the AAR Tenn-Tom Committee braced itself to oppose the next likely stage: requests for Congressional appropriations to start construction.[36] Because of the controversy surrounding the economic assumptions that underpinned the Corps's benefit/cost ratio, members of the AAR Tenn-Tom Committee turned their attention to the 3.25 percent interest rate used by the agency. By substituting the Water Resources Council's recommended 4.625 percent interest rate, the committee argued, the Tenn-Tom's benefit/cost ratio dropped from 1.24 to 1 to 0.93 to 1.[37] When the Corps announced a revised benefit/cost ratio of 1.6 to 1 in 1968, the AAR Tenn-Tom committee redoubled its efforts to oppose the appropriation of construction funds at the next set of hearings before the House and Senate Public Works Subcommittees.[38]

The Role of Local Promoters

The railroad's aggressive efforts led the waterway's advocates to realize that they needed an independent group of supporters lobbying for the Tenn-Tom. They launched this effort in 1958 with the creation of the Tennessee-Tombigbee Waterway Development Authority (TTWDA), a Congressionally-sanctioned multi-state compact, joined initially by Alabama and Mississippi, and eventually including Tennessee (1959), Kentucky (1962), and Florida (1967). The TTWDA had no purpose other than to promote the waterway. Its board of directors consisted of the governors of the member states plus five appointees from each. In its formation, the TTWDA avoided the burden and uncertainties of private financing by receiving its funding directly from the tax revenues of its member states.[39]

The TTWDA established its headquarters in Columbus, Mississippi, roughly at the midpoint of the proposed waterway. Alabama's Director of Conservation, William H. Drinkard, was appointed administrator, and the manager of Columbus's Chamber of Commerce, Glover Wilkins, filled the assistant administrator post. Wilkins, who was then 46 years old, knew the planned waterway best, having promoted the Tenn-Tom on behalf of the Chamber of Commerce since 1947. He succeeded Drinkard in 1962 and remained at the helm of the TTWDA for nearly 23 years. He made a career out of promoting the waterway, and was extraordinarily good at what he did. A smooth-talking native Mississippian, Wilkins embodied old-fashioned Southern charm, which was enhanced by his large stature and prematurely gray hair. Diplomacy and public relations came naturally to him, and he proved instrumental in organizing and maintaining a broad coalition of supporters. Wilkins thrived on the high-level political hobnobbing demanded by his post and in soothing the egos of elected and appointed officials. The key ingredient in his success, however, was his unfailing and passionate belief in the waterway and what he was convinced it would do to uplift his native South.[40]

The TTWDA was, in fact, typical of promotional organizations that have historically solidified local support behind major Corps of Engineers navigation projects. Throughout the twentieth century, the business and commercial interests of a region to be served by a project have financed these privately organized groups, sometimes with financial assistance from local government agencies. The TTWDA differed from the standard local support group in its state sponsorship, its multiple-state membership, and its Congressional sanction. The advantages of this arrangement extended beyond financial security, for what really counted in a national political fight was the ability to marshal the support of the states' governors, legislatures, U.S. Congressional delegations, and business interests in systematic and interlocking lobbying efforts. No Corps project had ever enjoyed the backing of such a sophisticated, powerful, and well-financed organization, and the TTWDA proved vital to the Tenn-Tom's ultimate completion.[41]

The TTWDA promoted the project in a number of ways. Public relations became a cornerstone of concern, and the TTWDA developed an effective program of publicizing and popularizing the waterway, which included close contact with the media. Its sound financial backing enabled it to support a full-time staff and sponsor their widespread travel, to hire senior technical consultants, and to publish a wide array of promotional literature. By helping stimulate and maintain local enthusiasm for the project, the TTWDA in turn gave the area's Congressional delegations political currency in advancing the waterway. It also played a key organizational role in lobbying Congress and the various presidential administrations, and was instrumental in lobbying the Alabama and Mississippi state legislatures to pass the necessary bills and bonds to meet the nonfederal financial requirements.

In stimulating discussion of the waterway among the local population, the TTWDA helped shape the issues of the debate by presenting the Tenn-Tom as the best means to revitalize the impoverished region. The waterway promised prosperity: during construction, there would be new jobs and increased demand for the region's goods and services; after opening, the waterway would entice industrial development in adjacent counties. The widespread belief that federal expenditures—*any* federal expenditures—bolster a local economy underpinned the Tenn-Tom's popularity, and the TTWDA fueled these expectations.

The TTWDA also conducted special studies, both to strengthen its own lobbying and support-generating activities and to help the Corps of Engineers in its political struggles. Because economic arguments drew much Congressional attention, the TTWDA retained Joseph R. Hartley, professor of transportation at Indiana University, as its economic consultant. Hartley testified and provided unsolicited economic assessments to the Congressional Appropriations committees and to the Bureau of the Budget (BoB; subsequently renamed the Office of Management and Budget). His assessments unabashedly praised the Tenn-Tom, and maintained that the Corps conservatively underestimated project benefits.[42]

In 1969, the TTWDA submitted a reanalysis of the Tenn-Tom benefit/cost ratio to the Public Works Subcommittee of the House Appropriations Committee. In his testimony before the subcommittee in May 1970, Glover Wilkins stressed the waterway's economic benefits. Although Wilkins explained that the principal benefits would result from savings in bulk commodity shipping costs, he argued that the worth of the project also rested upon benefits to be accrued in the areas of recreation, area redevelopment, economic development, land enhancement, flood control, and improvements in water quality and quantity. Some of these benefits, especially those of a regional nature, could not be used in the Corps's official economic justification. And Wilkins's claim for flood control benefits was categorically not part of the Corps's plans for the waterway; if anything, the Tenn-Tom could be expected to exacerbate flooding in several areas. Wilkins did not stop with economic benefits, however. He pleaded that the waterway would also provide a "psychological stimulus" to the people of this depressed area: if people *believed* the Tenn-Tom would bring economic development, then they were more likely to make that prophesy come true.[43]

By the late 1960s, the TTWDA had developed a lobbying routine that mirrored Congress's appropriations cycle. As the Authority's head, Wilkins traveled to Washington about twelve times a year, often accompanied by members of his staff and/or board of directors. Every autumn, he brought his board members—from one state at a time—to Capitol Hill to meet with the Congressional delegations from their states. As he explained: "We tell them what we want, and we urge them to make their feelings known to the Executive Branch."[44] Wilkins and his members met with key staff members of the Office of Management and Budget every September or October to support the Corps's budgetary requests. Then, generally in February, the TTWDA administrator would attend Congressional appropriations hearings on the Corps's civil works program to learn the status of the agency overall and, in particular, the status of the Tenn-Tom. Cherry-blossom time (April or May) brought Wilkins and his members back to Washington to testify at the public hearings and to lobby Congressional members.[45]

In addition to lining up witnesses for the hearings and soliciting written statements to be submitted for the record, TTWDA staff also helped draft testimony and statements for those people wanting such help. In May 1970, for example, the House Public Works Subcommittee held hearings in which the initial construction funds for the Tenn-Tom were debated. Wilkins and his staff drafted a statement for Representative John J.

Duncan. In sending the statement to the Tennessee Congressman, Wilkins said he hoped it met with his approval, but added, "please feel free to change it up in anyway or, if you prefer, we would be more than happy to re-write it entirely with suggestions from you."[46] The Authority continued to offer its assistance in writing "suggested statements" for waterway supporters testifying before Congressional appropriations hearings throughout the decade.[47]

Wilkins, in his quest to cultivate good relations with the Bureau of the Budget, Congress, the Corps, and other government agencies, consistently pushed the limits of appropriate behavior. In a candid 1975 interview, Wilkins described his work as a lobbyist: "We do quite a bit of entertaining for the decision makers," he said of himself and his staff, but quickly added, "I don't think that decisions are made on entertaining and things like that nearly as much as ordinary lay people think that it is." What was important, he said, was having lunch with a Senator or Representative and talking about "nonbusiness" matters, "so that when you do ask him for money it kind of pays off a little bit better." Even better yet, Wilkins confessed, was getting Congressional members and their spouses to attend an Authority-sponsored retreat. "This is one of the reasons that we bring our wives to a lot of our meetings," he explained, "because if we can get a Congressman and his wife there with our people and our wives at some place where everybody can relax, we get a better chance to sell him a bill of goods . . ."[48]

The appearance of conflict of interest was not something of which Wilkins was highly conscious, and the maintenance of "distance" (at least the perception of such in public) was generally done by government officials. For example, Wilkins urged BoB budget examiner Donald G. Waldon to attend the Authority's February 1970 quarterly meeting at a lush resort in Point Clear, Alabama, following the President's recommendation that construction funds be appropriated for Tenn-Tom.[49] Waldon evaluated water projects at the bureau and was the TTWDA's main personal contact there. Like Wilkins, he was a native of Mississippi, and the two men became quite friendly. Nevertheless, Waldon was careful not to accept travel funds or anything else from the Authority. He refused Wilkins's invitation, stating: "As we have discussed on several occasions, the Budget

Bureau, as an institution, is overly sensitive to possible activities that could reflect on the conduct of its staff."[50]

Wilkins's enthusiasm often brought him into conflict with one of the project's key supporters in the House of Representatives, Mississippi's Jamie Whitten. Whitten took pride in his ability to read the political landscape in the House, and he rarely hesitated to share his recommendations for the Tenn-Tom with Wilkins. But the two men did not always agree. For example, Wilkins decided that he should arrange a meeting between the deputy director of the Bureau of the Budget, Caspar Weinberger, and elected officials in favor of the waterway sometime during November 1970. Wilkins believed that previous meetings with BoB officials proved advantageous and would do so again, especially as waterway advocates sought the highest possible appropriation.[51] Whitten, however, thought that "we should leave well enough alone until we got the first shovel in the ground." He felt fortunate that the Bureau had approved the initial $1 million for construction earlier that year, and he believed that undue publicity for the project might "rock the boat" and delay construction. Upon learning of Wilkins's meeting with Weinberger in November and the accompanying press coverage, Whitten told the TTWDA administrator: "As we have said many times, there are many differences of opinion as to exactly what should be done; but I certainly would do nothing else which might give the Bureau of the Budget an excuse to hold up funds and thereby delay a start. These thoughts I have tried to get over to you on several visits but don't seem to have succeeded."[52]

Wilkins, of course, did his best to appease Whitten, who was perhaps the most crucial member of the House in the Tenn-Tom appropriations process. "I have tried as best I know how to comply with your wishes and recommendations," Wilkins said in a long, apologetic letter, adding that he would "try even harder in the future."[53]

Although the Corps, the Congressional supporters, and the TTWDA shared the same ultimate objective, their relationship was not always straightforward. Members of the TTWDA often depended on the Corps for information, and the Authority staff sought to develop and maintain a close working relationship with Corps employees and political supporters on Capitol Hill and elsewhere in Washington. The Corps leadership, however, recognized that political necessity required the agency to remain at arm's length from the Authority. They were two distinct organizations and needed to maintain that appearance, despite their frequent cooperation and interdependence. The Corps continued to work with the TTWDA—in fact, it often relied on the Authority for many things—but made an effort to keep the blatant conflict of interest infractions to a minimum.[54]

The Role of the White House and Capitol Hill

A major breakthrough for Tenn-Tom proponents occurred in December 1967 when Senator John Sparkman of Alabama convinced President Lyndon Johnson to meet with the leading Congressional supporters of the waterway to talk about resuming preconstruction planning, which had been stopped in 1951. In light of his reelection campaign and the Democrats' need to retain support in the South, the President agreed to request preconstruction planning funds for the Tenn-Tom, making it the only new water project to be so funded that year.[55]

It was no coincidence that the Tenn-Tom reemerged at this time. The political power of Mississippi and Alabama's Congressional delegations had grown steadily since World War II, due primarily to the seniority rule in Congress's committee system and the longevity of the two states' elected officials.[56] The most powerful Congressional supporters of the waterway included: Jamie L. Whitten, Democratic Representative from Mississippi, Chairman of the Subcommittee on Agriculture, Environmental and Consumer Protection of the Appropriations Committee and second ranking majority party member of the full committee; Bob Jones, Democratic Representative from Alabama, Chairman of the Subcommittee on Economic Development of the Public Works Committee and second ranking majority party member of the full committee; Senator John C. Stennis, a Mississippi Democrat, Chairman of the Armed Services Committee, and, within the Appropriations Committee, Chairman of the Subcommittee on Public Works and the Atomic Energy Commission; James O. Eastland, also a Democratic Mississippi Senator, Chairman of the Judiciary Committee, and President Pro Tempore of the Senate; and Democratic Senator John J. Sparkman of Alabama who chaired the Committee on Banking, Housing and Urban Development and the Joint Committee on Defense Production, as well as serving as Majority Whip. Jack Edwards, a Republican Representative from Alabama, also provided important Congressional backing for the project. A native of Alabama who had practiced law in Mobile for ten years, Edwards was elected to the House of Representatives from Alabama's First Congressional District in November 1964. Because the First District encompassed the port of Mobile, which stood to benefit more from Tenn-Tom traffic than any other city, Edwards became a staunch supporter of the waterway. His 1968 assignment to the House Appropriations Committee further enhanced his effectiveness.[57]

When Richard M. Nixon assumed the Presidency in 1969, Edwards—one of the few Southern Republicans in Congress at the time—became coordinator for the House of Representatives' effort to obtain the initial construction funds for the waterway, as supporters of the Tenn-Tom effectively linked the waterway to the G.O.P.'s "Southern Strategy." The Republican party recognized its growing appeal to Southern voters during the 1950s and 1960s, as economically conservative Democrats increasingly cast their votes with the Republicans. The Southern Strategy became a key presidential campaign component for the G.O.P. after Barry Goldwater swept the Deep South in 1964. Nixon looked to the South for support in 1968, but Alabama Governor George C. Wallace's independent presidential candidacy limited Nixon's success and very nearly threw the election into the House of Representatives. Determined to capture the forty-five electoral votes won by Wallace in 1968 (Wallace had defeated Nixon in both Alabama and Mississippi[58]), Nixon's political strategists took an early interest in the South as they prepared for the 1972 election. The political popularity of the Tenn-Tom among senior Southern Congressional members thus elevated the waterway's importance in Nixon's quest to boost his support, especially among conservative Democrats.

By backing the Tenn-Tom, Nixon could demonstrate that he was not neglecting the South. Virginia Hooper, a member of the Mississippi Executive Committee of the Republican National Committee, made that clear in January 1970 to Harry Dent, deputy counsel to the President: "I can think of no project which could prove to the South that President Nixon really does 'care', like the inclusion of this very worthy project in his Budget. This would be another item to put the lie on those who say the President is all talk and no action."[59]

Nixon's support of the Tenn-Tom exacted its costs, however, because the waterway, aside from its dubious economic merits, conflicted with the values and objectives of the emerging environmental movement. Always politically astute, Nixon rode the wave of the new environmental enthusiasm by signing the National Environmental Policy Act on 1 January 1970, making it his first official act of the decade. He used the occasion to proclaim his administration's commitment to rectifying environmental damage. Environmentalists were further encouraged to believe the administration sincerely supported their cause when the President stopped the Cross-Florida Barge Canal in early 1971. Ironically, by recommending construction of the controversial Tenn-Tom, Nixon endangered whatever support he had gained from the environmental community with the Florida decision.[60]

Nixon's campaign strategists realized that the political ramifications of the Cross-Florida Barge Canal and the Tenn-Tom differed substantially, largely because Florida did not sit squarely within the Deep South political block. Stopping the Cross-Florida Barge Canal seemed politically safe; Nixon stood to please and hoped to gain the support of the state's growing environmental interests, with minimal danger of alienating Southern political interests. The decision over the Tenn-Tom posed a different set of choices, however. Local and regional environmental interests appeared to lack the strength, influence, and organization in Mississippi and Alabama that they enjoyed in Florida. Opposition to the waterway therefore offered the Nixon Administration little political support from the relatively thin ranks of Southern environmentalists and risked substantial political fallout from the many Southern interests backing the project.[61] The architects of Nixon's Southern Strategy carefully weighed these factors. Well aware of the growing political clout they could bring to bear on behalf of the project, members of the multi-state TTWDA met with Bureau of the Budget Director Robert P. Mayo in October 1969 to request that Nixon's budget for fiscal year 1971 include funds for the waterway's construction.[62]

The Nixon Administration took the risk. It included $1 million for Tenn-Tom construction in the fiscal year 1971 budget, even though not all of the President's senior officials supported that decision. In a letter stamped SENSITIVE, Robert E. Jordan III, Special Assistant to the Secretary of the Army for Civil Functions, wrote to the then Director of the Office of Management and Budget George Shultz that some of the Corps's activities were of questionable value. "I have found in my assessment of the water resources needs of the Nation that major new navigation projects are of low priority," he stated, and then listed the Tennessee-Tombigbee Waterway as one of the most egregious examples within the President's budget. Although the Tenn-Tom would probably push down rail rates, he explained, "no significant economic gain to the Nation as a whole, commensurate with the project costs, appears likely as a result of [its] construction." Jordan recognized, however, that the Tenn-Tom engendered strong political support within the affected region and "that commitments may have been made in the past which will make it difficult to assign [the waterway] a low priority."[63]

Despite the Nixon Administration's desire to reorient national priorities, therefore, the political expediency of supporting the Tenn-Tom was irresistible, and the President's endorsement remained firm. The Congressional Appropriations committees considered the President's budget and left the $1 million request intact. Nixon signed the 1971 public works appropriation bill in October 1970.[64]

The appropriation of construction funds represented a watershed for the Tenn-Tom and its supporters. As Wilkins told the Mississippi State Board of Water Commissioners in December 1970: "The primary interest of the Tennessee-Tombigbee Waterway Development [Authority] in the past has, of necessity, been seeking and obtaining monies from the Federal government for feasibility studies, advanced planning and engineering, and now construction. With the contemplated construction start our attention must now also be turned to development of the planned use of the Waterway and how this planning can be carried out."[65]

Entering a New Phase

For over two hundred years, the Tennessee-Tombigbee Waterway had been an engineering dream and a political quagmire. Aside from the seductive engineering challenges of building such an immense and complicated waterway, the enormous cost that it would entail and the uncertainty of the waterway's use by commercial shippers meant that the route to final completion would never be smooth. Competition from Southern railroads and from barge traffic on the Mississippi River dogged the estimates of benefits. Although the Mississippi was longer, it provided a route unencumbered by locks and wide enough to handle tows of over forty barges (five times the size of tows that would be able to negotiate the Tenn-Tom). Added to the uncertain economic advantages of the project was the fact that it traversed a poor and lightly populated region of the South, a factor that did not help in the high-stakes national competition for federal public works projects.

Clearly, the signing of the 1971 appropriations represented a time of celebration for those who had worked so hard to make the Tenn-Tom a reality. The fruits of decades of engineering and economic studies and of political lobbying seemed at hand. It may have even appeared to proponents that the questioning of the economic justification had been lain to rest. It was a brief moment of calm, however. As President Nixon had already discovered, a storm of protest was brewing, in the form of a new environmental movement in the United States. In the meantime, as the next two chapters describe, the construction of the Tenn-Tom proceeded. The forests were cleared, the channels dug, and the water rerouted to flow to the Gulf.

CHAPTER II **Design**

AS WITH MOST MAJOR PUBLIC WORKS PROJECTS, the planning and designing of the Tennessee-Tombigbee Waterway required the balancing of complex challenges, some technical, some political. Because nearly four decades passed from the time Tenn-Tom was authorized to the time it opened to traffic, its ongoing design clearly illustrates that interplay of technical and political considerations. Although all large engineering projects necessarily take into account their own particular blend of technical, economic, legal, political, and social concerns, the significant shifts in public values represented by the environmental movement that flowered during the middle of the Tenn-Tom's lengthy gestation makes that waterway especially suggestive of the societal impacts on public works. As this chapter will address, those elements of the project designed prior to the late 1960s were markedly different from those design aspects produced after 1970 with regard to their environmental sensitivity.

At its most basic level, the Tenn-Tom's design was primarily influenced by the lay of the land. In conceptualizing the project, Corps planners segmented the waterway into three sections—*river, canal,* and *divide*—each reflecting the engineering approach taken to address the distinct geographical characteristics found along the path of the project.

The *river* section encompassed the southern portion of the waterway, which followed the widest stretches of the Tombigbee River. Here engineers sought to make the Tombigbee navigable by constructing a series of four locks and dams, and by straightening, widening, and deepening the natural river channel. The *canal* section paralleled the upper reaches of the Tombigbee, which Corps designers determined was too narrow and winding to provide an economical navigation channel. Their solution was to build a separate canal (containing five locks) adjacent to the river. The northern, or *divide,* section presented the greatest earthmoving requirement, for here the project crossed the westernmost extremity of the Appalachian mountain chain in the hilly region of northeastern Mississippi, which formed the geographic divide separating the Tennessee and Tombigbee watersheds. Linking these rivers called for the excavation of a 27-mile-long cut through the hills and the construction of a single, high-lift lock (the only high-lift lock, at 84 feet, on the project).[1]

Organization

To understand the engineering history of the Tenn-Tom, it is important to know something about the agency responsible for its development. Established in 1802, the U.S. Army Corps of Engineers has acquired two principal missions, one that supports construction related to the military and the other devoted to civilian water resources development. For its domestic civil works activities, the primary lines of authority flow from the headquarters in Washington, D.C., through the nine division offices, and to the thirty-seven district offices. The actual planning and design of projects takes place at the district level, under guidelines established by headquarters and under review of the appropriate division. Thus, at its top, the Corps looks like a military organization, while at its regional offices, it appears more like a decentralized civilian agency.[2]

The geographical boundaries of Corps districts are defined by regional watersheds, and because most water resources projects modify a particular drainage system, they usually reside within the jurisdiction of a single Corps district. The Tenn-Tom, however, connected two watersheds, and in so doing, it cut across the agency's Nashville and Mobile districts. This posed an organizational challenge that led senior agency officials to consider establishing a special, temporary district just to build the waterway. They also contemplated giving the entire project to the Mobile District, where the largest portion of the project rested. In 1946, the Corps's leadership chose a compromise solution: the Nashville District would design and construct the divide and canal sections, while the Mobile District would design and construct the river section and would serve as the lead office for the entire project. Although only 25 miles of the waterway fell within the geographical boundaries of the Nashville District, Nashville gained responsibility for some 85 miles of the project. This division of work made Nashville the recipient of about 60 percent of the construction funds.[3]

Dividing the Tenn-Tom between Nashville and Mobile created another organizational problem because the two districts also reported to different divisions, the Ohio River Division and the South Atlantic Division, respectively. To maintain the military chain of command, the Corps gave the South Atlantic Division the lead, with responsibility for preparing the annual budget document and testifying before the Congressional appropriations committees. In 1967, the agency's leaders reapportioned the jurisdictional responsibility to resemble more closely the geographical boundaries of the two districts. Under the revised arrangement, Nashville would design and construct the divide section, including the Bay Springs Lock and Dam, and Mobile would handle the rest. Overall responsibility for the waterway remained with the Mobile District.[4]

The design work was done in many stages over a number of years. Corps personnel at the Mobile and Nashville districts designed nearly everything "in-house" and submitted their work for review and approval to officials at the South Atlantic Division and headquarters. Engineers from the districts, divisions, and headquarters usually explored major design concepts in depth at special conferences. The culture of engineering within the Corps was thoroughly bureaucratic; the engineers who planned and designed the Tenn-Tom remained faceless to the outside world. It was the organization that designed and built water projects, not individual people.[5]

In June 1960, Mobile District completed the basic plan for the Tenn-Tom, which the Chief of Engineers approved in April 1962. This document focused on the overall engineering concept, and as such contained no detailed design. It dealt with such broad considerations as how many locks and dams the waterway would have, the general location of those structures, rough channel alignments, and the like.[6]

Each major segment of the waterway—such as the locks and dams—then received their own individual general design memorandum, or GDM. GDMs articulated the definite location and orientation of navigation structures and such things as the number and dimensions of spillway gates. Because most of the GDMs were prepared after 1970 and the enactment of the National Environmental Policy Act, alternative designs were evaluated to determine the soundest approach from engineering, economic, and environmental perspectives. At this stage of the design process, pertinent state and federal agencies were invited to comment on the project specifications. Corps planners then considered any existing or potential problems raised in the agency critiques, and evaluated their recommendations. These agency views were both summarized in the printed design document and printed verbatim in the appendices.[7]

Further refinement and specificity of the design for the discrete elements of each GDM led to feature design memoranda, or FDMs. Mobile and Nashville district personnel wrote FDMs for such items as lock gates and operating systems, spillways, river bend cutoffs, and navigation channels. Following approval by Corps division and headquarters officials, the districts prepared "Plans and Specifications," which constituted the detailed instructions needed by construction firms to bid on the project components and to build them.

River Section

The waterway's river section begins near Demopolis, Alabama, where the Black Warrior River empties into the Tombigbee. From that point, the Tenn-Tom extends north on the Tombigbee, widening, deepening, and straightening about 149 miles of the river,[8] and traversing a countryside characterized by broad, flat pasture land, broken by gently sloping hills. The Tombigbee is a geologically old river whose meandering course has etched an expansive flood plain varying in width between 1 and 2 miles, and containing extensive stands of bottomland hardwood forests. The river itself rarely narrows to less than 200 feet along this stretch of its run, and while its banks are mostly low, through the centuries its twisting path has occasionally pressed against highlands, in the process carving bluffs rising 20 to 30 feet in height. To overcome the 117-foot change in elevation from the southern to northern end of the river section, the Corps constructed four conventional locks and dams: the Gainesville Lock had a lift of 36 feet, and the locks at Aliceville, Columbus, and Aberdeen each overcame height differentials of 27 feet.

The river section was so named because the navigation channel of this part of the waterway remained within the banks of the Tombigbee. Where the river fell short of the minimum channel dimensions of 9 feet of depth and 300 feet of width, the Corps made up the difference by dredging. The agency also widened several river bends to accommodate the expected barge traffic. Canalization of the Tombigbee in this section of the Tenn-Tom reflected the traditional Corps of Engineers approach to navigation projects on the Arkansas, Ohio, and Upper Mississippi rivers and elsewhere, where the engineers confined the navigation channel to the bounds of the natural river.[9]

Like many rivers in the Southeast, the Tombigbee twisted and turned like a snake. Boats often had to ply two miles of river channel to advance one mile as the crow flies. To shorten the watercourse and to reduce the number of curves, the Corps called for the construction of several cutoffs, which were by-pass canals dug across the necks of river loops. Once severed, these former river loops—called *bendways* by the agency's engineers—became artificially-made oxbow lakes. The Corps eventually constructed 35 cutoffs, which shortened the waterway by 40 miles, while isolating 71 miles of extraneous river.[10]

Because barge traffic would be diverted from the bendways to the cutoff channel, the army engineers initially planned to allow these severed loops to silt over. They realized, however, that if the bendways remained connected to the navigational channel, their shallow backwaters and broad standing timber areas could provide important fish and wildlife habitat, as well as recreational boating and fishing reserves free of commercial barge traffic. The fate of these amputated river loops therefore became a matter of environmental concern. Without human intervention,

Rattlesnake Bend cutoff in the Tenn-Tom river section. Dredges were used to deepen, widen, and straighten the main stem of the Tombigbee River in the southern reaches of the waterway. *(Courtesy of U.S. Army Corps of Engineers, Mobile District.)*

the long-term existence of the bendways was doubtful. River-suspended sediments would eventually fill in the ends of these loops, thereby creating true oxbow lakes. The principal current, along with its scouring properties, would seek the main channel, while the water that passed through the bendways would move much more slowly and deposit more of its suspended sediments.[11] The eventual elimination of these stretches of the river would threaten fish populations in the Tombigbee, and so the U.S. Fish and Wildlife Service and several environmental groups pressured the Corps to preserve these vestiges of the old river by keeping their openings dredged, which the agency ultimately did.[12]

Aside from the bendways and the conversion of the free-flowing Tombigbee into a series of still-water impoundments, the river section's most significant environmental alteration involved the discarding of dredged material. The Corps's standard method of disposal involved spreading the dredged material (which typically consisted of about 80 percent "carriage" water and 20 percent solid material) along the river banks. The pumped water would then wash back into the river along with a certain amount of suspended material, significantly increasing the turbidity downstream. The heavier particles remained on or near the river banks. In years past, dredged material was used to fill wetlands, but by the 1970s wetlands were widely recognized as important elements of a river valley's ecology. Destruction of wetlands and of bottomland hardwoods would adversely affect the river's water quality, increase flooding, and rob the region of prime fish and wildlife habitat. It also degraded a river's general aesthetic qualities. From an engineering standpoint, this method of disposal created long-term maintenance problems because the solid materials eventually eroded back into the river. The tremendous amount of dredging required in the Tenn-Tom's river section made the traditional approach all the more environmentally controversial.[13]

The general design completed in 1960 envisioned spewing the dredged material along the banks of the Tombigbee. By 1971, Corps planners realized this procedure was unacceptable for environmental reasons, yet they included no specific plan in their environmental impact statement for how the material would be discharged, promising only to devote significant attention to determining the optimal methods for disposal.[14]

The Corps eventually developed a new disposal concept. Rather than spreading it along the river banks, the dredged material was pumped several hundreds of feet to upland areas, where it was contained in delineated disposal sites. A continuous dike, typically nine feet tall and ten feet wide at the crest and built from material dug from the interior of the disposal site, encircled each repository. The sites were internally divided by an earthen wall into two cells. The primary cell—the larger of the two—was designed to accommodate all the dredged material during construction, while the secondary cell provided storage space for fifty years of subsequent maintenance dredging. Contractors pumped dredged material into one end of the primary cell, where it was allowed to stand—or *pond*—before the water overflowed at the other end into the secondary cell. During the time the water stood in the primary cell, the coarser suspended materials settled, which caused the primary cell to receive the bulk of the solid material. Water entering the secondary cell typically contained suspended material of a much finer nature. The water was allowed to pond in this secondary cell and release more of its suspended material before it was then passed back into the river. This method substantially improved the quality of the water reentering the Tombigbee, while sparing hundreds of acres of wetlands that would have otherwise been sacrificed.[15]

The disposal of dredged material—even when incorporating the upland, cell method—generated widespread concern on aesthetic grounds because of the initial barren, scarred appearance it produced. Corps designers masked this problem by hiding the disposal sites behind uncleared buffer zones. Initially, agency officials considered a buffer zone of 500 feet between the disposal sites and the river, but they subsequently scaled that down to 300 feet for channel frontage and 150 feet for the other sides—dimensions eventually approved by the U.S. Fish and Wildlife Service, Environmental Protection Agency, and state agencies from Mississippi and Alabama.[16]

The buffer zones provided both visual screens for the disposal areas and seed sources for the natural revegetation of the sites. The interiors of the disposal cells were initially cleared of trees and brush, because the dredged materials would eventually kill most trees left standing. By the mid-1970s, however, Corps biologists argued successfully to limit deforestation within the disposal areas in order to retain wildlife habitat and accelerate revegetation. The decaying trees would help rebuild the soil and would provide habitat for arboreal animals and birds. Moreover, the design change offered significant cost savings, as the Corps could forgo contracts for land-clearing.[17] In all, there were over one hundred disposal areas in the river section, some of them covering 150 to 180 acres.[18]

All ten locks on the waterway were of standard Corps design with chamber dimensions of 110 by 600 feet. All but two, Lock D in the canal section and Bay Springs Lock in the divide section, were of the gravity wall design. This meant that the structural strength of the lock—notably its ability to resist sliding and overturning—came from the overall weight of the massive concrete walls. Each segment of the lock walls at every structure acted as an independent monolith.[19]

Gainesville Lock and Dam was the first navigational structure on the Tenn-Tom. The construction site, which encompassed a 250-foot-wide section of the Tombigbee River bordered by 30-foot-high bluffs, sat some 53 miles upriver from the existing Demopolis Lock and Dam and about one mile northeast of the town of Gainesville in extreme west-central Alabama. An earth dam across the old river channel created a reservoir covering 7,200 acres and extending 45 miles upriver to the Aliceville Lock and Dam site. To reduce the navigational distance by about 4.23 miles, the Corps dug a 9,200-foot-long canal through a large meander near the dam site. The 36-foot-lift lock formed the downstream end of that cutoff canal.[20]

The Federal Water Pollution Control Administration (FWPCA), the predecessor of the U.S. Environmental Protection Agency, reviewed the Corps's draft design memorandum for Gainesville Lock and Dam in 1968. Regional Director John R. Thoman expressed the FWPCA's concern about the possible effects of the structure on water quality, especially in lowering the levels of dissolved oxygen. He recommended the installation of "reaeration devices . . . to raise the dissolved oxygen to as near saturation value as practicable."[21] Although the Corps followed federal guidelines by publishing Thoman's letter in its final general design memorandum, agency officials dismissed the recommendation, stating: "Since water quality is not an authorized purpose of the Tennessee-Tombigbee Waterway and no

The Gainesville Lock and Spillway, the first navigational structure built on the Tenn-Tom. This 36-foot lift lock straddles the Tombigbee River near Gainesville, Alabama. *(Courtesy of U.S. Army Corps of Engineers, Mobile District.)*

tangible benefits were furnished to establish the economic justification of such improvements, there is no basis for the inclusion of reaeration devices as part of the features proposed by FWPCA."[22]

The subsequent passage of the National Environmental Policy Act (NEPA) and the mounting environmental criticism of the Tenn-Tom led Corps planners to reverse themselves on the water quality issue. As recommended by the FWPCA, the spillway at Gainesville Lock and Dam received special attention, given its status as the major structural element affecting water quality. Water impounded behind the dam was likely to become deficient in dissolved oxygen, and the original spillway design—which consisted of a fixed-crest spillway that passed water over a smooth sheet of concrete—did little to improve the quality of water leaving the reservoir. To increase dissolved oxygen in the water downstream of Gainesville, Corps personnel redesigned the project's spillway to include a lip—or "reaeration ramp"—about eight feet below the spillway crest which would create a 30-foot-high waterfall that directed the cascading water to hit a deflector bucket and splash back into the air. Maximizing turbulence and air-to-water interface in this way greatly increased reaeration. Corps planners were happy to discover that the incorporation of these "environmental aspects" at Gainesville did not boost the cost of construction.[23]

Pressured by private environmental organizations and various conservation agencies within the federal government to lessen the environmental damage caused by their water projects, Corps officials looked for ways to blunt criticism of the Tenn-Tom short of abandoning the project altogether. Previously, agency designers were reactive, altering the Gainesville dam spillway only in response to outside pressure. What they now sought was a self-initiated design change made strictly for environmental reasons, which they could hoist up their public relations flagpole as a symbol of their caring and good intentions. The Columbus Lock and Dam provided just that opportunity.

As originally designed, Columbus Dam threatened Plymouth Bluff, a rich Cretaceous fossil bed about 7 miles northeast of Columbus, Mississippi. The dam would raise the water level at Plymouth Bluff by 29 feet and would cover more than half of the exposed Eutaw Formation and two fossiliferous strata embedded in Selma Chalk. Geologists and students in the earth sciences had studied the area for over a hundred years, and the Mississippi State College for Women maintained a lodge at the top of the bluff for research and recreation. The Corps described the impact of flooding the site in its environmental impact statement: "The bluff would lose its attraction for visiting study groups from other areas as well as most of its value as a teaching-research site for local universities and its natural beauty and recreational usefulness would be drastically reduced." Having stated this, the agency promised to evaluate design alternatives that would avoid harm to this resource.[24]

Geological investigations for the Columbus Lock and Dam undertaken in August 1971, however, were confined to the original placement near Plymouth Bluff. The proposed site was found to be underlain by the Eutaw Formation, which produced excellent foundation conditions.[25] Yet Corps officials felt a strong need to demonstrate their agency's environmental sensitivity. Plymouth Bluff, the most popular geological formation within the entire region, seemed eminently worth saving, both for its own sake as well as for its public relations value. And because the detailed design for Columbus Lock and Dam had yet to be prepared, relocating the structure would cost the Corps little from either an engineering or an economic standpoint.[26]

By November 1971, designers had specified that the lock and dam be moved 3,800 feet upstream, thus sparing the unique paleontological site from flooding.[27] Corps officials pointed to the relocation of Columbus Lock and Dam as proof of their environmental awareness, but resiting the dam was not really as significant from an engineering and design point of view as many of the waterway polemicists implied.[28] The preliminary stage of planning made the transfer a relatively easy task. As Corps planner James D. Wall reported in spring 1971, "there is no good foundation data yet available and the finally selected site could be moved perhaps a mile or two if better foundation conditions would obtain."[29]

While the Corps continued to defend the Tenn-Tom on environmental grounds throughout the project's construction, the agency also found itself vulnerable to economic criticism, as the waterway's cost began rising far above official projections. Within the river section, designers paid special attention to the expenses associated with keeping the area's abundant groundwater from damaging the various construction sites.

Alluvial floodplains, consisting as they do of materials washed down from the surrounding highlands, usually contain highly pervious sands and gravels. These water-bearing soils allow easy horizontal movement of subsurface water. Such was the case in the Tombigbee River floodplain, where the Corps planned to erect its massive navigation locks. To build these structures in the dry, which was the fastest, safest, and most economical manner, engineers had to prevent the seemingly ubiquitous groundwater from entering the excavated construction sites. Initially, the engineers followed the standard method of "dewatering," which called for a series of deep wells encircling the construction site, which itself was contained within a coffer cell formed by an unbroken wall of interlocking metal sheets driven into the ground or by deep trenches back-filled with impervious material such as compacted clay. Sumps and pumps within the protected construction site would remove rain water and any ground-

Plymouth Bluff, overlooking the Tombigbee River about seven miles northeast of Columbus, Mississippi. This rich Cretaceous fossil bed was spared inundation when Corps of Engineers officials relocated the Columbus Dam site 3,800 feet upstream. *(Courtesy of U.S. Army Corps of Engineers, Mobile District.)*

water that had managed to seep in. Because an entire excavation could be lost if the pumps failed during critical periods of construction, emergency auxiliary sources of power had to be provided to compensate for electrical outages. Occasionally, the sheet metal walls would fail along their interlocked seams. This dewatering system proved time-consuming and expensive to install, maintain, and operate, especially with the heavy pumping requirements that were required during the three to four years it took to build each lock.[30]

The high costs associated with the traditional dewatering method led designers to look for alternative solutions. The relatively new concept of earth-filled slurry trench construction drew their attention. The use of slurry methods as an engineering solution for underground construction originated in Italy and France during the early twentieth century. Prior to World War II, the use of slurry methods in the United States extended only to water-well and oil-well drillers, who filled their shafts with a slurry mixture to prevent them from collapsing. Gradually during the 1950s and

1960s, contractors began to use the slurry method in underground construction within urban areas—primarily for large building foundations, subways, underground parking facilities, and large underground utilities tunnels. Even by the early 1970s, slurry trench construction in the United States was rarely used in dewatering large construction sites outside cities.[31]

Gainesville Lock and Dam became the first structure on the Tenn-Tom to receive detailed design. Although the slurry trench concept was suggested as early as March 1970, designers rejected it in favor of the traditional approach.[32] One year later, however, pressure to reduce mounting costs sent the engineers back to the drawing board in search of viable alternatives. George H. Mittendorf of the South Atlantic Division championed the slurry trench method, which, he argued, could reduce "the cost of the cutoff walls by about 35%" at the Gainesville Lock.[33] Despite some hesitancy,[34] the Corps tried the slurry trench method of construction and found it to be both efficient and cost effective. The agency consequently added the slurry trench concept to the design documents for the remaining locks in the river section. By October 1975, Corps officials estimated that the new technique had saved the government $4,379,500.[35]

The procedure was quite simple. Large backhoes and draglines dug 3 to 5-foot-wide trenches down to the impervious Eutaw clay (usually 30 to 40 feet deep, but at some locations extending down 70 feet). While the trench was being excavated, a mixture of water and bentonite slurry (bentonite is a gray powder made from natural clays containing montmorillonite) was pumped in. The dissolved bentonite would adhere to the trench walls forming an impermeable film, which served to keep the walls from collapsing. Contractors then mixed the excavated material with natural clay, sand, gravel, and bentonite slurry and pushed it back into the trench to displace the water and slurry solution and to form a permanent impermeable membrane. The displaced water and slurry mixture was pumped into a holding pond where it was recycled. The process proved far less costly than the traditional methods of dewatering, but designers and contractors shared reservations about its effectiveness because it had not been used in a large-scale project in the United States. The performance of the slurry trench at Aliceville Lock and Dam surpassed all expectations, however, which led to its incorporation into the waterway's remaining structures.[36]

Canal Section

The waterway's canal section began just south of Amory, Mississippi, and extended 46 miles north through the wide flood plain of the upper Tombigbee River to the Bay Springs Lock and Dam in Mackeys Creek.[37] This run of the Tombigbee was narrower and more serpentine than its downstream reaches. Early in their planning, Corps designers determined that it would be uneconomical to widen and straighten the river in this section. Instead, they proposed building a lateral "perched canal" for this stretch of the waterway which would parallel the upper reaches of the Tombigbee's East Fork. It would be perched in the sense that it would sit up from the river on the hillsides of the east flood plain. The navigation channel would consist of an excavated canal between parallel levees. The west levee would separate the canal and the river; the east levee would prevent streams and creeks originating in the uplands from entering the canal. This local drainage would first be fed into an excavated ditch running alongside of the canal, before it would be channeled under the waterway through reinforced concrete culverts. Water from storms exceeding the capacity of the culverts would be allowed into the canal by way of concrete spillways. Excess water within the canal would in turn be discharged into the river by means of similar spillways. Five locks—identified as A, B, C, D, and E—would overcome the 140-foot difference in elevation from end to end of the canal section. Three of the locks—A, D, and E—had 30-foot lifts, while the lifts at Locks B and C were five feet shorter.[38]

The perched canal concept represented the traditional design approach for these conditions. During the early 1970s, however, Corps officials began to question this plan as they prepared the project's detailed designs.[39] From a hydrologic standpoint, it became clear to the agency's engineers that maintenance of the perched canal would be extremely demanding, especially for those areas where the culverts passed under the navigable channel. Small floods would tend to silt up the culverts, while big floods threatened to block them with debris, thereby submerging large areas above the waterway and even penetrating the canal itself. In addition, the relatively narrow confines of the parallel levees subjected the canal to potential surge problems from the release of water from each lockage at Bay Springs. Moreover, this design made the canal almost entirely dependent on the Tennessee River for its water supply, because all local drainage would be passed under the waterway. Quite apart from engineering considerations, the perched canal posed troublesome aesthetic problems, as it would form an unsightly scar on the land—a straight ditch stretching for miles above the Tombigbee River.[40]

Arthur M. (Bud) Cronenberg, a tough-minded, forty-year veteran of the Corps and chief of the Mobile District's Hydraulic Structures Design Branch, became the leading proponent for the alternative and novel "chain-of-lakes" design concept. He was the first person to present the idea, and after he retired in 1971, several of his colleagues elaborated and refined it.[41] The chain-of-lakes followed essentially the same alignment as the perched canal design. The levee would remain on the west bank—the riverside of the canal. This dike would be built from material excavated for the navigation channel. The east bank levee, however, would be eliminated, thus creating a series of reservoirs with their eastern boundary determined by the natural lay of the land. Near the levee, maintenance dredging would ensure a navigable channel, while the rest of the impounded water would serve as a series of lakes for aesthetic and recreational purposes. The streams and creeks coming off the hills would thus feed directly into the waterway. This water would be passed into the natural river to the west by means of strategically located spillways. The five locks—whose locations and lifts remained essentially unchanged from the perched canal design—would be incorporated into earthen dams that tied the levee into higher ground. Thus, the project would not require the elaborate system of culverts running under the channel, and the broad surface area of the lakes would absorb the surges produced by the lockages at Bay Springs.[42]

By 1973, designers at the Mobile District had become convinced of the superiority of the chain-of-lakes concept. The engineers argued that the chain-of-lakes "will provide numerous environmental and ecological benefits not available in the authorized plan and would eliminate the need for a drainage ditch. The appearance of the waterway will be more natural and the turbidity of the water resulting from traffic will be less. The lakes will provide a desirable habitat for fishes and waterfowl and provide various recreational benefits." Although this plan nearly doubled the land to be flooded (8,300 rather than 4,300 acres), the chain-of-lakes option did not necessitate the acquisition of additional property; the perched canal required the same amount of acreage because the land outside the levees would be subjected to periodic flooding.[43]

Despite the engineering advantages of the chain-of-lakes design, senior Corps officials were leery of making such substantial modifications to the waterway without soliciting Congressional approval. As they recognized, the perched canal remained "the authorized scheme discussed in previous reports." While the Corps enjoyed the discretionary authority to make modest changes to its projects without the approval of Congress, an alteration of this magnitude risked exceeding the legal limits of that authority. On the other hand, the Corps was under intense pressure (most specifically in the form of a lawsuit brought against it by the Environmental Defense Fund) to demonstrate it was not violating the mandate of the National Environmental Policy Act, and, as the agency spokespersons were eager to emphasize, "the chain-of-lakes approach was environmentally more acceptable and more beneficial than the perched canal approach."[44]

Mobile District engineers continued to lobby their superiors on behalf of the chain-of-lakes design.[45] They maintained that the perched canal had no major advantages over the chain-of-lakes. Construction costs for the two designs varied little. The major structures, canal excavation, and land acquisition were the same. Land clearing requirements were less for the perched canal, while the chain-of-lakes needed only a single levee. The cost of the drainage ditches and culverts in the first design were offset by the cost of the spillway structures in the second design.[46] These arguments ultimately prevailed, and the Corps officially adopted the chain-of-lakes design in 1974. The designers had favored the chain-of-lakes concept primarily because it offered a better engineering solution. Corps planners and public relations officials, on the other hand, supported the change because it added significant opportunities for flat water recreation, in addition to being more aesthetically pleasing.[47]

Like the resiting of Columbus Lock and Dam to preserve Plymouth Bluff, the redesign of the canal section became a banner of the Corps's environmental sensitivity. Agency spokespersons and other Tenn-Tom enthusiasts hailed this change as a triumph of environmental considerations over those of engineering. Yet the canal section had been used as an example of environmental sensitivity even under the old design. The river section involved radical alterations to the Tombigbee's riverine ecology, and the divide section required a massive earth moving effort. The canal section, in comparison, appeared far more benign, whether under the perched design or chain-of-lakes design. Because the excavated material in the canal section would be used to build the levees, no wetlands would be lost to disposal sites. As Glover Wilkins of the Tennessee-Tombigbee Waterway Development Authority boasted to a *Time/Life* photographer who was filming the proposed waterway route in spring 1970, in the canal section "we will leave the rambling river stream and hardwood forest intact while we go up on the hillside to build the canal."[48] When coaching Representative Thomas G. Abernethy of Mississippi in January 1971 about dealing with environmental criticism of the Tenn-Tom, Wilkins wrote: "Another bit of phraseology we are using is that our canal will be a 'dry cut' meaning we will not disturb the wet lands [*sic*]. I wish you would consider using this also as it has a good connotation."[49]

The abandonment of the perched canal for the chain-of-lakes represented a major design change. Although less grand in scale, Lock D also led Corps officials to consider a significant deviation from the original design concept. Unlike the chain-of-lakes, however, political considerations outweighed the favored engineering approach at Lock D, thus exemplifying the shifting technical and political complexities of civil engineering design.

The issue at Lock D involved unusually poor foundation conditions and high artesian pressures. Foundation conditions throughout the canal section were less than ideal for the construction of the large navigation locks, but the Corps's initial surveys and subsurface testing supported the designers' belief that low-lift locks of the conventional gravity design could be built at all five sites. The ground at the Lock D site, however, was particularly swampy. When technicians began taking core samples from

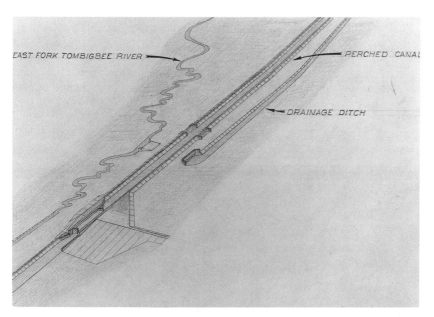

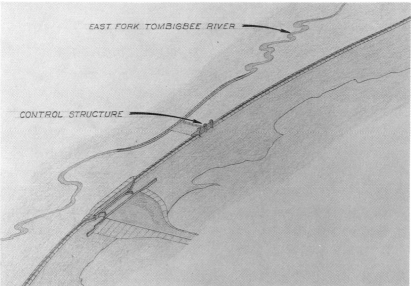

The canal section proved the most conceptually challenging segment of the Tenn-Tom. The combination of practical engineering concerns—notably how to handle the local run-off during and after storms—and aesthetic considerations led Corps officials to abandon the original "perched canal" design (above) in favor of the innovative "chain-of-lakes" design (below). These drawings were prepared by engineers at the Mobile District in 1973 to help explain the two design concepts to members of Congress. *(Courtesy of U.S. Army Corps of Engineers, Mobile District.)*

the area in 1975, they found seams of soft clay and groundwater under high artesian pressure, which together provided inadequate support for bearing the weight of a gravity wall lock.[50]

Agency engineers began investigating alternative design approaches, and settled on two principal plans. The original overall plan for the canal section called for lifts of 25 feet at Lock C and 30 feet at Locks D and E. The first design alternative included lifts of 25 feet at Lock C, 25 feet at Lock D, and 35 feet at Lock E, while the second alternative called for lifts of 25 feet at Lock C, the elimination of Lock D, and 60 feet at Lock E.[51] Although the Corps's designers believed these to be the most rational engineering solutions, the agency's legal counsel advised against making any major changes in the overall design of the canal section—whether that be changing the heights of the locks or completely eliminating Lock D—out of concern that it would "invite litigation."[52]

This sensitivity to the legal implications of further altering the waterway's design concept came as the result of a second lawsuit filed to stop the Tenn-Tom. Of the several charges brought against the Corps by the Environmental Defense Fund and the Louisville and Nashville Railroad, the plaintiffs challenged the Corps's discretionary authority to adopt the chain-of-lakes design without Congressional authorization. "As the District considers the possibility of eliminating Lock D in the Tenn-Tom Waterway or otherwise altering the canal section," Corps attorney David H. Webb advised the engineering staff, "foremost thought must be directed to the guiding principle that ultimate engineering (technical) decisions must be made *subsequent* to economic, environmental, or other studies rather than studies being performed because of an engineering decision. Furthermore, the District must avoid the *appearance* of studies being undertaken as a consequence of predetermined decisions."[53]

Corps officials from the various divisions and branches within Mobile District continued to study the design alternatives. In January 1977, the district's planning division summarized the pros and cons of the various options. Elimination of Lock D posed significant losses for recreation in the canal section. Without Lock D, the canal section would be missing its largest planned recreation site, the Beaver Lake Recreation Area. One less lock meant one fewer lake, and this reduced the amount of real estate required for the project by 1,350 acres. However, the absence of Lock D would require the excavation of an additional 10 million cubic yards for the navigation channel. Navigational benefits stood to improve without Lock D: passage through the waterway would be reduced by the time originally allotted for locking through Lock D—an estimated savings of $46.6 million over the fifty-year life of the project. Annual operation and maintenance costs would be reduced as salaries for personnel to operate Lock D could be eliminated as could the maintenance costs of the facility. Actual construction costs would be $1,776,000 less for building the project without Lock D. Finally, without Lock D, Lock E could be resited to a more favorable location where the structure could be built on sandstone.[54]

Mobile District's engineers reported that all three plans were "feasible as to engineering integrity and safety," that their costs were "nearly equal," that the "environmental impacts of each plan differ but none should pose insurmountable problems," and that the elimination of Lock D would result in fewer recreational benefits than the other alternatives.[55] They called for further studies, and emphasized that the elimination of Lock D would require approval from the Secretary of the Army and any alteration from the General Design Memorandum would require a supplement to the environmental impact statement. They also suggested that any major changes might require Congressional authorization.[56]

Despite the strong engineering and economic arguments in favor of eliminating Lock D, political and legal considerations led senior Corps officials to retain the lock and pursue the otherwise less than ideal technical solution. Mobile District therefore adopted a special rigid U-frame—or "bathtub"—design for Lock D which allowed the entire structure to float. To produce this rigid structure, the army engineers designed a thick, heavily reinforced concrete floor. This change lowered the bearing pressure from four tons per square foot for the standard gravity design to about two tons per square foot because the entire floor would distribute the weight of the lock. In a gravity design, only the walls carry the structure's weight to the supporting earth. The bathtub design also guarded against differential settlement. This design cost more than the others, but agency decision makers—fearful that project opponents would demand an environmental assessment *and* Congressional authorization for any major change—determined that it was worth the extra expense.[57]

Divide Section

The most dramatic element of the Tenn-Tom was the divide section, which extended north from Bay Springs Lock and Dam, through a 27-mile-long cut in the ridge separating the Tennessee and Tombigbee river basins, to an arm of Pickwick Lake on the Tennessee River. The lock at Bay Springs, with a lift of 84 feet, was the waterway's largest. The divide cut, which pierced the 570-foot-high ridge, averaged 75 feet in depth and reached a maximum depth of 175 feet. In all, over 160 million cubic yards of earth were excavated from the Mississippi hills; enough dirt, claimed the U.S. General Accounting Office, "to build a two-lane highway from the Earth to the Moon."[58]

This earthmoving challenge threatened to be the most costly aspect of the entire project, and led Corps designers to search thoroughly for alternative construction techniques. In retrospect, the most extraordinary proposal involved the use of nuclear explosives. The concept of excavating a canal by means of simultaneous detonation of a row of buried nuclear explosives sprang from the U.S. Atomic Energy Commission's (AEC) Project Plowshare program, which was created in 1957 to explore the possible civilian applications of nuclear power, such as large-scale excavations for use in highway, harbor, and canal building. The incentive for Plowshare had come a year earlier when Egypt seized the Suez Canal, which motivated Harold Brown of the AEC's Lawrence Radiation Laboratory (LRL) to propose that nuclear explosives be used to construct an alternative canal across Israel. Although support for Brown's specific proposal evaporated with the resolution of the Suez crisis, the general concept intrigued AEC officials enough for them to establish a special program, headed by the LRL at Livermore, California (the University of California ran the laboratory for the AEC). Soon thereafter, Major General W. C. Potter, Governor of the Panama Canal Zone, asked the LRL to assess the practicability of constructing a sea-level canal through the Isthmus of Panama using nuclear explosives.[59]

These were the days before Americans began to protest vigorously against the use of nuclear energy for domestic projects. Project Plowshare had an enthusiastic and devoted following among the engineering community, many of whom worked diligently to generate high public expectations and support for the program. The physicist Edward Teller, for example, claimed that "we're going to work miracles."[60] The U.S. Army Corps of Engineers demonstrated its interest in the civil engineering prospects of Plowshare in 1962, when it established the Nuclear Cratering Group to assist the LRL in developing excavation technologies and techniques.[61] As Captain Louis J. Circeo, Jr., of the Corps's Nuclear Cratering Group told the National Academy of Sciences' Highway Research Board, of all the various industrial applications associated with nuclear explosives, "large-scale excavation is one of the most promising."[62] He said that "because of the economic and time savings possible on many projects through the use of nuclear explosives, engineers should be aware of the possibilities of this method."[63]

Such arguments captured the attention of the Tennessee-Tombigbee Waterway Development Authority (TTWDA), whose members were eager to explore any construction alternative that might make the waterway economically feasible. Recognizing that Plowshare advocates were looking for practical and high-visibility applications for nuclear excavation, they raised the possibility of using the Tenn-Tom's divide cut as a test project. Mississippi's governor, who was a member of the TTWDA board of directors, assigned the state's Industrial and Technological Research Commission to broach this idea in 1960.[64] Officials at the LRL responded enthusiastically,[65] and in September they sent two of their senior scientists—Edward Teller and Gerald W. Johnson—to Mississippi to inspect the proposed construction site and meet with state and Corps of Engineers officials. Teller and Johnson were duly impressed, and recommended that the Research Commission and TTWDA fund a feasibility study, toward which the LRL would provide technical assistance at no charge as part of its Plowshare responsibilities. Lacking the necessary discretionary funds, Lieutenant Governor Paul Johnson convinced the engineering and construction firm Brown and Root of Houston, Texas, to undertake the study at no cost, with the prospect that it might be engaged on any future work.[66]

Lieutenant Colonel Ernest Graves, Jr., director of what later became Nuclear Cratering Group, assisted with the feasibility study and provided one of the first cost estimates in September 1960. According to Grave's preliminary figures, nuclear excavation offered a modest $400,000 savings out of a total project cost then placed at $227 million.[67] Nevertheless, the radical engineering solution continued to generate interest. A 1963 *U.S. News & World Report* article observed that "the first use of a nuclear explosion for peacetime construction may be to help build a canal in the U.S. South, instead of a harbor in Alaska or another Panama Canal." The story claimed that officials from Mississippi, Alabama, Tennessee, and Kentucky had launched a campaign "to 'sell' nuclear excavation of part of a planned 253-mile waterway to connect the Tennessee River with Alabama's Tombigbee River."[68]

The Limited Nuclear Test Ban Treaty of 1963, which outlawed atmospheric testing, did much to change the political climate for Project Plowshare proposals, because explosives developed under the guise of peaceful use could easily be perceived as having (if not actually having) military applications. Nevertheless, the program continued, although with less public fanfare and essentially no real prospects for implementation. Tenn-Tom promoters did what they could to sustain interest in using the waterway as a pilot project for constructive uses of nuclear explosives. As TTWDA administrator Glover Wilkins wrote AEC chairman Glenn T. Seaborg in February 1964, "the potential study of using nuclear devices on the divide cut of the Tennessee-Tombigbee Waterway is particularly appropriate at this time due to consideration being given to building a new canal across the isthmus in Central America," especially in light of the concurrence among policy makers that "uses of nuclear devices for peaceful excavation will have to be first tried within the Continental United States before attempting any project outside."[69] Mississippi's governor assured Seaborg "that evaluation of nuclear explosives in the construction of civil works will be welcomed rather than opposed" by the state's citizens.[70]

The technical evaluators of the Tenn-Tom feasibility study, however, developed serious reservations about the proposal, among them the impact of fallout on the region's dairy cows, the potential seismic hazard, the need to evacuate large numbers of people, and the possibility of contaminating the aquifers that served as the principal source of water for communities in northeastern Mississippi and northwestern Alabama. Moreover, it was not even certain that nuclear excavation would reduce construction costs.[71]

Following his conference at the LRL in March 1967, Lieutenant General William Cassidy, the Corps's Chief of Engineers, ruled out all future consideration of using nuclear explosives to build the Tenn-Tom.[72] South Atlantic Division Engineer Major General George H. Walker testified before the House Appropriations Subcommittee on Public Works that to use nuclear excavation to build the Tenn-Tom, the Corps "would have had to select another and longer route because of the geology of the proposed route. This longer route made the nuclear method more expensive, and therefore we did not adopt that."[73]

Ultimately, the Corps pursued a conservative civil engineering approach for the divide cut. Because the expense of large-scale excavations depends primarily on the types of material to be removed, designers selected a route that avoided solid rock, passing instead through an area whose soils were characterized by fairly consistent erodible silty-sand, made up mostly of the Eutaw Formation. Contractors could use giant earth-moving equipment—of the type often employed in open-pit mining—to dig the material, thus reducing the unit cost of excavation. This type of soil, however, required gentle slopes to avoid erosion, and a staircase of earthen benches had to be built. This design requirement increased the total amount of material to be excavated: at the deepest sections of the project, the cut would be nearly a half-mile wide at the top.[74]

The amount of earth to be removed from the divide cut varied depending on the alignment, slope design, and other factors. December 1973 estimates placed excavation of the 27-mile-long divide cut at 161 million cubic yards, up from the 1970 estimate of 144 million.[75] Reducing this amount became a major design challenge because the volume directly influenced the costs, which were typically calculated on a per cubic yard basis. Lowering the excavation requirements would also lessen the problem of disposal—a problem related to real estate as well as environmental considerations, because of the land needed for dumping. Safety and future maintenance costs (especially from slides and sloughing) had to be weighed against excavation costs and the environmental consequences of displacing enormous quantities of earth.[76]

The 1966 General Design Memorandum used $0.27 per cubic yard as the unit cost for excavating the divide cut. This figure translated the estimated 144 million cubic yards of common excavation to a cost of $38,928,000. Subsequent environmental legislation necessitated the additional care and treatment of this material to minimize environmental damage. By May 1973, members of Corps headquarters worried that "the cost of spoil area treatment on the Tenn-Tom Project could generate near prohibitive costs." Their concern stemmed from the agency's recent experience on the Trinity River navigation project, where "the estimated cost of extra haul to areas not directly along the banks, acquisition of land for spoil areas and landscape treatment to satisfy environmental requirements was in excess of $150,000,000." Reduction of required excavation would thus help lower construction costs at several levels. In considering ways to achieve such reductions, designers explored the engineering pros and cons of limiting the depth of the divide cut by digging a summit level canal with a set of locks (each 40 to 60 feet high) at either end. They sought to learn how much excavation would be saved in such a plan, how much the two additional locks would cost, and how much water would have to be pumped up to the canal from the Tennessee River's Pickwick Lake.[77]

Corps officials concluded that despite the reduction in excavation required by the summit level canal, the additional two locks would increase construction costs by $32 million over the original design, while operation and maintenance of the summit level canal and two locks would cost more and would include the extra expense of pumping water. Moreover, project benefits would be substantially cut by this plan because of the increased lockage time imposed upon barge traffic. These poor economic factors doomed this design alternative.[78]

Engineers remained confident they could reduce the needed amount of excavation, despite the various technical challenges. They concurred, however, "that the most important single factor in the excavation design was the dewatering procedure. Excavation of the divide cut will be in the dry, not by dredging."[79] When, in 1974, Nashville District engineers proposed using explosives in the divide cut, however, South Atlantic Division executives worried about the legal ramifications. The 1971 environmental impact statement did not address the effects of blasting. To be safe, the division staff reasoned, the Corps should file a supplemental statement assessing the environmental impact of explosions in the divide cut. Engineering Division Chief James D. Wall warned that without such a supplement "further court action could occur should blasting be undertaken." Nevertheless, Wall believed that filing a supplemental statement "would, in all probability, create considerable confusion at this time." He therefore instructed the district engineers to proceed with design and construction without preparing a supplement "under the assumption that any normal construction procedure is covered by the scope of the EIS."[80] A year later, however, division officials decided the agency should forgo the use of explosives in excavating the divide cut.[81]

Stabilization of the cut slopes became one of the major points of contention between engineers at the Nashville and Mobile districts. Nashville engineers wanted to protect the slopes with riprap (loosely placed stone) along the entire course of the divide cut. This represented a more conservative engineering approach. Nashville's designers argued that it was simpler and more efficient to place the rock "in the dry," than to come in later to repair sections and be forced to lay the rock from floating platforms. Mobile District, however, had adopted a wait-and-see approach along the river and canal sections—leaving cut slopes unprotected, and later placing riprap on those sections that had eroded—and favored the same for the divide section. Although the point was debated, Nashville District was not overruled, and the divide cut became the only section of the waterway to be lined fully with riprap.[82]

Disposing of the huge amounts of earth from the divide cut posed one of the most environmentally controversial aspects of the project, and placed the Corps under extreme pressure to produce a viable and environmentally acceptable solution. As early as 1967, James J. Tozzi, a senior staff analyst with the office of the Secretary of the Army, addressed the "scenic consideration" of dumping the excavated material from the divide cut. In criticizing the General Design Memorandum, Tozzi stated that the

Corps's plan to pile the excavated earth adjacent to the waterway would "decrease the attractiveness of the area by creating either large mounds of debris or a large expanse of land without vegetation." He said that the agency had "deposited the excavation materials of the Cross Florida Barge Canal near its banks" and was "now in the process of correcting the situation by spending additional funds to improve the attractiveness of the area." He urged a close evaluation of the divide cut plans.[83]

Despite Tozzi's recommendation, the 1971 environmental impact statement paid scant attention to the problem, stating only that the excavated material would be "wasted adjacent to the cuts." The report added, however, that final designs for the disposal areas were still under consideration, and the agency anticipated making significant changes for environmental reasons. The disposal alternatives included "use of some of the material as fill for roads and highways, rehabilitation of blighted areas, such as abandoned gravel pits and eroded gullies, filling and site preparation of industrial and commercial sites," and even the dumping of excavated material in Pickwick Lake to cover mercury-laden sediments.[84]

The Environmental Defense Fund, in its lawsuit to halt construction of the waterway, argued that the Corps should not be allowed to proceed until it disclosed its final plans for the earth dug from the divide cut. Corps officials, on the other hand, told the court that to provide "the best environmental protection and to achieve the most environmental enhancement, placement of spoil must be dealt with on an area by area basis . . ."[85] It was not until July 1974, after the conclusion of the court case, that the agency announced its plans for disposing of the excavated material: the estimated 160 million cubic yards of earth would be dumped in 51 adjacent valleys. Following the displacement of this material, the Corps promised to "revegetate" the poor soil—poor because it came from deep underground and thus was highly acidic and low in organic content—and eventually give the property to the Mississippi State Park Commission for use as park and game management lands. The Corps would compensate for the loss of wildlife habitat by building ponds within the disposal areas.[86]

In addition to this enormous excavation, the divide section contained the Tenn-Tom's most impressive navigational structure, the Bay Springs

Lock and Dam northeast of Tupelo, Mississippi. The 120-foot-high and 2,750-foot-long rockfill dam created a 7-mile-long reservoir (which covered 6,700 acres and created some 133 miles of shoreline) that remained at the same elevation as Pickwick Lake. The divide cut joined the two bodies of water. The rock foundation at the Bay Springs site allowed engineers to design a high-lift lock, which a soil foundation would not support. At 84 feet, the Bay Springs Lock became the United States' third highest lock east of the Mississippi River.[87]

Bay Springs Lock emptied into the artificial channel of the canal section, which led designers to worry about the downstream surges that would occur each time the lock emptied its 45 million gallons of water. This potential problem was exacerbated by the fact that the Corps wanted to design the lock to fill and empty quickly, so as to lessen the time ships spent in passage, thus maximizing the economic benefits claimed for the waterway. In this respect, Bay Springs Lock represented a textbook case of civil engineering design: how to balance the competing goals of safety, speed, and cost.[88]

Nashville District designers remained uncertain about the downstream turbulence caused by the rapid emptying of the lock chamber. In 1974, they commissioned the Corps's research entity—the Waterways Experiment Station (WES) in Vicksburg, Mississippi—to build a hydraulic model to test the operational effects of the lock discharge system. Over the next two years, WES carried out the model test and found that Nashville District's original design produced dangerous downstream surges. It therefore recommended modifying the design of the lock discharge system and altering the geometry of the lock approach to diffuse the surge and slow the swelling of the water surface.[89]

The Waterways Experiment Station also built and tested a 1:25 scale model of Bay Springs Lock to study the longitudinal floor culvert filling and emptying system proposed by the Nashville District. This system featured tuning-fork shaped culverts in the floor of the lock chamber. Safety, reliability, and speed of operation were primary considerations. Safety issues posed a greater technical challenge in high-lift locks than in low-lift locks, as greater potential existed for dangerous surges within the lock chamber during filling.[90] The Tenn-Tom's other locks used side port filling and emptying systems, which consisted of longitudinal culverts in each lock wall, connected by short ports to the sides of the lock chambers. This configuration allowed water to enter and leave the lock chamber through wall ports arranged symmetrically along the chamber. It took 8.5 minutes, for example, to empty the 36-foot-lift Gainesville Lock. Although efficient for low-lift locks, this system held less appeal for high-lift locks like Bay Springs, where the 84-foot lift would take about 30 minutes to fill using this system.[91]

The speed of filling and emptying Bay Springs Lock represented a significant economic issue because navigation savings represented the lion's share of benefits claimed for the Tenn-Tom, and those savings were measured in part by the annual tonnage passing through the system. Several factors already restricted the maximum potential tonnage, including tow sizes limited to eight barges and the numerous curves in the watercourse. Reducing the lockage time was important because the more traffic that could be passed through the system (at least in theory), the more benefits the Corps could claim for the project. The quest for the highest possible benefit/cost ratio led designers to streamline whatever traffic bottlenecks they could, and the Bay Springs Lock became an obvious target. In 1976, Corps economists calculated that a four-minute increase in the lockage time would reduce by over $4.5 million the annual navigation benefits (assuming maximum traffic on the waterway).[92] Bay Springs, therefore, became the limiting factor in the Tenn-Tom's traffic capacity, and agency engineers found great economic incentive to design the fastest hydraulic filling and emptying system possible within safety limits.

Balancing speed and safety, however, was no simple matter, especially when it came to designing a hydraulic filling and emptying system for a lock with an 84-foot lift, the equivalent of a seven-story building. The side port filling and emptying system, the Corps's standard design for its low-lift locks, would create dangerous turbulence within the Bay Springs Lock chamber if the filling time was cut below 30 minutes. Nashville District engineers strongly favored the bottom longitudinal filling and emptying system, even though the district had yet to build one. The design concept was not new to the Corps, however, and Nashville designers patterned the Bay Springs Lock hydraulic system after the filling and emptying system

The Bay Springs Lock, the largest of the Tenn-Tom's ten locks. With a lift of 84 feet, this structure would release 45 million gallons of water with each downstream lockage. *(Courtesy of U.S. Army Corps of Engineers, Mobile District.)*

on the Bankhead Lock on Alabama's Black Warrior River. The Waterways Experiment Station conducted model tests and found the system the most favorable yet developed "from a technical standpoint." Prototypes for the other principal design alternative, the bottom lateral system, indicated an additional 4 minutes in filling and emptying times over the longitudinal system, which for Bay Springs boasted a filling time of 8.5 minutes and an emptying time of 12 minutes. (The emptying time was kept slower to reduce downstream surges.)[93]

The Bay Springs filling and emptying system, which represented the state-of-the-art in hydraulic design, could be operated with safety so quickly because the flow of water was evenly distributed throughout the lock chamber. Two 14-foot by 14-foot culverts channeled water to and from the lock chamber. The huge valves, two for filling and two for emptying, weighed 49,700 pounds each. Even with only one valve working, the flow of water entering or leaving the chamber would be evenly distributed.

Environmental considerations, most notably water quality, further challenged the designers of Bay Springs Lock and Dam. In the original design, most of the water released downstream during lockages would have been withdrawn from the lower layers of Bay Springs Lake. Through physical and mathematical modeling of the system, however, engineers determined that this water would have been significantly cooler than the natural thermal regime of the Tombigbee's tributaries in the spring and low in dissolved oxygen content in the summer and early fall. These conditions would have a long-term detrimental effect on the river's aquatic organisms. Corps designers solved this problem by placing a submerged weir in the lock approach, which allowed lock operators to withdraw water from different layers of the lake, depending on water conditions.[94]

From Drawing Board to Building Site

As the world's largest engineering organization, the Corps has long prided itself as a leader in the field of water resources development. When it came to designing navigation systems, the agency considered itself second to none. Those skills were put to the test with the Tennessee-Tombigbee Waterway. With society increasingly questioning the value of environmentally destructive projects like the Tenn-Tom, the Corps's engineers worked to address those concerns by redesigning elements of the waterway to reduce their ecological impact. To be fair, the design alterations did make a difference: relocating Columbus Lock and Dam saved Plymouth Bluff from inundation; installing reaeration ramps at the spillways improved water quality; realigning disposal sites pardoned wetlands otherwise sentenced to be destroyed.

Despite these design changes, however, building the Tenn-Tom necessarily resulted in profound environmental damage, and this unavoidable impact continued to fuel the opponents' outrage. To add insult to injury, construction proved vastly more expensive than originally estimated. As the next chapter indicates, the technical challenge of building the Corps's largest-ever water project was exacerbated by the legal and political attempts to kill the waterway on environmental and economic grounds.

CHAPTER III **Construction**

THE TENNESSEE-TOMBIGBEE WATERWAY followed the typical arrangement for Corps civil works projects: agency personnel planned and designed nearly every aspect of the waterway, while private contractors actually built the project under the supervision and inspection of the Corps's district offices. In the case of the Tenn-Tom, the Mobile and Nashville districts divided the construction into scores of individual segments. The Corps advertised each segment by providing detailed descriptions of the work to be done, the materials and equipment to be supplied, the schedule to be met, the technical and financial requirements of the contractors, and the deadline for receipt of bids. Once a district awarded a contract, the field offices of its construction division administered the contract. At the peak of construction, 75 prime contractors, 1,200 subcontractors, and their nearly 3,000 workers toiled to complete the project. The complexity of the organization necessitated a clear chain of command. If problems and significant changes in project design could not be resolved at the field offices, they were elevated within the agency as far as necessary—that was, to the District Engineer, the Division Engineer, and, if need be, to Corps Headquarters.[1]

Even under the best of circumstances, management of a large public works project involves juggling people, schedules, equipment, and tasks. Scores of widely varying activities must be closely monitored and coordinated to keep a project on schedule and within budget. Tasks are intricately choreographed, like elements of a dance. For example, if the canal was to run under an existing railroad line, construction crews could not begin to dig until a bridge had been built and the railroad re-routed over it. Any delay in erecting a single bridge could create a bottleneck for the entire project. The same domino effect operated in numerous other discrete—and seemingly inconsequential—components of the waterway. Long before construction started in a certain area, field surveys, soil sampling, and model studies had to be completed so that engineers could refine their designs. From the finished designs, plans and specifications had to be completed (and those documents reviewed and approved by several offices within the Corps bureaucracy) prior to the project being put out to bid. Real estate transactions had to be completed before construction could start. Often, several construction contracts were carried out simultaneously, from clearing and grubbing to dredging, from the relocation of highway bridges and pipelines to the building of locks and dams.

Such overlapping contracts are not unusual on large water projects, but this was no ordinary waterway. It consisted of ten locks and dams, *each* the equivalent of a major Corps project. Because the ten structures, as well as the massive channel excavations, were started at different times, yet concurrently pursued, the management challenge transcended nearly all other Corps civil works endeavors. The fact that the 234-mile-long waterway cut across state lines and across Corps of Engineers district boundaries magnified the difficulty all the more by increasing the number of interested and/or responsible parties.[2]

Construction Sequencing

In both the design and construction of the waterway, Corps planners had decided to work consecutively from south to north—beginning in Demopolis, Alabama, with the river section, then on to the canal section, and finally to the divide section. This represented the traditional sequencing of navigation projects. For the engineers and contractors, this approach offered an orderly and rational progression. By avoiding undue logistical and organizational complications, it made the project easier to manage. From the viewpoint of programming, which was linked to the seemingly whimsical tides of Congressional appropriations, it was also simpler. Moreover, such an approach allowed the Corps to open sections of the project gradually to commercial navigation, which made at least part of the project functional should construction be interrupted or stopped for any reason.

The Corps followed this strategy until 1973 when congressional proponents began advocating an alternate approach based upon political considerations. When Major General Daniel A. Raymond, the South Atlantic Division Engineer, presented the annual Tenn-Tom budget request to the House Appropriations Subcommittee on Public Works in March, Representative Jamie L. Whitten asked him about the possibility of compressing the construction schedule for the waterway. Whitten specifically raised the issue of starting and completing the divide section earlier than originally planned. Because the divide cut was the most time consuming element of the Tenn-Tom, Whitten thought the Corps should abandon its original scheme and start working on that section simultaneously with its work on the southern end. Coincidentally, the divide cut sat in Whitten's congressional district in northeastern Mississippi. Persuading the Corps to start construction of the divide cut early would bring the benefits of federal expenditures all that much sooner to his constituents. But Whitten's motivation encompassed more than immediacy of constituent politics. He knew that the faster the project was built, the harder it would be to stop. Raymond resisted the congressman's suggestion, holding fast to the traditional engineering approach of progressing sequentially, and this led to a heated exchange between the two men; but finally Raymond succumbed. As a result, Whitten made the rare proposal to add funds beyond the publicly announced Corps capability (that is, the maximum amount of money the agency itself estimated it could spend on the project during the fiscal year) submitted in the President's budget.[3]

Raymond provided Corps headquarters with a detailed background report on the implications of Whitten's request, which called for a formal reply. Raymond thought the waterway could be completed two years ahead of schedule (then slated for 1983) if the Corps accepted certain "concepts and risks." Those concepts and risks meant developing the tightest possible schedule; relying on full and regular annual appropriations; awarding overlapping contracts; and controlling cost increases.[4]

Brigadier General James L. Kelly, Deputy Director of Civil Works, provided the Public Works Subcommittee with the Corps's response in April. "We have reanalyzed our planning, construction and funding schedules for the project," he reported, and claimed that "project completion could be physically accomplished by 30 June 1981," a savings of nearly two years. This accelerated schedule necessitated an earlier construction start on the divide cut, he said, and the Corps would have to assume "the availability of the required funds without fiscal constraint in future years." Although Kelly testified that increased funding would be needed during the earlier years of construction, he claimed the accelerated schedule would not increase the overall cost of the project (estimated in March 1973 to be $465 million).[5] Satisfied with this statement, the subcommittee added $3 million to the agency's fiscal year 1974 budget to start construction on the divide cut.[6]

Representative Tom Bevill of Alabama, Senator John Stennis of Mississippi, and other congressional allies joined Whitten in urging the Corps to hasten construction and to bring several sections of the waterway under simultaneous construction.[7] Agency leaders sympathized with this congressional directive and did their best to accommodate the revised schedule. Borrowing from its emergency military construction experience, the Corps incorporated a fast-track approach to building the Tenn-Tom—a method characterized by concurrent design and construction.[8] Speeding up the Tenn-Tom meant not only accelerating design and construction, but also quickening the pace of real estate acquisition, highway and railroad relocations, and cultural resources surveys. This change placed rigorous demands on the agency's programmers and stretched the Corps's standard project administration scheme to its limits.[9]

Construction Management

Knowing that the accelerated schedule demanded the best possible construction management the Corps could muster, agency leaders made the Tenn-Tom a top priority. This emphasis resulted in part from the close attention paid to the waterway by members of Congress, the Office of Management and Budget, other federal agencies, state agencies, the federal courts, the railroads, environmental organizations, the news media, and the general public. The Corps maintained its traditional decentralized management structure throughout construction—district, division, headquarters—although it faced the unusual circumstance of having two districts in two divisions share responsibilities.[10]

Because Mobile District held overall responsibility for the Tenn-Tom, programmers at Mobile served as the overall managers. Vernon S. Holmes headed the program office and John Jeffrey Tidmore assisted him directly with regard to the Tenn-Tom. Their responsibility as programmers included compiling budget data for presentation to Congress each year; monitoring expenditures and updating the estimated construction costs; developing the construction schedule; and allocating the funds to the various offices. In dealing with the Tenn-Tom, they had to know what was going on at every stage of the project, from surveying and testing to design work, from real estate acquisition to contract advertising and award, from construction to operations and maintenance. Few individuals knew more about the status and details of the waterway than Mobile District's programmers.[11]

In an effort to improve its oversight of the waterway, the South Atlantic Division instituted a series of quarterly meetings. Major General Carroll N. LeTellier, the South Atlantic Division Engineer from 1973 to 1976, created the position of Special Assistant to the Division Engineer in charge of coordinating all the efforts on the Tenn-Tom—including the quarterly meetings—and with direct access to the Division Engineer (thus bypassing the normal chain of command). LeTellier selected Robert L. Crisp, Jr., to fill this new position, which later carried the title of Tennessee-Tombigbee Coordinator. Crisp had served as the division's chief of the Geology, Soils and Materials Branch (later renamed the Geotechnical Branch) and was well known within the division for his blunt, hard-driving personality and his keen ability to complete difficult tasks and to motivate people, either through inspiration or intimidation. The quarterly meetings, which were not standard Corps procedure, offered all parties involved a new management tool. The South Atlantic Division used the meetings to avoid duplication of effort in the field and to make sure deadlines were met and budgets not exceeded. Although the Division Engineer or his assistant for civil works presided over the meetings, Crisp actually organized them and made sure any assignments that came from them were completed. The chiefs of the planning, engineering, real estate, construction, and programming offices from both the division and the Mobile and Nashville districts typically attended. The meetings informed members of the various Corps offices what was required of them, and they helped managers of the project to anticipate potential problems, while enhancing their overall perspective. The districts' programmers benefitted by being able to anticipate more accurately the funding needs of the various aspects of the waterway.[12]

Crisp organized the meetings so that each responsible officer within the Nashville and Mobile districts made a formal presentation addressing his office's work over the previous, present, and upcoming quarters and comparing schedules for the same periods. Unmet deadlines or budgets required explanations. This hot-seat approach proved extremely effective in motivating the officers because they spoke before the Division Engineer or his top assistant, the two District Engineers, and the heads of the various departments. It used embarrassment (and the threat of embarrassment) as a management tool. Crisp's role included doing most of the dirty work, such as asking the tough questions in front of the group.[13] Vernon Holmes recalled the process and the corresponding preparation it demanded as "a pain in the ass," although extremely effective. The meetings emphasized to employees at all levels that the Tenn-Tom held top priority within the agency.[14] As an added benefit, the Division Engineer usually attended the quarterly meeting that preceded his appearance before the congressional appropriations committees, so that he might receive an effective briefing on the project's status.[15]

Divide Cut Construction

Because the divide cut represented the biggest and therefore the most time-consuming element of the waterway, Corps officials explored ways to expedite its construction. One relatively simple solution, it seemed, would be to start digging sooner in the south end of the divide cut. Robert

Crisp observed that awarding contracts early in this section would offer several benefits, "primarily in that this more shallow excavation area would provide an excellent full scale test for development of construction procedures, know-how, erosion control methods, and shift some construction funds from the years of maximum funding to years of lower funding requirements."[16] And so, in April 1976, the Corps announced its intention to build a test trench, which, it explained, would enable it "to evaluate the effectiveness of dewatering methods, proposed soil stabilization methods, and possible unforeseen problems associated with construction of the waterway." The test trench was to consist of two sections, which together extended 6 miles. Almost 3 million cubic yards of earth would be excavated. Some of this material was to be used in building a permanent waterway side-slope on the west bank, while the excess would be placed in a disposal area and treated. The Corps hoped to evaluate the effectiveness of conventional excavation equipment, most notably draglines and scrapers. Deep wells and trenching were to be examined for their efficiency as a dewatering technique. Following construction, the agency would test for erosion (and the possible need to lay down a protective layer of stone, or riprap) by diverting Mackeys Creek through the completed trench.[17]

In a later attempt to accelerate construction, Corps officials grouped the last three major contracts on the divide cut into a single, jumbo contract. The combined contract consisted of digging over 11 miles of channel and building 174 drainage structures. It included cutting through the geographical divide separating the Tennessee and Tombigbee rivers, thus making it the deepest and widest excavation site on the entire waterway (up to 175 feet deep and 1,400 feet wide).[18] To tackle this enormous project, three of the nation's largest engineering/construction firms merged to form Tenn-Tom Constructors. This joint venture organization consisted of the Morrison-Knudsen Corporation of Boise, Idaho (which held a 55 percent interest in the firm and overall sponsorship), Brown and Root Inc. of Houston, Texas (35 percent), and Martin K. Eby Construction Company of Wichita, Kansas (10 percent). The Corps awarded the contract to Tenn-Tom Constructors in March 1979 and asked that construction begin the following month. The $271 million contract represented the most expensive single civil works contract ever awarded by the Corps for a project in the United States.[19]

Joint ventures on large public works were not new to either the construction industry or Morrison-Knudsen. Since Harry Morrison and Morris Knudsen founded the heavy-construction company in 1912, Morrison-Knudsen had built over 150 dams, in addition to numerous other projects such as bridges, highways, irrigation canals, airfields, hospitals, breweries, and military installations. In 1931, Morrison-Knudsen joined the joint venture firm Six Companies, Inc., which built the record-setting Hoover Dam, and it also worked jointly on other major projects, including the Grand Coulee and Bonneville dams on the Columbia River and the San Francisco-Oakland Bay Bridge.[20]

Tenn-Tom Constructors purchased all new equipment specifically for the Tenn-Tom contract, which allowed it to tailor the equipment to the job rather than fitting the job to an existing fleet of equipment. The job itself entailed digging an 11.3-mile section of the divide cut, in the process excavating over 95 million cubic yards of earth—roughly half the total moved to build the entire Panama Canal. The types of materials varied from alluvial silts (about 34 million cubic yards), to the clay and shale layers of the Eutaw formation (about 58 million cubic yards), to the laminated fine silts and clay of the McShan formation (about 3 million cubic yards), to the gravel, cobbles, and clay lenses of the Gordo formation (about 1 million cubic yards). The different properties of these materials necessitated different approaches for their excavation. The top 23 million cubic yards of alluvial silts formed a heavy, mud-like soil some 20 feet deep. Because the waterlogged material could not support heavy earthmoving equipment, Tenn-Tom Constructors dug a pit through this material, then removed the rest by carving the vertical face with two large 15-cubic yard hydraulic shovels. The material was hauled away by a 64-unit fleet of huge 50-ton, all-wheel-drive dump trucks. The remaining 11 million cubic yards of alluvium were loaded on dump trucks by six hydraulic backhoes, which ranged in capacity from 2.5 to 8 cubic yards.[21]

Thirty-three million cubic yards of the Eutaw formation (a sandy material ranging in thickness from 40 to 100 feet and far easier to handle than the saturated alluvial silts) were excavated by two horizontal and one

A convoy of 50-ton, all-wheel-drive dump trucks worked around the clock removing earth from the Tenn-Tom divide cut. Over 160 million cubic yards of dirt were excavated in the divide section of the waterway alone, nearly equalling the amount dug by the builders of the Panama Canal. *(Courtesy of Morrison Knudsen Corporation, Boise Idaho.)*

vertical Holland loaders. These machines were large scrapers equipped with conveyor belts that cast the material off to one side. The Holland loaders were simultaneously pushed and pulled by a pair of crawler tractors and were capable of filling a 50-ton dump truck in less than a minute. The Holland loader produced a continuous flow of material, which fed the steady procession of dump trucks. The other 25 million cubic yards of Eutaw formation were removed by twenty-two 24-cubic-yard, twin-engine scrapers. Hydraulic backhoes and 6-cubic-yard draglines extracted the 3 million cubic yards of McShan formation. Forty-eight bulldozers and a variety of pickup trucks, water trucks, cranes, graders, and other support vehicles and smaller pieces of equipment (such as air compressors, portable power generators, water pumps, and gas welders) rounded off the equipment fleet, which totaled 470 pieces—the largest inventory ever assembled by the mammoth Morrison-Knudsen Corporation.[22]

Dewatering the construction site and excavating the millions of cubic yards of earth posed continual challenges, as did the disposal of that great

The most dramatic element of the Tenn-Tom was the divide cut, which at its extreme was 175 feet deep and 1,400 feet wide. The excavated material filled 51 adjacent valleys. *(Courtesy of U.S. Army Corps of Engineers, Mobile District.)*

quantity of material. Because the divide cut comprised the most hilly terrain of the entire waterway, Corps planners had sited the disposal areas in the valleys adjacent to the project. The Tenn-Tom Constructors' contract area contained 16 disposal sites, the largest of which held 33 million cubic yards. To reach these dumping grounds, the contractors built and maintained a first-class system of haul roads. As the dump trucks and scrapers began filling the disposal areas, they found that the waterlogged alluvial silts could not support the weight of this heavy equipment if the silt was piled too deeply. Tenn-Tom Constructors solved the problem by alternating layers of alluvial material with the sandy soils of the Eutaw formation. Wherever possible, the contractors mixed the two materials together before dumping it. After filling the disposal sites, the contractors contoured and reseeded the area with native plants.[23]

As a general rule, the faster a contractor finishes a job, the more money he or she will make. Equipment not committed for lengthy periods can be moved to other jobs, as can key personnel. For the Corps's part, the controversy raging over the Tenn-Tom in Congress made the agency's leadership eager to see the project completed quickly. To accomplish this goal, Tenn-Tom Constructors used its field operations personnel for 19.5 hours per day, Monday through Friday, and 16 hours on Saturday, while the equipment department worked three 8-hour shifts per day, 6 days per week. To accommodate this schedule, an extensive lighting system was installed to illuminate the construction site and disposal areas.[24]

At the height of construction, Tenn-Tom Constructors maintained a work force of 935. Labor often caused the company trouble. When construction began in 1979, several other contractors were working at adjacent sites, and still others were building other federal projects in the region, such as the Tennessee Valley Authority's Yellow Creek nuclear power plant (later abandoned) and its new Pickwick Lock on the Tennessee River, all employing hundreds of local workers. Lee W. Miles, resident manager for Tenn-Tom Constructors, recalled that "the shortage of labor caused the workers to have a very poor attitude towards work, due to the fact that if they were fired, they could go to work the next day." By 1981, following the completion of other Tenn-Tom contracts, the labor "shortage," as Miles called it, eased, and along with it "the employees' attitude improved."[25]

Aside from digging the canal, dumping the excavated material, and building the drainage structures, Tenn-Tom Constructors faced another large task: lining the lower slopes on both sides of this 11.3-mile section of waterway with protective limestone riprap. The slopes averaged 90 feet from top to bottom and required nearly 900,000 tons of rock. The standard method of placing riprap called for extended-arm backhoes and cranes to lower the rock in place. Usually, this procedure lacked uniformity, and the stone surface had to be subsequently leveled, or dressed. Because the contract called for synthetic filter fabric to be laid under the rock surface to reduce subsurface erosion, care had to be taken not to puncture the fabric when placing the stone—a significant problem if the 150-pound rocks were dropped from a height greater than one foot, which was difficult under the standard method of operation.[26]

The "Rock Monster," designed for lining an 11.3-mile stretch of the Tenn-Tom divide cut with nearly 900,000 tons of protective limestone riprap. After it had finished its work on the Tenn-Tom, the $1 million machine was sold for scrap. *(Courtesy of Morrison Knudsen Corporation, Boise Idaho.)*

The scale of this task motivated Morrison-Knudsen to explore novel approaches to the project. At the company's headquarters in Boise, Idaho, engineers Charles Spencer and Samuel Chambers—with input from Pat Monnet, the company's general equipment superintendent for the Tenn-Tom contract—designed a new machine specifically for the placement of riprap. The riprap machine, known as the "Rock Monster," was designed, fabricated, and tested in Boise before being dismantled, shipped to the construction site in northeast Mississippi, and reassembled. The device consisted of a 145-foot-long main truss frame (which looked much like a bridge truss), which would span the width of the slopes. The truss was mounted at both ends on sets of crawler tracks. The machine was self-propelled and was powered by its own electrical generating unit. To assure that the truss ran parallel to the slopes, which varied in angle from 2:1 to 3:1, hydraulic jacks adjusted the height of the uphill end of the unit. The truss supported a tram system, upon which rode a "vibratory" or "feeder" hopper loaded with rock, which it placed on the slopes—the rocks being dropped from a low height and hitting a metal deflector to reduce impact (and possible damage to the filter fabric). A "shuttle" hopper was supplied with rock by front-end loaders at the base of the slope, and transferred the riprap to the feeder hopper. The machine made a swath of seven feet at each pass, and evenly distributed a blanket of 18-inch rock, which significantly reduced the amount of smoothing to be done later. The machine placed an average of 280 tons of rock per hour, more than double the installment rate by the traditional method.[27]

The Rock Monster proved to be a great success, and Spencer, Chambers, and Monnet took out a patent on their design. The machine's suitability for smaller scale projects was questionable, however, and Morrison-Knudsen disassembled the machine and sold it for scrap after it completed the Tenn-Tom contract, despite its $1 million cost in design and construction.

Dewatering was one of the major challenges in constructing the divide cut. The Eutaw-McShan and Gordo aquifers in northeastern Mississippi and west-central Alabama supplied water to most of the area's residents, who drew their water from wells.[28] The Tenn-Tom threatened to change the region's groundwater characteristics, but in 1970 the extent of its impact remained unknown. The initial plans called for dewatering the aquifers with deep wells so that the contractors could work in the dry. In some of the more shallow cuts, dewatering was to be accomplished using drainage trenches instead of deep wells.[29]

The Corps realized from the beginning that digging the divide cut would lower the area's water table and disturb some local water wells.[30] Anticipating this, the Nashville District initiated a domestic well inventory in 1970. By July 1971, technicians had located and measured 2,100 wells in an area extending three miles on either side of the Tenn-Tom's centerline at the northern and southern ends of the divide cut, and six miles on either side in the main portion. Investigators recorded such information as the wells' location, depth, type, and use. The Corps began monitoring groundwater levels and flows in 1972. Piezometers (instruments used to measure water pressure) and observation wells were installed at numerous strategic locations and readings were taken periodically. The agency con-

Aerial view of the partially completed divide cut, which was built by Tenn-Tom Constructors, a special joint venture consisting of the Morrison Knudsen Corporation, Brown and Root Inc., and Martin K. Eby Construction Company *(Courtesy of Morrison Knudsen Corporation, Boise, Idaho.)*

tracted the U.S. Geological Survey to assist in the groundwater monitoring.[31] In May 1976, the Corps received its first formal complaint about lowered water levels in a domestic well.[32] In the following years, farms and houses up to five miles away from the waterway experienced problems with their wells. By December 1977, 29 more complaints were filed, all of them stating that lowered water levels rendered their domestic wells useless. Many residents filed complaints with their county attorney and the Corps about the loss of the use of their private water wells in or near the vicinity of the Tenn-Tom.[33]

Geologists from the Nashville District investigated the complaints. They compared the state of the wells in question with data obtained from the 1970 well inventory, took into consideration patterns of local precipitation, checked data from the Corps's piezometers and observations wells, and weighed other pertinent data. The Real Estate Division then informed the complainant of the findings. As Nashville District geologist Marvin Simmons told the *Northeast Mississippi Daily Journal:* "A water well is not a permanent installation—you can't expect it to last forever. We don't want bad publicity, but, on the other hand, we can't afford to pay everybody."[34]

Corps officials projected that about 100 domestic wells in the divide cut section could be affected by construction activities, and that about 35 would actually be damaged. They estimated that correction of these damages would cost $230,000, although the agency's attorneys thought U.S. and Mississippi law might not hold the government liable for these damages. The Corps's leadership, however, thought it right—both morally and politically—for the agency to correct any damages resulting from construction of the Tenn-Tom, and they appealed to Representative Jamie Whitten for help.[35] By the time Tenn-Tom opened to traffic in 1985, Corps technicians had investigated 182 complaints relating to domestic water wells. They deemed only 42 of those wells to have lost water due to the effects of the waterway, and the government paid to have those wells drilled deeper.[36]

Environmental Considerations

Responsibility for minimizing the adverse environmental impact of the waterway rested primarily with the Corps, for the agency's engineers designed the project. Environmental concern did not stop with the design of the project, however; construction figured prominently. Because private contractors actually built the waterway, some critics worried that contractors would feel less constrained to follow environmental regulations than would a government agency. To minimize environmental disruption (in terms of erosion, turbidity, and the like), the Corps required all primary contractors to submit, along with their bids, a Contractor's Environmental Protection Plan and Erosion Control Plan. Agency supervisors then monitored the contractor's adherence to the provisions of those plans.[37]

The contractors' environmental protection plans included assurances that each craft superintendent would educate his or her workers on the environmental requirements of the job. They stated that the contractors would not "deface, destroy, or injure" trees or shrubs outside the limits of the construction site, and that if such actions occurred, the contractors would repair the damage. Ground and aerial photographs of the preconstruction site were to be provided the Corps supervising officer as a means of monitoring environmental impacts. Contractors promised not "to disturb, remove, or damage any item, article, or finding that appears to have any apparent historical or archeological value."[38]

To maintain water quality, the Corps prohibited contractors from dumping any sewage effluent, fuels, oils, acids, chemicals, insecticides, herbicides, or other harmful materials into streams. If contractors discarded waste materials in unauthorized areas, they were required to excavate the contaminated ground and replace it with suitable fill in order to prevent groundwater contamination. Contractors were also responsible for conducting training courses for their key personnel to ensure their knowledge of pollution standards, detection of pollution, and general procedures pertaining to environmental protection on the job site.[39]

Corps officials realized they would have to provide the contractors with incentive to maintain environmental quality, as such incentives were not ingrained in the construction industry. As Nashville District biologist Joe Cathey recalled, not a single contractor who worked on the Tenn-Tom held environmental quality as a priority. "Their goal was to make a profit," he said, "to get out of there as quickly as they can," and "to avoid spending money" on things they considered unnecessary. The Corps was forced, therefore, to incorporate specific environmental guidelines into its plans and specifications—general comments on environmental goals simply failed to produce results. Without set criteria, contractors generally did as little as they could to protect the environment. Written requirements, however, allowed Corps officials to monitor the work and apply penalties when the contractors fell short. The penalty imposed by government inspectors usually came in the form of work stoppages, not financial penalties. But construction shutdowns cost the contractors dearly, and the Corps found it an effective means of control.[40]

Despite all the efforts to limit environmental damage, building a waterway remained an inherently destructive as well as constructive act, necessarily involving severe scarring of the land, displacement of wildlife, and noise, air, and water pollution. In 1976, for example, contractors clearing land for an access road to the Columbus Lock and Dam site cleared a 135-

foot-wide swath, exceeding the Corps's plans and specifications for the project by 70 feet. Not only did the contractors decrease the "aesthetic value" of the land by ignoring the specifications, but their overly-wide corridor also damaged part of an old brick kiln associated with the extinct town of Plymouth, Mississippi. "Although nothing can be done to rectify this situation at Columbus Lock and Dam," Mobile District's planning chief Lawrence R. Green wrote the district's heads of construction and engineering, "precautions should be taken to insure that a similar situation does not occur on other projects." Green explained: "In many cases the appearance of the construction access roads influences the public's opinion of the Corps's sincerity toward the goal of minimizing environmental impacts. This factor . . . makes[s] it imperative that the aesthetic appearance of the roads receive special attention."[41]

The Corps remained under constant scrutiny with regard to the environmental impact of the waterway. When the agency gave notice that it intended to fill the reservoir behind the Gainesville Lock and Dam in the spring of 1978, the Assistant Chief of Fisheries for the Alabama Department of Conservation and Natural Resources, C. E. White, was livid. "We are shocked by the filling date of May, 1978," he wrote in a scathing letter to the Mobile District Engineer in December 1977. "No construction agency that had any concern for fishery resources and fishing would fill a reservoir of this magnitude after a date when the majority of the carnivorous species of fish had spawned," he continued. "As the water of the reservoir warms forage species will begin to spawn with a minimum number of carnivorous species to control their numbers. The result will be excessive numbers of fish which will grow at a very slow rate and will provide poor fishing when it should be providing very excellent fishing." He called upon the Corps to fill its reservoir in February instead, "so that carnivorous species can provide adequate young-of-the-year to control the forage fish which will spawn heavily during the summer of 1978."[42]

From Construction Site to Political Landscape

As the largest single civil works project ever undertaken by the Corps of Engineers, the Tenn-Tom represented an enormous construction effort. Yet the contractors who actually built the project relied, with but few exceptions, on traditional construction techniques and traditional heavy equipment. The Corps, too, relied for the most part on its standard operating procedures. Although a huge undertaking, the Tenn-Tom was easily broken down into manageable units—building ten locks simultaneously would have been a herculean task, but taken one at a time they were no more demanding than a typical Corps project. The fast-tracking of the Tenn-Tom begun in the mid-1970s was really much like the work done by the agency in its military construction, and presented its biggest challenge in the areas of coordination and management.

What separated the Tenn-Tom from all the Corps's previous engineering undertakings was the public scrutiny it underwent as a result of changing societal values and a new set of environmental laws, the most important of which was the National Environmental Policy Act. From a strictly technical standpoint, the Tenn-Tom was an impressive, if not an outstanding, project. Its real accomplishment, however, rested in the architecture of its political support and defense, which is the subject of the remaining pages of this book.

CHAPTER IV

The Environmental Debate

IN JANUARY 1969, the Tennessee-Tombigbee Waterway Development Authority (TTWDA) issued a promotional booklet proclaiming the myriad benefits of the Tenn-Tom. Anticipating possible objections, the TTWDA asserted that "many waterway projects are objected to by conservationists in our country on the grounds that they upset the world ecology, that is, the total pattern of relationship of all organisms. However, the Tennessee-Tombigbee Waterway has never raised objection from any conservationist."[1] Events of the next fifteen years belied that assertion, for the Tenn-Tom became one of the most environmentally controversial projects in the United States. Yet the TTWDA's optimism in 1969 was not entirely ill-founded. Despite the nearly continuous debate over the project's economic feasibility, few environmental criticisms had been raised. This changed dramatically during the next few years as the start of construction coincided precisely with the blossoming of the environmental movement.

Aside from its extensive destruction of riverine and wetland ecosystems and its joining of two entirely separate watersheds, the Tenn-Tom attracted substantial criticism from environmentalists because of its unprecedented size. Together with its questionable economic justification, the

Tenn-Tom came to symbolize federal misspending and environmental insensitivity. Opposition to the project presented the Corps of Engineers with serious problems, as the controversy generated national attention through its numerous debates within the Congress, the Executive Branch, the federal courts, and the news media.

Far from being an isolated affair, the Tenn-Tom dispute fitted squarely into the broad-based protest against the Corps's civil works program that had steadily gained momentum since the mid-1960s.[2] Charges of environmental insensitivity damaged the agency's public image and caused serious concern among its leadership. The Corps's approach to water resources management relied heavily on engineering solutions. While this approach was well received earlier in the century, by the 1960s it was increasingly at odds with the nation's growing environmental awareness. Moreover, many people raised questions about the huge government expenditures on traditional pork-barrel projects.

The five-state TTWDA and other waterway proponents defended the Tenn-Tom during the early 1970s by claiming that environmental opposition came almost entirely from outside the South. But such assertions were simply not true. Prior to the actual start of construction in 1972, the most probing and persistent environmental questioning of the Tenn-Tom came from concerned scientists and citizens residing in the Southeast, many of them faculty members and students at state colleges and universities. They were joined by private citizens in Mississippi, Alabama, Tennessee, and Kentucky who voiced their concern as individuals and as members of state and national conservation organizations and grass-roots citizens' action groups. The regional environmental protest was most vocally and effectively represented by the Committee for Leaving the Environment of America Natural, or "CLEAN," as it was most commonly known.[3]

The growing environmental movement of the late 1960s and early 1970s spawned scores of community-based organizations throughout the United States. These citizens' action groups focused primarily on local, grass-roots issues. CLEAN typified these new organizations. In 1969, Denzel E. Ferguson, Glenn H. Clemmer, and other members of Mississippi State University's biology department sought to bring together interested scientists and citizens to discuss and address the region's various environmental problems.[4] They selected the acronym "CLEAN" in February 1970, and officially founded the organization the following month with the expressed purpose of providing "a conduit through which concerned persons may channel their knowledge, abilities, and energies toward the betterment of our natural environment."[5] Membership in the Starkville, Mississippi-based organization was open "to any and all persons" interested in supporting environmental causes throughout the Southeast. Although CLEAN later claimed members from a number of states, faculty members and graduate students at Mississippi State University remained the largest contingent. CLEAN's original charter did not mention the Tennessee-Tombigbee Waterway; yet, within a month of its founding, CLEAN formed a special Tenn-Tom committee. Clemmer, an ichthyologist specializing in the taxonomy of fishes in Southeastern rivers, chaired the new committee. In taking on the Tenn-Tom, CLEAN members adopted a conservative approach, being careful not to appear unalterably opposed to either the waterway or regional economic development. Their widely distributed position paper addressed the ecological and aesthetic qualities of the free-flowing Tombigbee River, and asked only that a "careful" and thorough environmental study of the project's impact be made before construction was allowed to begin.[6]

Congress appropriated the initial construction funds for the Tenn-Tom in 1970, the same year President Richard Nixon signed into law the National Environmental Policy Act (NEPA). As one of the first major water projects to be built entirely under the auspices of NEPA, the Tenn-Tom became an important test case for the newly passed legislation. NEPA made the Tenn-Tom—like other federal projects—vulnerable to scrutiny and legal attack by environmental organizations and private individuals.[7] This resulted from the requirement that agencies prepare environmental impact statements (EIS), the adequacy of which could be challenged by lawsuits. Public review of the EIS gave citizens access to the information used in reaching the agency's decision, while public commentary allowed groups and individuals outside the federal government an opportunity to participate in the decision-making process and to place their concerns on the record. Legal recourse extended this ability further. NEPA, therefore, created a system of internal and external review of environmental considerations, forcing the Corps to consider the environmental effects of the waterway in unprecedented detail.

Tenn-Tom's Environmental Impact

By their very nature, inland navigation projects impose significant physical and biological changes in riverine environments by transforming flowing water systems of rivers into still water systems of impounded lakes. This process eliminates rapids and shoals, creates uniform river depths, slows the flow of water, and increases siltation. Ecologically significant wetlands and bottomland hardwoods are destroyed through flooding, the dumping of dredged and excavated materials, the widening of river curves, and the construction of river bend cutoffs. Subsequent navigation increases turbidity, waterborne pollutants, and bank erosion. Because the existence of large and varied fish and wildlife populations depend upon diversity in a river system, navigational goals often come into direct conflict with fish and wildlife values expressed by such agencies as the U.S. Fish and Wildlife Service.[8]

The Tenn-Tom required over 100,000 acres of forested and agricultural lands. About 40,000 acres would be flooded by the project, while the remaining lands would come under various uses, such as providing sites for the disposal of dredged material or providing buffer zones subject to periodic submersion. A large proportion (nearly 51,000 acres) of the land lost to the waterway was bottomland hardwood forests, which provided some of the richest and most threatened wildlife habitat in the Southeast. Upland hardwoods, mixed pine and hardwood forests, cropland, pasture land, and abandoned farm fields would also be sacrificed, but the destruction of the large continuous tracts of bottomland hardwoods constituted the greatest environmental casualty. The bluffs along the Tombigbee contained western Alabama's and eastern Mississippi's most spectacular and stratigraphically complete exposures of Upper Cretaceous sediments, and geologists worried that many of these outcroppings would be inundated by the waterway. Numerous historic and archaeological sites would also be destroyed. In creating some 40,000 acres of flat water surface, construction of the Tenn-Tom would eliminate the free-flowing nature of

170 miles of tributary streams and 140 miles of the main stem of the Tombigbee River. In so doing, most of the Tombigbee's abundant gravel bars (concentrated in the river bends) would be lost, and with them an important riverine habitat for mollusks. Straightening the river through the construction of "cutoffs" would create several miles of oxbow lakes which, in the absence of periodic dredging, would eventually silt over.[9]

The volume of excavated materials was enormous and its disposal posed a major environmental challenge. Overall, the waterway required digging more earth than was removed during the construction of the Panama Canal, and the dumping of this material threatened to transform the local topography. The Corps estimated displacing about 260 million cubic yards: 140 million cubic yards coming out of the divide section, 50 million cubic yards out of the canal section, and 70 million cubic yards out of the river section. The low organic content of the excavated material made revegetation difficult. Because the soils were essentially turned upside down within the divide section, some oxidation of minerals occurred when the subsoil was exposed to the atmosphere, resulting in the run-off of pollutants into the water system. Erosion was yet another serious issue associated with the massive excavations.[10]

The potential impact of these activities on plant and animal life was severe. The 115 species of fishes supported by the free-flowing stretch of the upper Tombigbee made it one of the richest rivers in North America in terms of native fish fauna.[11] The Fish and Wildlife Service attributed the Tombigbee's remarkable diversity of warm water fishes to its varied and excellent instream habitats, consisting of "deep pools, gravel and sand bars, swiftly flowing waters, undercut banks, and submerged trees."[12] R. Dale Caldwell, an assistant professor of biology at Texas A&M University and a specialist in the fishes of the Upper Tombigbee River, warned the Corps in 1971 that the slack-water conditions created by the Tenn-Tom would endanger at least three species of fish in the Tombigbee River.[13]

Environmentalists especially feared the ecological consequences of mixing the waters of the Tennessee and Tombigbee. Because the joining of two distinct river systems rarely occurs in nature, scientists were uncertain of the ultimate consequences of the Tenn-Tom link. Intermingling the fishes and other organisms of the two rivers might result in the complete loss of related species because of competition. Interbreeding might also take place, producing, in Caldwell's words, "one large, highly variable 'hybrid' species." He warned that "a great deal of research is needed before we begin mixing distinct faunas on a wholesale basis. It will be much better to know what happens under controlled conditions than to produce a Frankenstein-type experiment which is completely uncontrollable."[14]

To make their point about the unpredictable impact of blending the faunas of the two separate river systems, environmentalists cited the example of the Great Lakes, where the commercial fishing of trout and whitefish was destroyed by the introduction and subsequent proliferation of the lamprey, which entered the Great Lakes through the Welland Can-al constructed during the 1930s.[15] On the Tenn-Tom, biologists feared that the Eurasian water milfoil—an aquatic weed known for its rapid growth—might be introduced into the Tombigbee River system from the Tennessee River, where the exotic plant had established itself in several Tennessee Valley Authority lakes, including Pickwick. In addition, they noted the different chemistry of the two rivers and raised the question of what biological changes in the river might be triggered by changes in the chemical composition of the waters once they were mixed.[16]

Joining the two rivers threatened to pollute the Upper Tombigbee with the pesticides and heavy metals contaminating Pickwick Lake on the Tennessee River. The heavy metals in the lake included an abnormally high concentration of mercury, the environmental hazards of which became a national issue in 1970. In response to the high levels of mercury detected in some fish, Tennessee Valley Authority officials had already prohibited fishing at Pickwick Lake. Corps planners even studied the possibility of depositing several million cubic yards of excavated earth from the Tenn-Tom into Pickwick Lake "to cover mercury-laden sediments."[17]

The waterway also endangered the region's groundwater resources. The possible elevation of existing water tables was a potential problem in areas where the river was to be dammed. The opposite problem existed in the divide cut section where the deep excavation would intersect groundwater aquifers and thus disrupt their flow and, by acting as a drain, lower the water table. A subsiding water table posed a special problem for the seven adjoining northeast Mississippi counties that relied on groundwater for their drinking supply.

The possible effect of the project on the quality of the area's surface water raised still another issue. During construction, increased silting and turbidity would degrade water quality. The impoundments created by the waterway's ten locks and dams would trap the silt and nutrients that normally emptied into Mobile Bay, nurturing the bay's fisheries and tidal wetlands, and instead would cause excessive plant growth in the Tenn-Tom's sluggish waters. Disruption of the area's natural drainage patterns would also affect water quality. Pollution from boats using the waterway, as well as from residential, commercial, and industrial development around the project, further threatened water quality.

The Tenn-Tom would unavoidably alter the upper reaches of the Tombigbee River, which remained the last major free-flowing river in the entire Mobile Basin drainage system.[18] This fact was important from aesthetic, recreational, and philosophical viewpoints. Free-flowing rivers, unencumbered by human modifications such as impoundments, diversions, channelization, and riprapping, were becoming increasingly rare in the United States. As a diminishing public resource, their value increased as did the number of river enthusiasts such as canoeists, stream fishermen, hikers, bird watchers, and nature photographers. Throughout the country, preservation of wild and scenic rivers attracted growing sup-port.[19] Although not all of the Tombigbee remained free-flowing—its lower half (some 220 miles) had been impounded and channelized since 1915 as part of the Black Warrior-Tombigbee navigation project, and its upper 53 miles were channelized for flood control purposes in 1940—160 miles remained in their natural state.

No matter how much the Corps did to lessen the adverse environmental impact of the waterway—through fish and wildlife mitigation measures, design changes to improve water quality, realignment of the channel to preserve wetlands, and the like—construction of the locks and dams would inevitably destroy the free-flowing nature of the middle Tombigbee, replacing it with a slack water environment. For many environmental critics, the loss of the river epitomized their basic objection to the Tenn-Tom. This fact did much to shape the nature of the confrontation between the project opponents and proponents, for compromise on the part of the environmentalists meant to them the complete surrender of their cause. The environmentalists' goal of total victory held practical as well as philosophical significance, for without total victory, project opponents risked losing their visibility and thus their ability to attract the resources so vital to their continued existence and effectiveness.[20]

Organized Opposition

The Corps tried taking the offensive by maintaining that the Tenn-Tom would "enhance" the natural environment.[21] Waterway advocates repeatedly stressed that the Tenn-Tom would be the first major federal water project built under the policy guidelines of the National Environmental Policy Act, and thus the Corps would give high priority to environmental considerations.[22] Critics and supporters of the project, however, were not so naive as to overlook the fact that the authorization and planning of the Tenn-Tom long predated the passage of NEPA. Speaking before an environmental forum in April 1971, Colonel Harry A. Griffith, the Mobile District Engineer, addressed this issue by claiming that the waterway "is a transitional project. It is authorized, it is funded, and construction is due to begin. And here we are doing an eleventh hour job of superimposing the environment on this project."[23]

"Environmental enhancement" nevertheless became the banner phrase for the waterway's proponents. They emphasized the environmental "improvements" to be brought about by the project. Senator John Sparkman, for example, told Alabama television audiences that "the proposed Tennessee-Tombigbee link would take some of the traffic off the Mississippi River, strengthen Alabama's economy and improve the ecology of the area, just as waterway improvements under the Tennessee Valley Authority have done."[24] In response to pointed criticisms of the Tenn-Tom, Acting Mobile District Engineer Lieutenant Colonel Paul D. Sontag asserted: "We must consider the total environment and its enhancement for man as well as wildlife. Consequently, we believe that most of the changes that will occur will be beneficial."[25]

NEPA left the Corps of Engineers confronting a new set of rules.[26] Exactly how those rules would be applied was not clear in early 1970.

'Environmental organizations and concerned citizens demanded that the Corps complete its environmental study of the Tenn-Tom before starting construction. A letter-writing campaign that pressed this issue with the Corps, the Congress, and the Nixon Administration showed the spontaneity of early environmental opposition to the waterway.[27] At the University of Kentucky in Lexington, the environmental implications of the Tenn-Tom ignited the passions of John E. Cooper and Robert A. Kuehne. Cooper, the executive vice-president of the National Speleological Society and a graduate student in the university's zoology department, and Kuehne, an associate professor of zoology and a specialist in aquatic biology, wanted to consolidate the efforts of environmental organizations and individuals opposed to the waterway.

Cooper and Kuehne feared that the splintered nature of the opposition diluted its effectiveness. In July 1970, they mailed a *Dear Colleague* letter to over two hundred scientists throughout the Southeast urging them to write immediately to their Senators requesting that Tenn-Tom construction funds be withheld until after the Corps completed its EIS. Telling the scientists that the Corps had recently announced its intention to comply with NEPA and to incorporate environmental values into its benefit-cost calculations for all future public works projects, they asked rhetorically: "Where better to begin the application of environmental planning than with the Tenn-Tom project?"[28] The two biologists argued that government decisions could be reversed if Congress and the administration were "presented with logical reasons to do so," and they cited the Cross-Florida Barge Canal as an example of such change. "This," they claimed, "places a direct *responsibility* on the scientific community to supply information and informed opinion on a variety of significant issues."[29]

Cooper did not cloak his position from the Corps, yet he conceded that he was "willing to accept the environmental concern of the Corps of Engineers at face value." In return, he insisted that the agency face the possibility "that an objective environmental impact study by qualified scientists may find the construction of the Tennessee-Tombigbee Waterway should be precluded by ecological considerations." The Corps's treatment of the waterway construction as "a foregone conclusion," he argued, made it unlikely that the agency would undertake a complete and honest investigation of the project's environmental effects. "My attitude, and that of many of my colleagues," Cooper said, "is that it would be far more ecologically and economically reasonable to opt for an environmental study *before* land acquisition and construction begin, and we are consequently working towards those ends." As far as the environmentalists were concerned, "construction of the Waterway is not a cut-and-dried, accepted fact until an environmental study shows it to be ecologically feasible and establishes guidelines for safeguards."[30]

Cooper and Kuehne called a meeting in August 1970 at Mississippi State University to establish an environmental coalition of scientists and citizens from Alabama, Mississippi, Tennessee, and Kentucky. The organization was to be modeled on the Florida Defenders of the Environment, a group that had successfully opposed the Cross-Florida Barge Canal.[31] According to Cooper and Kuehne's press release, "one of the organization's first aims is to halt funding of the proposed Tennessee-Tombigbee Waterway until such time as an objective environmental impact study . . . has been completed and has shown the Waterway to be ecologically feasible." They added that the new organization intended to work closely with the Environmental Defense Fund (EDF).[32] Understanding the advantages of associating their regional environmental action group with an established and effective national organization, Cooper had informed EDF of their plans to form a specialized, anti-Tenn-Tom group, one that expected a tough battle, "what with the governors of a number of states straining at the barrel of pork."[33]

Nineteen people representing five regional environmental organizations and eleven Southern colleges and universities gathered at Starkville, Mississippi, to discuss possible courses of action.[34] Kuehne, who led the meeting, asserted that "it is obvious that the Corps considers Tenn-Tom construction a foregone conclusion and is only interested in environmental studies which might conceivably relate to modifications of construction plans as work progresses." He called for "a separate study of potential environmental impact, planned and implemented by an organization like Florida Defenders of the Environment, to include biology, hydrology, economics, history, archaeology, and other disciplines."[35]

In summer 1970, with only a handful of the newly-required environmental impact statements completed by all federal agencies, skepticism of the process was widespread among those environmentalists who understood the importance of establishing precedents. Kuehne's mistrust of the Corps and other federal agencies like the Fish and Wildlife Service (which he said were reluctant to "criticize each other's work to any extent") typified these broader concerns and led him to propose the preparation of an independent environmental study financed through voluntary efforts, membership dues, and private fund-raising drives.[36]

The main point of order, however, was to evaluate the wisdom of forming a new Southeast environmental action group dedicated to stopping the Tenn-Tom. Kuehne presented various organizational scenarios: becoming a chapter of a prominent group like the National Audubon Society; merging with the recently established Committee for Leaving the Environment of America Natural; or forming an independent organization of their own. Glenn Clemmer strongly advocated joining CLEAN, which, he said, had already formed a Tenn-Tom study group. Moreover, CLEAN had recently joined with EDF in a lawsuit to halt the U.S. Department of Agriculture from "further Mirex poisoning of the South in the so-called 'fire ant eradication program.'"[37] After debating the pros and cons of the three alternatives, the group voted unanimously to join CLEAN.[38]

This unanimity did not extend to the group's plan of action. John S. Ramsey, an associate professor in Auburn University's Department of Fisheries and Allied Aquacultures, feared that an independent environmental study of the Tenn-Tom might not be completed in time to inform the ultimate decision on the waterway. In yet another *Dear Colleague* letter to scientists in the Southeast, Ramsey stated that "if the federal government agencies are doing what may be the only environmental impact study on the Tennessee-Tombigbee Waterway project, they need to have all the unpublished or published facts we specialists can give them." He urged his colleagues to provide information to both the Corps and the Fish and Wildlife Service.[39]

Uniform support for the waterway among Alabama's and Mississippi's Congressional delegations, combined with the general political climate of the Southeast, prompted environmental critics to follow a conservative strategy. They insisted only that construction be postponed until after the EIS had been approved by the President's Council on Environmental Quality. As Clemmer explained: "We decided that we will at least serve as a watchdog agency, to make certain that whatever studies the Corps of Engineers undertakes will be done thoroughly, fairly and by independent scientists."[40] CLEAN chairman Boyd Gatlin elaborated on this point in a letter to Representative Thomas Abernethy, which stressed to the Mississippi congressman that CLEAN would "insist by any lawful means" that the EIS "be done properly and objectively and that its findings be promptly translated into policy." He emphasized, however, that:

> CLEAN is not opposed to the development of necessary waterways. We are not in favor of leaving every stream in its natural state, if the tampering can be adequately justified. Indeed, if the Tennessee-Tombigbee can be constructed so that its adverse effects can be reduced and then offset by real benefits to man's total environment, you will find no opposition from here.[41]

By September 1970, Corps officials gave their assurance that construction would not start until after the agency had filed its EIS.[42] With that public guarantee in place, project opponents readjusted their tactics. Many of them, like Ramsey, pressed to ensure that the environmental impact study be scientifically sound. This was the approach adopted by CLEAN, although it disappointed Cooper and Kuehne, who had pushed for a more activist role. Cooper complained that CLEAN had "dropped the ball in opting to look over the shoulder of the Corps in doing its environmental study instead of going ahead with a separate study of [its] own." He blamed this approach on CLEAN's lack of organization and "push."[43]

Most environmental organizations opposed to the Tenn-Tom increased their activities within the political arena. The 6,500-member Alabama Conservancy, for example, began a letter-writing campaign in February 1971. The Conservancy urged its members and friends to write their Senators and President Nixon, hoping that Nixon's decision to halt construction of the Cross-Florida Barge Canal would set a precedent for the Tenn-Tom. "In the letter to President Nixon," the readers were advised, "you may wish to praise him for the Cross-Florida Barge Canal decision, and to indicate that the same reasoning he used in that decision as to protection of natural, ecological values as opposed to a total focus on maximizing economic return also pertains to the Tennessee-Tombigbee Waterway."[44]

Under the presidency of Charles S. Prigmore, the Conservancy continued its active opposition to the Tenn-Tom through the spring of 1971. Prigmore and other members barraged Congressional members and government officials with letters detailing the severity of the waterway's environmental impact.[45] Representative Jack Edwards of Alabama responded sternly to the Conservancy's appeal, calling the Tenn-Tom "the first and only waterway project to be built with environmental enhancement a dominant feature in its construction." Edwards charged the Conservancy with "not serving the best interest of Alabama by raising hypothetical questions now," and said he would make an all-out effort in support of the waterway. By way of reminding the Conservancy of the give and take of environmental politics, he added: "I am supporting you on your Bankhead Wilderness and I in turn ask your support of the much needed Tennessee-Tombigbee project."[46] Prigmore acquiesced. In June, he told Edwards that "the Alabama Conservancy has given a great deal of thought to your telegram regarding the Tennessee-Tombigbee Waterway and we have decided not to pursue the issue further at this time." In turn, Prigmore requested Edwards's help "in getting a more acceptable bill in the House to establish a full 11,000-acre wilderness area."[47] The Conservancy's board of directors voted in July to discontinue the organization's direct involvement in the fight against Tenn-Tom. They thought they were already engaged in opposing too many other projects that were nearing completion to commit their limited resources—financial and political—to the Tenn-Tom campaign. Instead, individual members of the Conservancy interested in opposing the waterway agreed to work through CLEAN.[48]

The Corps's Response

In June 1970, the Corps's Mobile District completed its "Technical Studies Work Plan" for the Tenn-Tom environmental impact study.[49] At the agency's South Atlantic Division, Deputy Division Engineer Colonel R. P. Tabb criticized the study concept as too broad and ambitious. Above all else, Tabb emphasized that the agency's leaders did not want the environmental study to delay the start of construction any longer than absolutely necessary. He therefore ordered that "your greatest effort should be spent where we have the greatest chance to make project adjustments to better harmonize with the environment. Study in detail where the rock hits the water but don't try to chase every ripple to the shore and beyond." He also advised Mobile District to "avoid pure research and concentrate on improving the interaction of our engineering solutions with the environment."[50]

The Corps's Chief of Engineers, Lieutenant General Frederick J. Clarke, concurred. He told his field officers that, in light of "the magnitude of the Tennessee-Tombigbee project and its potential for controversy, I believe we should make every effort to expedite the environmental study."[51] At a meeting of Clarke's Environmental Advisory Board (EAB) in September 1970, board member Charles H. W. Foster and Director of Civil Works Major General Francis P. Koisch clashed in a heated debate over the waterway. EAB secretary Colonel Robert R. Werner described it in his summary of the meeting: "Foster questioned the rationale whereby this environmental research project only looked at means of minimizing undesirable impacts and did not consider whether the project was sound enough on environmental grounds to be constructed at all." Werner said Koisch "explained that while we were able to initiate the [environmental] study we were past the point where the Corps could stop the project."

Foster expressed what environmental skeptics had feared—that the army engineers were dead set on building the Tenn-Tom no matter what the environmental effects were.[52] Although the EAB members strongly opposed the waterway itself, they were asked only to comment on the proposed environmental impact study; and even there, the board's role was strictly advisory.[53] Corps officials listened but were unswayed. In November 1970, they publicly announced a three-phase, final environmental study plan.[54]

The first phase involved a five-month evaluation of the overall environmental effect of the waterway, which would culminate in an environmental impact statement for submission to the Council on Environmental Quality (CEQ)[55] by March 1971. The Corps reiterated its promise not to begin construction of the waterway until after this time. Moreover, the agency assured the public that "prior to construction of each of the major project elements, the environmental statement will be updated to present the latest environmental aspects."[56]

The study's second phase called for more detailed analyses during the engineering and design stages of the project "to assure that every opportunity to enhance the environment and to minimize detrimental effects is provided full consideration." These detailed studies, selected on the basis of phase I findings, would require "a sustained large-scale effort" on the part of Corps personnel and contractors.[57]

The third phase was less well-defined, as its scope would be determined during the latter part of the phase II study, which was to be completed in July 1974. Phase III environmental evaluations were to be conducted at a "reduced scale" after the waterway opened. Post-construction studies of water quality, groundwater, biological communities, erosion, and other areas studied during phase II would help Corps staff to compare the actual environmental effects of the waterway with the predicted effects, thereby providing the knowledge necessary to reduce the long-term adverse impacts through project alterations.[58]

Corps officials completed the draft environmental impact statement on 15 January 1971 and circulated it to 18 federal and state agencies, with the request that they file their comments within a month.[59] The Corps submitted the final EIS to the Council on Environmental Quality on 20 April 1971.[60] The document consisted of 32 pages of text, 19 pages of responses to comments made by other agencies, and a 58-page appendix of letters from the agencies. In it, the Corps claimed that it sought "to determine if there were any detrimental effects significant enough to forgo development of the project," but concluded that there were not. Moreover, the agency asserted that "the beneficial environmental impacts overshadowed the unavoidable adverse effects."[61]

CEQ Chairman Russell E. Train reviewed the EIS. He acknowledged that the document dealt only with the first of three proposed phases, with more detailed work to follow. Although he did not reject this approach, Train only tepidly approved the statement, stressing that "the nature and magnitude of the project give great importance to the remaining phases of your study." He asked that the Corps remain in close contact with CEQ throughout the design and construction of the waterway and that the Council "be informed of any environmental problems that arise."[62]

Critics universally found the final environmental impact statement unacceptable, and they altered their strategy accordingly. While previously they had urged that construction be postponed until after the EIS had been filed and accepted, they now objected to CEQ's approval of the EIS and asked that the report be redone before construction of the waterway could be started. The Corps's strategy of implementing a three-phase environmental study therefore backfired, for it left the agency vulnerable to charges that it had not completed a *full* environmental review of the project and that its EIS was inadequate. CLEAN's new chairman, James D. Williams—an ichthyologist and assistant professor of biology at Mississippi State College for Women—spoke for his organization's members when he reported to CEQ's chairman that spring:

> We are agreed that this study is inadequate and amounts to little more than a rewrite of previously published works. It represents very little in the way of original investigation, it is undocumented, and ignores a number of relevant questions. In short, we find almost nothing in it which would provide assurance the Corps is, indeed, concerned about the environment. In fact, the cavalier fashion in which the study disposes of complex problems in conservation is certain evidence that the Corps is only giving lip service to the legislation which established the Council on Environmental Quality.

Williams concluded by asserting: "I am sure you would agree that nothing could be more wasteful or more tragic than to have an impetuous beginning of the Tenn-Tom bring it to the kind of fate suffered by the Cross-Florida Barge Canal."[63] CEQ officials took no direct action, but they assured Williams that the "Council shares your concern that the environmental implications of the Waterway be carefully monitored, and we plan periodically to review the Corps' continued studies."[64]

Environmental critics quickly expanded their search for allies. Williams appealed to Environmental Protection Agency (EPA) Administrator William D. Ruckelshaus to "take immediate action" to ensure that the Corps prepare an adequate environmental impact statement before it was allowed to begin construction. "Long before a power shovel takes its first giant bite out of the countryside," said Williams, "is the proper time to determine the effects of lacerating the earth."[65] Acting Southeast Regional Administrator John R. Thoman told Williams that EPA shared CLEAN's concern with the environmental degradation that would result from construction of the waterway. EPA's views, Thoman said, were expressed in the agency's formal comments on the environmental impact statement. "We felt," he explained, "that the Corps had not adequately considered environmental effects," and that the proposal to examine environmental factors during (not before) the final design phase was unacceptable. "We cannot agree with this proposal," he said, "and feel that the total project should be evaluated before construction begins—not during design and construction. This project is already authorized, however, and this Agency has no authority to prevent its construction."[66]

Environmentalists confronted the Tenn-Tom in the news media, the White House, other federal agencies, and, later, the courts, but the Congress always remained the focal point, for it determined annual appropriations. In May 1971, Williams wrote to the members of the Public Works and Appropriations Committees in both chambers, urging them to look critically upon the Corps's request for additional construction funds. "We feel," Williams said, "that Congress should reevaluate this project and insist that the Corps adhere to both the letter and the spirit of the National Environmental Policy Act."[67] He also implored CLEAN members and interested environmentalists to "immediately WRITE, WIRE, OR TELEPHONE your Washington representatives," asking that the Corps be required to prepared a complete and adequate EIS before construction.[68]

Robert Kuehne testified before the House and Senate Appropriations Subcommittees on Public Works in May. He questioned the Tenn-Tom EIS and again drew parallels with the Cross-Florida Barge Canal.[69] Brent Blackwelder, then an analyst and lobbyist for the environmental organization Friends of the Earth, also testified; he called the Tenn-Tom "one of the most destructive projects proposed by the Corps," and urged Congress to withhold funding pending a full environmental study.[70] Kuehne followed up his testimony by writing the Corps's Chief of Engineers to request that the agency work with academic scientists in the Southeast in preparing a "comprehensive environmental impact survey of the Tenn-Tom."[71]

In addition to disillusioning CLEAN members and other regional Tenn-Tom opponents, CEQ's acceptance of the Corps's EIS had a sobering effect on the national environmental organizations. The action heightened the groups' interest in the huge waterway because it now—after decades of debate—threatened to become a reality. William M. Partington, director of the Environmental Information Center of the Florida Conservation Foundation, turned from the successful bid to stop the Cross-Florida Barge Canal to building an environmental coalition against the Tenn-Tom.[72] As he explained in a *Field and Stream* interview, leaders of Trout Unlimited, the National Audubon Society, and the Wilderness Society were all interested in "the upcoming citizen efforts on the Tenn-Tom."[73] The League of Conservation Voters also attempted to generate and organize grass-roots opposition to the project during the summer of 1971. In a letter to the organization's membership, League Chairman Marion Edey detailed the environmental problems associated with the waterway. "If you wish to help oppose this project," he said, "please write to President Nixon at the White House and ask him to abandon his support for the Tennessee-Tombigbee. Let him know how much you appreciated his opposition to the Cross-Florida Barge Canal, and ask him please not to break the precedent he set in Florida by supporting an equally destructive canal in Mississippi and Alabama."[74]

President Nixon Responds

To counter the Tenn-Tom critics, waterway advocates worked equally hard to nurture and promote the Nixon Administration's support, although they often lobbied from the privileged position of Washington insiders. The appropriation of construction funds in 1970 had stirred an immediate interest in launching the project in grand fashion, and waterway supporters began urging the President to attend a symbolic ground breaking ceremony. They understood, as Glover Wilkins claimed, that if Nixon attended the ceremony then "it will be *'his project'* and he is the one man who can pull it over the hill with no questions asked."[75] The Governors and Congressional delegations of Alabama, Mississippi, Tennessee, and Kentucky organized an effort to assure Nixon's participation in the ground breaking ceremony. Jack Edwards, a Republican Representative from Alabama, extended the initial invitation in June 1970 when he called the White House to request the President's appearance "anytime next spring." Presidential Assistant William E. Timmons advised Edwards that a decision could not be made until after the election.[76]

Kentucky Governor and TTWDA Chairman Louie B. Nunn, a Republican, assumed the leadership role in lobbying the President. In October 1970, he wrote Nixon estimating between fifty thousand to a hundred thousand people would attend the event, and offering to schedule the ceremony around the President's planned trip to dedicate the Arkansas River project in early June 1971. In a hand-written postscript, Nunn added: "This could be a 'real big thing' in a *key area* next year. Not Southern Strategy but *Southern Progress* good for 'the North.' Big regional opportunity."[77] When informed that the President could not schedule events that far in advance, Nunn reissued the invitation in December, stating that the ground breaking could be arranged at the President's convenience anytime in April or May. "I assure you this is a project of top priority with the people of the entire Southeast," he wrote, "and you will receive a tumultuous and warm reception."[78] In addition to appealing directly to the President, Nunn approached Nixon Special Assistant H. R. Haldeman. Nunn explained to Haldeman that President Johnson "used the project for political advantage" when he approved advance planning funds, and that Nixon, too, could generate substantial political support through his personal appearance at the ground breaking ceremony.[79]

The President's advisors differed on the issue. Timmons led those who encouraged Nixon's participation, based upon the political capital it would win him. Speaking at the ground breaking, attending the various publicized events, and holding a major press conference thereafter offered Nixon the opportunity to test the political waters of Alabama. Presidential advisor John C. Whitaker spoke for the staff members who believed the political risks of backing the Tenn-Tom outweighed the political benefits. The effect of Nixon's endorsement on the 1972 election was the central preoccupation. Nixon needed the support of the South, yet attending the ground breaking would place him on the home turf of his political rival,

George Wallace. Timmons expressed this concern to the Congressional promoters of the waterway. By late February 1971, he reported that Edwards "has assurances from Governor Wallace that George will not in any way use the occasion to embarrass the President. On the contrary, I'm told Wallace wants the President to receive full credit for the historic event."[80]

In March, Whitaker emphasized that the President might be subjected to considerable "environmental flak" in view of his earlier curtailment of the Cross-Florida Barge Canal.[81] Nixon decided later that month to forgo the ground breaking ceremony, citing his busy schedule as the reason he could not attend.[82] This decision disappointed leaders of the TTWDA who began working with Governor Wallace on alternate plans.[83] Nevertheless, project proponents and opponents continued their lobbying for and against the President's participation.

Council on Environmental Quality Chairman Russell E. Train recommended to Whitaker that Nixon oppose the waterway. Train reported that CEQ had only received a draft environmental impact statement for the Tenn-Tom which did not include comments from the other federal agencies. "It could prove an embarrassment and poor precedent to give endorsement," he advised, adding that the draft EIS "is undoubtedly going to be the subject of a suit against the Corps (and prove very embarrassing to the President if he's visible on this project)." Train said that the Corps failed to assess what will be done with the vast amounts of excavated material, as well as the environmental effects of that disposal. He was also concerned with the project's impact on groundwater and water quality, the destruction of paleontological and archaeological sites, the effects of mixing two river systems, and the loss of wildlife habitat. He added that "alternatives to the project are given little attention" and that "the economics of the project are extremely shaky." He concluded:

> Finally, the President has received broad support across the country for his decision halting further construction of the Cross-Florida Barge Canal, most of which support he could quickly lose by any public display on behalf of the Tennessee-Tombigbee. I do not think it is worth his jeopardizing this broad national support on an issue of only regional significance. There must be other events in the South he could attend without risking all he has gained with the Cross-Florida decision.[84]

Whitaker distributed Train's memorandum to Nixon's top domestic advisors in April to buttress his own recommendation that the President skip the ceremony.[85] As he told John D. Ehrlichman, the "Tennessee Tombigbee is a 'no-no' in my opinion."[86] The debate continued among the White House staff. In May, Nixon asked Train to report on the Corps's final environmental impact statement. "In our view," Train responded, "the Corps has taken the steps required under the [National Environmental Policy] Act before commencing the project and has shown an awareness of the need for continued environmental studies to avoid many of the possible adverse environmental impacts noted in the comments it has received from other agencies." Despite this compliance, however, he warned: "It may be anticipated that this will be a controversial project and that some conservation groups may join EPA in recommending that the project be reevaluated because of anticipated adverse effects on water quality, natural and scenic values, and a marginal benefit/cost ratio..."[87]

Wesley K. Sasaki, Assistant Chief of the Natural Resources Programs Division within the Office of Management and Budget, also advised against Nixon attending the Tenn-Tom ground breaking ceremony. In an April 1971 memorandum to senior OMB officials, Sasaki advised that "caution is in order." Even if the environmental impact statement deemed the project free of major harmful effects, unanticipated problems could always arise later. "No major controversy has yet arisen on the environmental aspects of the project," he said, "but conservation groups have historically been late in raising the issue on projects—often waiting until a project is under construction." He added that if Nixon "commits too strongly to the project now, the President could possibly be caught in the kind of environment vs development cross-fire that caused his cancellation of the Cross Florida Barge Canal."[88]

As Train explained, the filing of the final EIS placed the Corps in procedural compliance with the National Environmental Policy Act. The President still risked losing the support of the environmental community, but no legal constraints now prevented construction. Whitaker speculated: "There will be considerable flak and there may be a suit to enjoin construction—in fact, I predict the President's trip will attract the attention to ensure there is a suit." Nevertheless, Whitaker conceded that "the President is legally okay and, on balance, considering the pluses in Tennessee, Mississippi and Alabama, I'd go along with it."[89]

Project opponents continued their campaign against Nixon's support of the waterway. Joe Browder of Friends of the Earth wrote Whitaker that the recently filed environmental impact statement was "outrageous." "It may all boil down to simply better lobbying by the barge people," he observed, "but I can't believe that an honest economic and political analysis of this question would show that Tenn-Tom, and projects like it, are worth the political, environmental, or dollar cost. I hope you all don't get stuck having to push this one."[90] Fourteen members of CLEAN sent a telegram to the President in May urging him not to attend the ground breaking. "As professional ecologists and scientists we do not believe sufficient evidence has been presented to assure the public this project can be undertaken without the risk of doing serious damage to the environment," they wrote. "The exemplary position you took in regard to the Cross-Florida Barge Canal is firm evidence you have zealous regard for our natural resources. We ask you to insist that the Corps of Engineers complete their scientific studies of the environmental effects of the Tenn-Tom Waterway before construction is initiated."[91] CLEAN members also requested that Train "counsel the President and recommend to him that he defer any endorsement of an immediate beginning of the Tenn-Tom."[92] So, too, did the Alabama Conservancy speak out against the President's planned trip to Mobile, as did the Alabama League of Women Voters.[93] "If the project is going to hurt its surroundings, the cost of that damage should be included in the cost-benefit ratio used to justify construction," the Louisville *Courier-Journal* editorialized. "This might mean that the project might not prove economically feasible. But if the environmental damage is going to be extensive, it probably shouldn't be built in any case, economically feasible or not."[94]

In mid-May, Nixon decided to risk attending the ground breaking. At Whitaker's request, the Council on Environmental Quality prepared a list of questions and answers addressing the criticisms likely to arise from that decision. The perceived inconsistency between Nixon's halt of the Cross-Florida Barge Canal and his support of the Tenn-Tom generated particu-

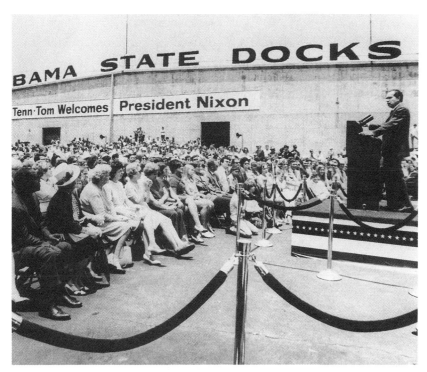

President Richard Nixon was the featured speaker at the May 25, 1971 ground breaking ceremony in Mobile Alabama, which symbolically launched construction of the Tennessee-Tombigbee Waterway. A subsequent court injunction postponed the actual start of construction for over a year. *(Courtesy of Glover Wilkins Tennessee-Tombigbee Waterway Archives, Archives and Museums Department, Mississippi University for Women, Columbus, Mississippi.)*

lar concern. The CEQ recommended that the administration's spokespersons tell the press that the two decisions were quite different: that the Corps filed an EIS for the Tenn-Tom but not for the Florida canal; that "although both projects were authorized by Congress long ago (Cross-Florida—1942; Tennessee-Tombigbee—1946), work has not begun on the Tennessee-Tombigbee project until after a thorough reexamination in light of modern values"; and that the Florida canal "would have destroyed a uniquely beautiful natural stream, one of a very few of its kind in the United States," while the Tenn-Tom threatened "no such natural treasures."[95]

Nixon visited Mobile on 25 May 1971 to dedicate the start of construction.[96] He was joined by Alabama Governor George Wallace, Mississippi Governor John Bell Williams, Kentucky Governor Louie Nunn, Florida

The Environmental Debate 105

Governor Reubin Askew, and Alabama Senators John Sparkman and Jim Allen. The President traveled to Birmingham after his speech in Mobile to address a gathering of editors from the Southeast. Prior to his brief statement to this audience, members of his cabinet discussed various domestic policy issues. Whitaker thought Nixon's appearance at the Tenn-Tom ground breaking made it "a black day for the environment" and he recommended to Presidential Assistant John Ehrlichman that the planned discussion of environmental issues by the Secretary of Interior be replaced by an overview of farm topics by the Secretary of Agriculture. "While the President is digging his ditch in Mobile and fouling up the environment," Whitaker quipped, "it would be inappropriate, in any way, to have an environmental briefing in Birmingham. There is just no way to win that one."[97] Ehrlichman agreed, and made the substitution.[98]

The press highlighted the political implications of Nixon's participation. The Louisville *Courier-Journal* observed that his attendance at the symbolic ground breaking was the President's first visit to Alabama since assuming office. "While the trip was ostensibly to speak at the ceremony," the paper reported, "it also was viewed as a bid for Southern support in 1972."[99] Political columnists Rowland Evans and Robert Novak speculated that Nixon's Alabama trip and his support of the Tennessee-Tombigbee Waterway represented an effort to persuade Wallace to stay out of the 1972 Presidential election.[100]

The political benefits envisioned by Nixon's aides never materialized. Wallace ran for president again in 1972 and, as Whitaker and Train had predicted, the subsequent litigation and injunction against the Tenn-Tom embarrassed the administration. Environmental issues in general were bruising the Nixon White House, which led Ehrlichman to send Attorney General John N. Mitchell a confidential memorandum stating: "As you know, our batting average has been dismal on successfully defending major new governmental construction projects against injunctions by environmentalists armed with Section 102 of the National Environmental Policy Act . . ."[101]

Toward a Legal Confrontation

The enactment of the National Environmental Policy Act in January 1970 bolstered the confidence of environmentalists around the country, filling them with hope that the new law would reorient the programs and activities of federal agencies. Critics of the Tennessee-Tombigbee Waterway considered their faith in NEPA to be well-founded, especially in light of what they considered to be the self-evident and overwhelming environmental damage that would accompany construction of that particular project. Consequently, members of CLEAN and like-minded opponents focused their energies on ensuring that a full environmental impact statement be completed and reviewed, believing that the project would never pass muster under the Act. When the Corps's EIS won approval in April 1971, environmentalists began questioning the adequacy of the EIS and asked that the Corps postpone construction until it revised its environmental evaluation of the project. That was not to be. Along with its gala ground breaking ceremony at the end of May, the Corps invited bids to start construction that summer.

Environmentalists felt stymied and betrayed. Their faith and confidence in the ability of the National Environmental Policy Act to reorient agencies like the Corps of Engineers was shaken. Yet the Council on Environmental Quality's green light for the Tenn-Tom EIS and President Nixon's public kickoff of the project served to arouse the interest of the Environmental Defense Fund (EDF). The organization of environmentally concerned attorneys and scientists had already succeeded in winning court-ordered injunctions against the Corps's Gillham Dam and Cross-Florida Barge Canal, despite the fact that both projects had been under construction for several years prior to the passage of NEPA. The Tenn-Tom seemed an ideal project to challenge next, given its enormous scope, its potential adverse environmental impact, and the fact that any construction would take place entirely under the aegis of NEPA. The waterway therefore fitted squarely into EDF's strategy of litigating carefully selected projects to establish legal precedents under the new Act.[102]

The local group, CLEAN, had successfully delayed construction of the Tenn-Tom by insisting that the Corps first complete an environmental assessment. When administrative review of the environmental impact statement failed to stop the waterway altogether, however, CLEAN officials began to abandon their conservative approach in favor of bolder, more confrontational means. The Environmental Defense Fund's interest in extending and strengthening the National Environmental Policy Act through selected lawsuits proved opportune, therefore, and representatives from EDF and CLEAN began exploring their common objectives in spring 1971.[103]

CHAPTER V   **In Court–Act I**

THE ENVIRONMENTAL LITIGATION spawned by the National Environmental Policy Act did much to fuse the interests of regional and national environmental organizations, in large part because using the courts cost money, and most regional groups had fewer resources than their national counterparts. Such was the case for the Committee for Leaving the Environment of America Natural (CLEAN) and the Environmental Defense Fund (EDF), although it was not the Tennessee-Tombigbee Waterway that first brought the two organizations together. Nearly a year before the Council on Environmental Quality accepted the Tenn-Tom environmental impact statement, CLEAN and EDF had joined forces to combat the U.S. Department of Agriculture's (USDA) proposed fire ant eradication program in the Southeast.

It was CLEAN's first big involvement with any issue. To rid the Southeast of the imported fire ant, the Department of Agriculture proposed spraying some 120 million acres in nine states with the biocide Mirex, a chlorinated hydrocarbon—a relative of DDT—added to a mixture of corncob grits and soybean oil. Three applications were planned per annum, with each application equalling 1.25 pounds of Mirex bait per acre. This $200 million program was projected to last for twelve years. To stop the indiscriminate application of the alleged carcinogen (all surfaces were to be treated—ponds, streams, land, and cities), the Environmental Defense Fund intended to bring suit against the USDA. Members of EDF contacted CLEAN officials about joining them as co-plaintiffs in the suit, which CLEAN did. In August 1970, EDF and CLEAN filed a complaint for injunction in the U.S. District Court for the District of Columbia.[1]

Despite the lawsuit, the courts did not decide the fate of the proposed fire ant eradication campaign. That was done by the state of Georgia when it withdrew its support. Because the program depended upon the cooperation of the nine southern states, USDA stopped promoting it after Georgia pulled out. Even though the litigation became moot, EDF and CLEAN achieved their objective of clarifying and making public the environmental problems associated with the indiscriminate spraying of Mirex.[2] A lawsuit against the Corps of Engineers to enjoin construction of the Tenn-Tom, it was reasoned, might achieve the same results—bringing to national attention the adverse environmental impact of the waterway and, by implication, other large-scale navigation projects.

EDF's confidence in tackling the Tenn-Tom had been strengthened by its recent success in stopping another Corps project, the Gillham Dam on the Cossatot River in southwestern Arkansas.[3] The Gillham Dam suit became the first fully-litigated National Environmental Policy Act (NEPA) case, and EDF was anxious to build on its momentum. In opposing the Tenn-Tom, EDF retained the same attorney who had secured the preliminary injunction against Gillham Dam, Richard S. Arnold of Texarkana, Arkansas. Arnold had earned a reputation as an extremely talented and articulate litigator. He graduated with high honors from Yale University in 1957 and with highest honors from Harvard Law School in 1960, after which he worked as a law clerk for Supreme Court Justice William J. Brennan, Jr. He subsequently practiced law in the District of Columbia for four years, before returning to Arkansas where he was made a partner in his family's successful law firm, Arnold and Arnold.[4] The Gillham Dam case taught the Arkansas attorney a great deal about the Corps's organization, its operating and planning procedures, its regulations, and its manner of justifying civil works projects. It was on the basis of this accumulated knowledge and Arnold's abundant talent that EDF selected him to litigate its suit against the Corps on Tenn-Tom.

EDF and CLEAN formed their coalition to stop the Tenn-Tom in the spring of 1971. James Williams joined as a third plaintiff. The CLEAN chairman was a native of Alabama and a resident of Mississippi, and his participation in the suit helped shortcircuit the possible criticism that opposition to the waterway came only from out-of-state instigators. Jon T. (Rick) Brown assisted Arnold as an attorney for the plaintiffs. Brown, who was based in Washington, D.C., was a young, hard-working attorney with EDF who had had some experience working on the Cross-Florida Barge Canal litigation.[5] Williams helped Arnold and Brown by inviting his environmental colleagues to testify on the inadequacies of the Corps's environmental impact statement.[6]

The plaintiffs filed their complaint in the U.S. District Court for the District of Columbia on 14 July 1971. EDF explained that "the plaintiffs filed the suit on behalf of themselves, their members, and people who have enjoyed the Tombigbee River and wish to preserve its integrity as a natural eco-system."[7] Response to the lawsuit varied, depending in large part on where one stood with regard to the project prior to the litigation. However, not all project proponents were unsympathetic to the suit. The Louisville *Times* expressed this sentiment when it called the environmental lawsuit "a welcome intervention." "If nothing else," the newspaper editorialized, "it may delay construction (due to start this month) long enough for a thorough airing of problems not satisfactorily answered by the Army Corps of Engineers in its report recommending the project." The editors continued, stating that the *Times*, "as many readers are aware, has been an enthusiastic supporter of the proposed canal in the past . . . Recent developments, however, have given us second thoughts." They argued that answers to the environmental questions raised by the project should be found before construction began, not pursued concurrently with construction as Corps officials were proposing.[8]

The litigation extended the environmental criticism that had surrounded the Tenn-Tom from the time Congress first appropriated construction funds in 1970. There were few surprises in the lawsuit. In describing the consequences of building the waterway, the plaintiffs mentioned the flooding of 24,000 acres of land; the periodic inundation of an additional 46,000 acres of land; the destruction of 170 miles of free-flowing tributary streams and 140 miles of the Tombigbee River, and the loss of associated recreational, aesthetic, and ecological resources; the elimination of prime fish and wildlife habitat; the loss of important archaeological, historical, and paleontological sites; the deleterious effect on the area's groundwater; and the varied ecological problems associated with joining the waters of two distinct river systems.[9]

The plaintiffs maintained that the Corps was proceeding without Congressional authority, and they requested that the court issue a permanent injunction against construction. Like most lawsuits against the government, the Tenn-Tom complaint contained multiple "causes of action." The plaintiffs understood from the beginning that not all causes would be heard by the court; they simply hoped that one or more would be.[10] Of the seven charges, the principal one claimed the Corps violated the National Environmental Policy Act. The army engineers were also accused of violating the Fish and Wildlife Coordination Act of 1934, the benefit/cost section of the Rivers and Harbors Act of 1936, the Historic Sites Act of 1960, the Environmental Quality Improvement Act of 1970, the Fifth and Ninth Amendments to the Constitution, and the Public-Trust Doctrine.

In challenging the project's economic justification, the plaintiffs claimed that the Corps had exaggerated the benefit/cost ratio. "The economic claims of the defendants are arbitrary, capricious, irrational, and in violation of standards and guidelines favored by the Office of Management and Budget," they charged, ridiculing as "ludicrously low" the 3.25 percent interest rate used to calculate the benefit/cost ratio. The government's interagency Water Resources Council, for example, had recommended use of a 5.25 percent interest rate for federal water projects. The plaintiffs quoted the Office of Management and Budget's suggested 10 percent interest rate, which would have produced a negative benefit/cost ratio for the Tenn-Tom. They also argued "that environmental costs have never been weighed in determining whether benefits outweigh costs."[11]

In order to prevent the Corps from proceeding prior to a full court hearing, the plaintiffs filed a motion for preliminary injunction on 21 July 1971.[12] Judge John Lewis Smith, Jr., was assigned the lawsuit, and he scheduled a hearing on the motion for 15 and 16 September. The Corps agreed to refrain from signing a construction contract until at least 24 September.[13]

In formulating their arguments, the plaintiffs benefitted from the voluntary assistance of numerous university faculty members opposed either to the waterway, or to the manner in which the Corps evaluated the project's environmental impact and planned for the waterway's construction. Paul Roberts, an assistant professor of economics at the University of Florida who had recently challenged the economic underpinnings of the Cross-Florida Barge Canal in a similar lawsuit, helped the plaintiffs frame their economic critique of the Tenn-Tom.[14]

John E. Cooper, then an assistant professor of biology at the Community College of Baltimore, and his wife Martha R. Cooper compiled scientific data about the Tombigbee River basin and helped organize opposition to the project among academic biologists.[15] The Coopers provided lead attorney Richard Arnold with a detailed critique of the Tenn-Tom environmental impact statement, along with an appendix of solicited evaluations from academic scientists sympathetic to their cause.[16] The Coopers challenged the Corps's assertion that environmental losses could be mitigated by "advanced engineering and design work."[17] They questioned such reliance on technological solutions, claiming that "each 'fix,' in the classic pattern, precipitates a number of problems, and each of these requires a 'fix,' each of which causes additional problems, and so on, *ad infinitum*."[18] In addressing the agency's claim that ongoing studies would assist it in determining the effects of the divide cut on the groundwater hydrology, thereby allowing it to take counteracting measures, the Coopers quipped: "There's that old engineering mentality again—we can fix anything we bust." "There should be a law," they added, "requiring engineers to learn something about living systems."[19] The Coopers also provided specific technical rebuttals to many points raised by the Corps. Such voluntary assistance by the scientific community proved invaluable to the Environmental Defense Fund in this case, as it did in most of EDF's early lawsuits.

In August, Arnold informed EDF chairman Dennis Puleston that the Corps intended to make two basic arguments. The first dealt with the 3.25 percent interest rate, which the agency insisted was set not by itself, but by the Water Resources Council "pursuant to statutory authority." Arnold recommended that this issue be researched thoroughly, for he believed the Corps could be effectively challenged on the legality of that point. The Corps's second argument sought to establish that a preliminary injunction was inappropriate because "irreparable injury" would not occur from the proposed initial construction, given that the Tombigbee River would not be impounded. That would come later, two or three years down the road. Arnold thought this argument fundamentally unsound. "If, as we contend, the project has not been preceded by the necessary studies required by law, any alteration of the natural environment will cause an irreparable injury," he said. "The fact that the injury will not be in the form of an impoundment of waters should not be controlling."[20]

Defense Strategy

Like most federal agencies, the Corps was prohibited from representing itself in court, being defended instead by the Department of Justice. Justice Department attorneys, however, necessarily relied on the technical assistance of Corps personnel in preparing for the case.[21] They also had to contend with interested parties beyond the named defendants. Tennessee-Tombigbee Waterway Development Authority (TTWDA) officials naturally worried about the litigation, which threatened to delay, if not stop altogether, the project they had worked so long and hard to create. On the day the Environmental Defense Fund filed its suit, TTWDA administrator Glover Wilkins told his board of directors: "We have half expected this and have been concerned about it for some time." He reassured the board that the plaintiffs did not "have a leg to stand on in filing this suit." He defended the environmental impact study and added that "in attacking our benefit:cost ratio, I personally feel the Corps of Engineers know[s] a lot more about this than a group of biological environmentalists."[22]

Wilkins traveled to Washington, D.C., where he solicited assistance from the waterway's Congressional supporters. He feared that the Corps and the Department of Justice might not do everything in their power to win the suit, and he hoped his Congressional allies would help ensure that the two agencies would put forth their very best effort.[23] He met with key Congressional members and their staffs and with officials from the Corps and the Department of Justice. Wilkins thought the Corps would not receive the best possible defense if the lawsuit was assigned to the Land and Natural Resources Division of the Justice Department—the division that typically represented the Corps on matters of civil works. He urged Mississippi Senator James O. Eastland, chairman of the Judiciary Committee, to ask the Attorney General to transfer the case within the Department of Justice to the Civil Division's Special Litigation Counsel, whose attorneys tended to have more seniority and actual courtroom experience. Eastland complied, and the Justice Department responded to his intervention by reassigning the case and placing it under the direction of veteran litigator, Irwin Goldbloom.[24]

While Eastland lobbied the Justice Department, Republican Representative Jack Edwards of Alabama focused on the Nixon White House. Edwards asked the President to coordinate the positions of the Council on Environmental Quality and the Environmental Protection Agency, which, he argued, stood on opposite sides of the issue—the CEQ having cleared the Tenn-Tom and the EPA having condemned it. He urged Nixon to direct CEQ's Russell Train to issue "an affirmative statement for use by the court in September."[25] The White House refused to take this step; in part because it wanted to distance itself from the emerging environmental controversy and in part because CEQ could do no more than attest to the fact that the Corps adhered to the mechanics of the environmental impact statement process. As White House staff assistant Richard M. Fairbanks explained, Edwards mistakenly assumed CEQ was pro and EPA con on the matter of the Tenn-Tom, while in truth "both are moderately skeptical."[26]

By the end of July, Wilkins could report to the TTWDA's board members that Congressional allies had succeeded in assuring that the lawsuit would "receive top attention and priority" from the Justice Department and the Corps. "Our friends in Congress," he said, "presented the urgency of the case and the President of the United States is aware of it and has ordered full-steam ahead in its defense." Wilkins explained that the TTWDA's own counsel, William G. Burgin, Jr., remained on retainer and that the Authority had also hired John H. Gullett, a Washington-based attorney, on a day-to-day basis. While Wilkins assured the board he was being financially prudent, he warned that "before this thing is over we could be committed for many thousands of dollars."[27]

Burgin, a Mississippi State Senator whose law practice was based in Columbus, had represented the TTWDA on various matters in the past—primarily in working within the state legislature to gain appropriations for the waterway (he chaired the Appropriations Committee)—and he played an active role in the early stages of the Tenn-Tom litigation.[28] In August, Burgin, Gullett, and Wilkins met with Corps and Justice officials to discuss the government's litigation strategy. The group debated the desirability of asking the Court to require the plaintiffs to post "a substantial bond." They agreed that a transfer of venue would also be "desirable," but they reached no consensus as to when it should be requested.[29]

Wilkins subsequently met with Gullett and Burgin to plan the Authority's strategy. The two attorneys favored entering the litigation as defendant-intervenors. Wilkins was more cautious, believing the TTWDA should intervene only with the approval of Justice and the Corps. They all expressed concern over the possibility of a "taxpayers suit" being brought against the Authority. Such a suit could be filed, Wilkins explained, "whether we win or lose and this could prove extremely embarrassing." The risk of spending state funds to defend the Tenn-Tom in court, he said, was the TTWDA's exposure to criticism from taxpayers opposed to the waterway who had "just as much right to the money to defend their cause."[30]

Wilkins stressed to his board members that although the legal battles would be determined in the courts, "I think all of us have a big job in winning the battle of public opinion."[31] He told whoever would listen that the litigation was an "attack by a handful of academicians in the Columbus area and the eastern based Environmental Defense Fund."[32] In August 1971, the TTWDA released its own "Environmental Policy Statement."[33] This pamphlet sought to counter the public attacks on the Tenn-Tom by compiling pro-waterway articles, Congressional testimony, and Corps statements. Wilkins described it as "an information statement that can be used by friendly newspaper editors, Congressmen and other civic organizations in our battle to bring factual information to the public in contrast to all the 'hogwash' the environmentalists are putting out."[34]

Wilkins and his staff also met with editors and publishers in the Southeast to explain what the Corps was doing to minimize the adverse environmental effects of the waterway. Wilkins then held "a luncheon for the Washington press corps who write for our southeastern newspapers." Pleased with the results of these meetings and with "the good publicity," he nevertheless reported to his board members that this all-out effort to win the lawsuit was "costing us a pile of money." He explained that the TTWDA might soon go broke, if it continued spending at this rate, but added "if we were to lose this battle all would be lost and it wouldn't matter if we had money left over or not!"[35]

The Corps's Mobile District closely coordinated its efforts with the Justice Department. John W. Rushing, chief of Mobile's Environmental Studies Section, provided a point-by-point rebuttal of the major environmental criticisms brought against the waterway. Above all else, Rushing stressed that the Corps's assessment revealed "no potential environmental problems for which present science and technology cannot be expected to develop satisfactory solutions." He addressed the complaint that the Corps should complete all of its environmental studies before it decided to build the waterway by stating: "It is understandable that today's public, so acutely aware of environmental concerns, should wish that a thorough environmental study be completed before initiation of construction of a project as large in scope as the Tennessee-Tombigbee Waterway. But I believe that those who are clamoring for pre-construction completion of all three phases of the Tenn-Tom environmental study have a basic misunderstanding of the three-part study by the Corps."[36]

Corps planners had contributed to this misunderstanding by employing the concept of a "three phase" study. Mobile District had had little experience preparing environmental impact statements before it completed the Tenn-Tom document, and district officials believed the three-phase approach would convey the Corps's commitment to environmental values. Phase one in the plan consisted of completing the environmental impact statement. Corps planners proposed to go beyond this requirement to continue their environmental studies during construction so that project designs might be modified further to help minimize environmental damage; they also said they would undertake environmental studies during operation and maintenance to identify and correct problems under those conditions. Because these additional studies were labeled *phases*, however, and because the Corps had only completed the first phase in 1971, critics found it easy to accuse the agency of failing to study thoroughly the environmental effects of the waterway as required by NEPA. The language of the environmental impact statement had made the project vulnerable to legal attack and had planted doubts in the public's mind about the Corps's sincerity on environmental issues.[37]

Judge Smith heard the testimony of the plaintiffs' six witnesses on 15 September 1971. The defense produced six witnesses on the following day.[38] One week later, Smith granted the preliminary injunction, thus postponing the start of construction pending a court hearing on a permanent injunction. He required the plaintiffs to post a token bond of one dollar, despite protests to the contrary by the Corps. In issuing the preliminary injunction, Smith ruled that the plaintiffs "made a substantial showing" that the "defendants have not fully complied with the requirements of the National Environmental Policy Act."[39] Smith's law clerk later confided with Arnold that the judge had been predisposed to deny the requested preliminary injunction, but that he was persuaded by the plaintiffs' scientific witnesses. Arnold found this interpretation persuasive for, as he told EDF chairman Dennis Puleston, Smith "is a rather conservative judge and has a number of friends highly placed in the Corps of Engineers."[40]

Contrary to what many waterway advocates claimed, the primary purpose of the plaintiffs was not to postpone the project. Arnold adamantly emphasized his desire for an early trial to EDF officials: "I would hope that we could complete discovery with reasonable speed and bring the case to trial on the merits by the end of the year 1971."[41] To Senator Gaylord Nelson, himself an outspoken critic of Tenn-Tom, Arnold was even more to the point, claiming that the plaintiffs "intend to cooperate with counsel for the defendants in bringing the matter on for trial as quickly as possible. We have not brought this suit for purpose of delay, but rather in a good-faith effort to obtain an adjudication on this important project."[42]

The Question of Venue

Upon learning of Judge Smith's decision, Glover Wilkins called an emergency meeting of the TTWDA's board of directors to determine the Authority's future course of action.[43] The board members agreed that their vital interests were at stake, and that the TTWDA should therefore join the suit as a defendant-intervenor. The Authority retained Burgin's Columbus, Mississippi, law partner, Hunter M. Gholson, to file a petition to intervene and to handle the TTWDA's direct involvement in the litigation.[44] The Tombigbee River Valley Water Management District, the Tenn-Tom's official state sponsor from Mississippi, also filed as an intervenor, and hired Tupelo attorney Fred M. Bush, Jr., to represent it.

The TTWDA had already played an active role in the litigation through its behind-the-scenes maneuvering. It became even more aggressive in its official capacity as a defendant-intervenor. Its first major ploy was to push for a transfer of venue. The plaintiffs had filed the case with the District Court in Washington, D.C., for two main reasons: they believed that the District Courts in Mississippi and Alabama would be partial to the waterway; and the District and Appellate courts in the District of Columbia were thought to provide a more sympathetic forum for those groups bringing suit against the federal government.[45] Although the validity of this belief is debatable, the defendants shared that perspective. As Wilkins told his board of directors, moving the litigation to the Southeast where it would be "reported on by correspondents familiar with the area will make quite a difference." Few people in Washington, including judges, knew what the Tombigbee looked like, he said. "They picture it as a pristine, crystal-clean, fast-flowing trout stream or some such thing when it is in reality a log-jammed, stagnant, muddy or else flooding river!"[46]

The Corps's legal counsel favored joining the TTWDA in requesting a transfer of venue.[47] Justice Department officials, however, were reluctant. Attorneys representing the TTWDA believed that a motion for transfer of venue would be greatly strengthened if all the defendants and defendant-intervenors acted in concert. Accordingly, John Gullett lobbied Justice's L. Patrick Gray and Irwin Goldbloom. Gray, who was Assistant Attorney

General for the Civil Division, initially sided with Gullett. Goldbloom, the lead Justice attorney assigned to the Tenn-Tom litigation, did not. Goldbloom favored the defendant-intervenors filing for a transfer of venue, and he agreed that the government would not counter such a motion. However, he strongly opposed the government joining in the request to relocate the case. Such a move, he argued, would give the appearance of "judge shopping." Gray concurred and informed Gullett and the TTWDA that, while the Justice Department supported the intervenors filing for a transfer of venue, the government would not join them.[48] Following this decision, Richard Appleton of Corps Headquarters's Office of Counsel telephoned Assistant Mobile District Counsel Alfred Holmes to tell him, according to Holmes's notes, that Mobile District should assist the intervenors in their motion but should furnish such information "orally or by informal lists mailed direct to the intervenors so that the Corps of Engineers will not be put into any official position or any interpretation of speaking for the Department of Justice."[49]

Meanwhile, Gullett continued to lobby the Justice Department on other matters. He asked if Gray or another top Justice Department official would work through the White House to obtain a letter of support from CEQ Chairman Russell Train on behalf of the Tenn-Tom environmental impact statement. Gullett also asked Gray to assign another experienced attorney to assist Goldbloom. Four days later, Gray responded that he would put two more attorneys on the case. An endorsement from Train, however, was another matter. Gullett reported that John Whitaker, President Nixon's staff director for environmental and natural resources issues, told Gray that the Corps "would have a snowball's chance in hell of getting such a letter" from CEQ.[50]

EDF's Arnold met informally with Justice Department attorney A. Theodore Giattina, Goldbloom's principal assistant, in January 1972 to discuss the request for a transfer of venue. He learned that Giattina had recommended asking for a transfer to the U.S. District Court for the Northern District of Mississippi rather than to the U.S. District Court for the Southern District of Alabama located in Mobile. Giattina wanted the federal government to join the motion because he thought it would increase the likelihood of the transfer being granted. He was overruled, however, by his superiors, Gray and Goldbloom. As Arnold later told the EDF leadership, senior Justice Department officials "apparently are of the opinion that the motion will probably be denied, and that the Government's chances for victory at the trial would be impaired if they officially told Judge Smith that they wanted to get out of his Court."[51]

The defendant-intervenors therefore filed their motion alone in December 1971. Despite the recommendations of the Justice Department, they requested that the trial be held in the Mobile Division of the U.S. District Court for the Southern District of Alabama. They argued that the case should be heard in Mobile because most of the witnesses expected to testify lived in either Mississippi or Alabama, as did the thousands of people to be affected directly by the waterway. Moreover, they claimed that the bulk of the Corps's documents dealing with the project were in Mobile, and bringing such records to Washington, D.C., would be inordinately expensive. They pointed out that none of the plaintiffs resided in Washington, and charged the plaintiffs with having "shopped around for a forum to their liking."[52] When the plaintiffs then filed an objection to the venue change, Justice Department officials reversed their earlier decision and countered by filing their own motion for transfer of venue to the U.S. District Court for the Northern District of Mississippi.[53]

Judge Smith heard oral arguments for and against both motions in January 1972, after which he granted the transfer of venue requested by the federal defendants.[54] Rumors circulated that Mississippi Senator James Eastland influenced Smith's decision. The Judiciary Committee chairman had called Smith's preliminary injunction a "case of blatant judicial tyranny." "It is deplorable," he said, "that a Federal judge has, with one stroke of a pen, thrown a roadblock in the path of this great and envisionary project."[55] The media widely reported that Eastland had said Smith would never receive a circuit court appointment because of the temporary injunction he issued against the waterway.[56] Senator John Sparkman of Alabama also attempted to influence the court's decision. Three days before Smith was scheduled to hear arguments for and against the motions, Sparkman issued a press release strongly supporting the transfer of venue.[57]

The plaintiffs, at Arnold's recommendation, did not appeal the transfer of venue. Nevertheless, they were surprised and disappointed. Arnold thought Judge Smith's decision was "extraordinary," given that the District Court in Washington had already taken testimony for two days and that the TTWDA had voluntarily injected itself into the case *after* the preliminary injunction had been awarded and had then requested the case be transferred. Fifteen years later, Arnold, who was then a federal judge in Arkansas, still found the venue change puzzling. "I think that you will look a long time," he said, "before you find an instance of a transfer of venue in a situation like that."[58]

The Trial

Chief Judge William C. Keady of the U.S. District Court for the Northern District of Mississippi presided over the transferred lawsuit. The defendants were pleased to have Keady hear the case, for they believed he would be sympathetic to their position. Keady was a native of the area, a former Mississippi trial lawyer, and a close personal friend of Senator John Stennis.[59] The defendants and plaintiffs alike were well aware that courts tended to be sensitive to their regional political climates. As Nashville District Engineer Colonel William F. Brandes said: "In the Tenn-Tom case, many people (especially the environmentalists) consider that the suit was decided when it moved to Mississippi."[60] Wilkins told his board that he was "delighted" with the venue change, in large part because he felt in Mississippi "the court will have a better understanding of the area involved by the project as well as the people who have supported this Waterway for so many years."[61]

Keady scheduled the trial to begin in Aberdeen, Mississippi, in June 1972. In the pre-trial hearing in April, Keady settled a dispute over discovery procedures in favor of the Corps. The plaintiffs had requested that the army engineers make available to them any and all documents pertaining to environmental aspects of the Tenn-Tom. Attorneys representing the Corps vigorously objected to this broad request, claiming executive privilege in denying access to many of the requested materials. Keady agreed with most of the agency's arguments. He ruled that the Corps did not have to disclose inter-agency records, documents pertaining to future appropriations requests, or material compiled in preparing the case for trial. With regard to inter-agency correspondence, Keady stated: "It is essential to safeguard the free flow of ideas and views within the agency and for government agents not to be inhibited in the free and unlimited expression of views out of fear that their tentative thinking is subject to being later exposed in litigation, even though the ultimate decision reached by the agency may have small relevance to the first thoughts on the subject."[62]

Keady's decision put the plaintiffs at an extreme disadvantage. One of the key elements of their case was that the Corps's environmental impact statement was not used in the agency's decision making process as required by NEPA. The plaintiffs intended to argue that the decision to proceed with the project had already been made prior to the preparation of the EIS, and that the Corps was not about to reverse that decision. To document this, however, the plaintiffs needed access to the agency's files. Without such access, they were severely handicapped.[63]

The April hearing also handled a motion by the defendants to either dismiss the case, issue a partial summary judgment, or define and limit the issues to be tried.[64] Keady ruled that the court would hear only the plaintiffs' first cause of action—that the Corps violated the National Environmental Policy Act. He dismissed five of the plaintiffs' other causes of action, while EDF voluntarily withdrew the Public-Trust Doctrine violation.[65] Keady thus ruled that the plaintiffs' testimony on the Tenn-Tom's benefit/cost ratio and the economic effects of the waterway were inadmissible. Such economic analyses, he said, had to be settled between the Corps of Engineers and the Congress; his court would not address the substantive question of whether the project was good or bad—for economic reasons, or any other—but would focus instead on determining whether the agency had or had not complied with the procedural requirements of NEPA.[66] Economic criticism of the Tenn-Tom had been an important element of the lawsuit, as it had been in earlier EDF lawsuits to stop construction of the Corps's Cross-Florida Barge Canal and Gillham Dam, and the elimination of economic issues therefore proved a serious blow to the project's detractors.[67] If Keady's ruling came as a profound disappointment to the plaintiffs, it was received with an equal measure of delight by the defendants. Hunter Gholson expressed such sentiments to the TTWDA, stating: "We believe that the order is a great step toward our ultimate victory and that the posture of the case is now simpler and more favorable to us than its status under previous pleadings and the order of Judge Smith in Washington."[68]

The June trial lasted eight days. Most of the evidence was of a scientific or technical nature, presented by a procession of expert witnesses from both sides. Keady later spoke of the irony of this situation, given his own meager scientific background—limited to a high school biology course and a college course in geology.[69]

The defendants had hired a private systems analysis consulting firm, Meta Systems, to provide an independent evaluation of the methodology and background data used by the Corps in preparing the environmental impact statement. The Cambridge, Massachusetts-based firm consisted of several Harvard University faculty members from different disciplines, and three of them—Robert P. Burdin, Harold Thomas, and Joseph Harrington—testified at the June trial. Burdin played an additional advisory role for the defendants by anticipating the plaintiffs' line of argument, suggesting counterarguments, and recommending additional expert witnesses.[70]

The plaintiffs sought to obtain the best possible witnesses—well-known experts in their respective fields—rather than obtain witnesses only from the Southeast in an attempt to impress Keady and the Southern audience.[71] They produced expert witnesses from several biological subdisciplines, as well as from engineering, geochemistry, and sociology. Many of these people had already gained valuable experience dealing with the Corps on other water development projects. David Stansbery, director of the Ohio State Museum, and Jack Burch of the University of Michigan Museum of Zoology each contributed two weeks of their time. In addition to testifying, they sat in the courtroom during the entire trial and advised the plaintiffs' attorneys in the evenings.[72]

Project supporters bused in by the TTWDA filled the courtroom throughout the trial. Keady allowed these spectators to wear and carry pro-Tenn-Tom pins and bumperstickers into his court.[73] Nashville District Engineer Colonel William Brandes, who attended the entire hearing, wrote later about the tenor of the trial. "Despite the obvious sentiments of the populace," he observed, "the mood was more jovial than tense." He said that "audible comments from the spectators were frequent. On hearing one witness describing the clear, pristine beauty of the free-flowing Tombigbee River, one overall-clad man said loudly and incredulously, 'Is that *our* Tombigbee River he's talking about?'"[74]

In August 1972, Keady ruled in favor of the defendants, dissolving the preliminary injunction and dismissing the complaint with prejudice.[75] The plaintiffs appealed Keady's decision to the Fifth Circuit Court of Appeals in New Orleans. In an attempt to prohibit the commencement of construction until the case was heard, the plaintiffs also filed a Motion for Injunction Pending Appeal, which Keady denied, stating that the plaintiffs failed to show that the absence of an injunction would lead to irreparable injury and claiming that an injunction would harm the public interest because of substantial construction cost increases.[76]

The plaintiffs—now technically "plaintiffs-appellants"—quickly filed another Motion for Injunction Pending Appeal, this time with the U.S. Court of Appeals for the Fifth Circuit. In arguing against Keady's decision made two days earlier, the plaintiffs presented several cogent points. They challenged the Corps's decision making process, which, because of the proposed three-stage environmental study, skewed the procedural requirements of NEPA. "The decision to build the Tennessee-Tombigbee Waterway had already been made at the time that the EIS was being prepared," they argued. "The EIS was simply a paper formality, something to get through before issuing a notice to prospective bidders on the first contract." They also challenged the Corps's claim that no irreparable injury would be done to the environment by construction at the southern end of the waterway because the waters of the Tennessee and Tombigbee rivers would not be mixed and because there would be no impoundment of the Tombigbee River. "It is not a particular mixing of two river systems, or a particular impoundment of the Tombigbee River, against which NEPA protects the interests that plaintiffs represent here," they stated. "The purpose of the statute, and the theory of plaintiffs' claim, is rather that *no substantial environmental disruption through federal action should take place until NEPA has been complied with.*"[77]

The Fifth Circuit Court of Appeals denied the motion in September

1972.[78] Rather than challenge this denial—which was issued without an opinion—the plaintiffs' attorneys instead focused on preparing their brief and arguments for the actual trial.[79]

Oral arguments were presented to the Fifth Circuit Court at New Orleans in May 1973. The Circuit Judges hearing the appeal consisted of Irving L. Goldberg of Dallas, Texas; Charles Clark of Jackson, Mississippi; and Paul H. Roney of St. Petersburg, Florida. The appellants argued that the Corps had failed to comply with the National Environmental Policy Act for all the reasons raised in the previous trial. Their most serious challenge to the District Court's ruling, however, was based on Keady's refusal to hear economic objections to the waterway or any other substantive arguments challenging the Corps's decision to build the project. The appellants told the court that the economic objections actually outweighed the environmental objections. They accused the Corps of using an unrealistically low 3.25 percent interest rate in calculating the project's benefit/cost ratio, and they charged the agency with failing to include the costs of environmental damages associated with constructing the waterway, such as the loss of stream fishing.[80]

The Corps retorted that it had, in fact, complied with NEPA, and that its environmental impact statement adequately identified "all significant and probably adverse environmental effects." It also defended the three-phase environmental study of the Tenn-Tom as fulfilling the spirit and the purpose of NEPA. Economic determinations, the Corps argued, were political decisions made by the Congress and the President, and were therefore not subject to judicial review.[81]

The Fifth Circuit Court took nearly a year to issue its decision. In April 1974, it ruled in favor of the Corps. Judge Clark wrote the opinion, which held that the environmental impact statement complied with the requirements of NEPA. It also stated that, "although the court had power to review substantive agency decisions whether to proceed with environmentally controversial projects, Congress itself, by appropriating funds for the project in question after having studied and debated its environmental impact, had supplanted the Corps' recommendation that the waterway be built and had thereby precluded such review."[82] This ruling that substantive agency decisions were subject to review in federal court was the only major difference between the opinions issued by the Court of Appeals and the District Court. The appeals court rejected the appellants' claim that environmental impact statements were required to be "detailed." "It is entirely unreasonable," the court said, "to think that Congress intended for an impact statement to document every particle of knowledge that an agency might compile in considering the proposed action."[83]

The affirmation of the District Court's ruling left the Tenn-Tom opponents in a quandary. One option was to file a Petition for Certiorari with the Supreme Court. Although some attorneys advising EDF favored this course,[84] Arnold did not. He doubted that "the Supreme Court could be prevailed upon to grant review and reverse this judgment." This was unlikely, he contended, because Judge Clark took "great pains to emphasize that his opinion is limited to the peculiar situation before him."[85]

EDF chairman Dennis Puleston told Arnold that the Fifth Circuit's decision was "a great disappointment to us, of course, but not altogether unexpected when we consider the way other water resource cases have been going recently." Puleston said the EDF's Executive Committee would consider the possibility of appealing the case to the Supreme Court. He thanked Arnold for putting up "a splendid fight" against "some of the worst environmental insults on record."[86] By mid-June 1974, EDF's directors decided to turn their energies elsewhere.[87]

Reassessment

The litigation delayed the start of construction by a year, but it is uncertain whether the lawsuit postponed the ultimate completion of the waterway. Judge Smith's preliminary injunction pertained only to actual construction work. However, many other aspects of the waterway were allowed to continue—aspects that were absolutely critical to the advancement of the project. These activities included planning, design, and real estate acquisition, and the Corps pursued them all vigorously. In fact, some agency engineers felt desperately behind schedule in 1971 and found the construction delay to be a benefit in disguise, as it gave them time to catch up with their design work.[88]

Nashville District's Colonel Brandes thought the litigation made the Tenn-Tom a better project. "The Corps writes honest reports," he claimed, and "it operates meticulously within the guidelines it is furnished." However, he reasoned that "no one works so well that they can't be made to do better by some hostile scrutiny and criticism. Many features of the Tenn-Tom project were reappraised with painstaking attention to environmental effects simply because the project was under attack, and a number of beneficial changes were made."[89]

Brandes admitted that the Corps's decision to describe the Tenn-Tom EIS as the first part of a three-phase process was "unfortunate." He explained: "It was an honest but naive admission that all the environmental effects could not be ascertained in advance. The plaintiffs jumped on this point gleefully and often, asserting that this was *prima facie* evidence the statement was preliminary, incomplete and incompatible with NEPA's 'detailed' requirement." He added that, although the agency "survived that line of attack," it was "certain that if the Corps had it to do over, they would never have mentioned the three-phase method."[90]

The effects of the Tenn-Tom lawsuit extended beyond the waterway. Even before the litigation ended, legal counsel within Corps headquarters sought to draw lessons for the entire agency. In April 1972, Mobile District Counsel Alfred Holmes outlined these lessons to a Corps-wide environmental law seminar. He stressed the need for "close cooperation" with Justice Department attorneys, especially the need for Corps engineers "to educate fully the trial attorneys" on the technical facts of the project. He emphasized the problems caused by the Corps's terminology of "three phase" environmental studies, and recommended agency officials avoid using language in their environmental impact statements which "might be construed to mean the Corps has not identified the environmental impacts." Holmes also noted the importance of the change of venue to a District Court outside Washington, D.C., and the desirability of well organized intervenors joining the lawsuit, both for their legal advice and for their ability to work the local news media. Finally, he advised Corps attorneys to study the interrogatories filed by the Environmental Defense Fund for their insights into the nature of environmental litigation being brought against Corps projects.[91]

The Tenn-Tom's opponents also assessed the failed court case. Although successful in winning a preliminary injunction in 1971, the subsequent transfer of venue to Mississippi, the severe restrictions placed on their discovery, and the dismissal of their complaint on the project's economics made a hard task all the more difficult. But losing the lawsuit did not stop the antagonists. Rather, they considered their options and turned to fight the waterway in other arenas.

CHAPTER VI **The Environmentalist-Railroad Coalition**

ENVIRONMENTALISTS WERE THE most visible and politically active critics of the Tennessee-Tombigbee Waterway during the early 1970s, but they were not alone. Because the barge and rail industries were vigorous economic competitors, the American railroad industry had a long history of opposing federal funding for new or expanded inland navigation projects, and the proposed connection of the Tennessee and Tombigbee rivers had not escaped its attention. Throughout the twentieth century, vocal opposition (mainly in the form of economic criticism) from the railroad industry had served as a counterpoint to the Tenn-Tom's enthusiasts.

As long as Congress refused to appropriate construction funds for the waterway, the railroads considered their efforts sufficient. That changed dramatically in 1970, however, after the Corps received $1 million to start building the Tenn-Tom.[1] The railroads' complacency about the untenability of the waterway had been shattered, the result of changing political circumstances in the House and Senate, which no longer regarded the Tenn-Tom's questionable economic underpinnings as a fatal flaw.

In March 1970, James G. Tangerose of the Association of American Railroads (AAR) urged the Association's senior management to consider its options.[2] Tangerose suggested that the railroad industry put forth three people to testify before the forthcoming House and Senate appropriations hearings: a representative from a railroad in the immediate vicinity of the Tenn-Tom, such as the Louisville and Nashville (L&N) Railroad; an independent economist; and Tangerose as the AAR representative. William C. Wagner, manager of commerce at the L&N Railroad, urged his colleagues to consider the "possibility of bringing in on our side some of the groups which are interested in environmental conservation, such as the Sierra Club and the Audubon Society." The AAR executives decided that the environmental groups would be approached, not to testify before Congress along with the railroads, but to voice separately their environmental objections to the project.[3]

Finding a Common Ground

Tangerose contacted the Sierra Club and was surprised to learn that it had no active chapter in either Alabama or Mississippi.[4] E. Leo Koester, L&N's Manager of Public Communications, attempted to establish relations with regional environmental groups in the Southeast, but he reported that "there has been no progress made in securing a witness to oppose this newest ditch as a threat to conservation and/or ecology."[5]

Meanwhile, the railroads planned their testimony. In April 1970, John F. Smith, Jr., L&N's commerce attorney, wrote to officials of the Gulf, Mobile and Ohio Railroad, the St. Louis-San Francisco Railway, and the Illinois Central Railroad about that testimony, and informing them that the L&N was "contemplating approaching an ecologist and a Washington public relations firm for the purpose of getting our position before the public." Such a campaign would be costly, and he asked "whether or not your railroad would share in the expense of going all-out in an uphill fight against the Tennessee-Tombigbee appropriation." He argued that it was "an opportune time to make a fight on ecological grounds."[6]

When Tangerose, James B. Bennett, Jr. (head of the Department of Marketing and Transportation at the University of Tennessee at Knoxville), and Brooks H. Gordon (Assistant Manager for Commerce at the L&N Railroad), testified in May before the House and Senate subcommittees on public works, the environmentalists "found" the railroads.[7] Gordon reported back to L&N executives that representatives from several environmental organizations approached them after the hearings to ask if the railroads would be willing to join them in testifying against other Corps of Engineers projects. Gordon stressed that "it seems that the railroads have now found themselves with a new ally in fighting unwarranted waterway projects, the conservation and ecology groups. In working in cooperation with these groups the railroads now have, I believe, the best climate in years for stopping, or at least slowing down, some of these projects."[8]

L&N's leadership welcomed this information and continued its efforts to encourage opposition to the Tenn-Tom among environmental organizations. L&N Senior Vice President W. Gavin Whitsett approached National Audubon Society President (and former Secretary of the Army) Elvis J. Stahr. "I know it might appear that I am, as a railroad officer, attempting to obstruct our competitors, the barge lines," Whitsett said, "but we are not stupid enough to think the public would go to bat over the railroads' problems, but that the public would only become interested if they could be informed how much the public has at stake." He asked only that the Audubon Society examine the project to determine if it might have an interest in the adverse environmental impact of the waterway.[9]

Not all the railroad officials believed that environmental arguments could stop navigation projects. N. E. White, General Marketing Manager for the Illinois Central Railroad, told Tangerose that his company stood behind the AAR's opposition to the Tenn-Tom. He stressed, however, that "in our opinion the only way you can stop this or any other waterway project is to show a benefit-cost ratio of less than 1 to 1."[10]

This emphasis on economic opposition to the Tenn-Tom remained central to the railroads' position, yet there was a certain appeal to the environmental criticism, an attraction that grew even stronger when economic arguments failed to avert Congressional appropriations for construction. Railroad officials recognized the expanding popular and political momentum of the environmental movement. By June 1970, the anti-Tenn-Tom activities of CLEAN and the Alabama Conservancy had come to the attention of Leo Koester at the L&N. Koester sent newspaper clippings describing the activities of these two groups to Claude Ryan, L&N's regional sales manager in Birmingham, Alabama, and pointed out that CLEAN and the Alabama Conservancy "have an approach that is much more in the spirit of the times." Moreover, he wrote, "in matters of conservation, our railroad and the industry as a whole is comparatively on the side of the angels."[11]

Koester cultivated relations with environmental critics Robert Kuehne and John Cooper at the University of Kentucky in Lexington. Kuehne testified against the project at the May 1970 appropriations hearings, while Cooper submitted letters of opposition for the record. Koester provided them newspaper clippings and some minor reimbursement for attempting to generate concern among environmentalists in the Southeast and to establish an organization committed to stopping the Tenn-Tom.[12] Cooper acknowledged this assistance in July 1970. "Those Waterway boys are convinced they've got the game won," he told Koester, "but maybe we can throw a few shin-bruisers in their path. Who knows what might happen? I'm sure the FDE [Florida Defenders of the Environment] had no idea they would get as far as they have, what with millions already invested in the Cross-Florida Canal."[13] Cooper thanked Koester for recommending the political handbook, *Ecotactics*,[14] but told him he already owned that tract "and about every other book published on the conservation-population-environment area." Cooper closed by passing along Kuehne's suggestion that the Tennessee-Tombigbee Waterway Development Authority be investigated for using state monies to "suborn" federal funds. Kuehne had mentioned "the possibility of putting Nader's Raiders onto this."[15]

Koester believed strongly in the import of bringing together grass-roots opponents of Corps projects from around the country, and he tried to facilitate the development of such a coalition. The recent injunction against the Cross-Florida Barge Canal made this a natural case to emulate, and Koester worked to bring Bill Partington of the Florida Defenders of the Environment and Cooper together.[16]

In August, Acting Committee Chairman Brooks Gordon called a meeting of the 12-member AAR Tennessee-Tombigbee Waterway Project Committee "to discuss the developments of the past year, and to plan further action for the forthcoming year."[17] Committee members discussed the recent decision of the AAR's Board of Directors to eliminate the Waterways Section in September, when its long-time director, James Tangerose, planned to leave the AAR. The Waterways Section was but one of several divisions within the AAR scheduled to be eliminated in the aftermath of Penn Central Railroad's 1970 decision to file for bankruptcy; Penn Central's withdrawal from the AAR had reduced the association's budget by 27 percent.[18] Despite this news and the fact that Tangerose had assumed the lead role in nurturing the Tenn-Tom committee, the membership voted to maintain the Tennessee-Tombigbee Waterway Project Committee and to continue to oppose publicly the proposed navigation project. As Gordon explained to the executives of the four railroads represented on the committee, the railroads "have recently been joined on the 'field of battle' by the exceptional campaigns being conducted by those who are concerned by the damage that projects like the Tennessee-Tombigbee do to the environment." As a result, he said, "the climate has never been better for the railroads' point of view." Gordon recommended that the committee meet more frequently, that it maintain regular contact with AAR officials, and that it increase its expenditures "for communication and legislative intelligence."[19]

Of all the committee's members, the L&N Railroad maintained the greatest interest in the waterway. Its executives made numerous inquiries to federal agencies about the environmental implications of the Tenn-Tom. L&N Vice President-Law Philip M. Lanier, for example, wrote as an individual citizen to Russell Train in May 1971 to ask that the Council on Environmental Quality "demand that the Corps make—*in advance of construction*—a proper environmental study." Lanier asserted that the agency has "never confronted the questions raised by ecologists."[20]

Throughout this period, the L&N maintained a financial and advisory relationship with CLEAN, which was orchestrated by Leo Koester. Through frequent telephone conversations and correspondence with the CLEAN leaders, Koester—a well-respected public relations veteran—advised how they could best present their environmental case to the public and Congress. On several occasions, he sent them drafts of letters that might be broadly circulated under the CLEAN letterhead. He facilitated the development of a network of Tenn-Tom opponents and supplied

CLEAN with newspaper clippings and large batches of reprints for mass mailings. Although modest, the L&N's monetary contributions helped the financially strapped environmental opponents cover direct expenses such as postage, photoduplication, printing, and telephone.[21]

L&N's minor financial contributions to CLEAN became a subject of reconsideration when CLEAN joined the Environmental Defense Fund (EDF) in filing suit against the Corps in July 1971. As lead attorney Richard Arnold explained to EDF Executive Director Roderick A. Cameron: "Jim Williams [of CLEAN] and I discussed the railroad-contribution matter at some length Monday night, and we are agreed that no further contributions will be accepted from this quarter, at least until the question is brought wholly out in the open." Arnold thought this would not cause CLEAN any "undue financial hardship" as L&N's contributions to CLEAN had "been quite small, totalling perhaps $250.00 in the past."[22]

When the L&N later made an unsolicited $500 donation to the Environmental Defense Fund in August, Cameron returned the check with a letter to Lanier thanking the railroad for its gift. "After some reflection," Cameron explained, "we have concluded that it would be unwise for us to accept your contribution to our work at the present time." He described the suit filed against the Corps, noting that environmental issues were at the heart of the litigation, but adding that EDF analysts had also recalculated the Corps's benefit/cost ratio for the project and found it seriously lacking. Cameron contended that "the conclusion is inescapable that such a blatant subsidy of one mode of transportation over another will cause an unjustified handicap against the disfavored mode of transportation, in this case, the railroads." Based on this economic assessment, he invited the L&N to join as an additional party to the lawsuit. Cameron believed that the benefit in such a move would be to dispel the popular perception that the pursuit of environmentally responsible actions "conflict[s] with society's economic interests." He concluded:

> Although we would be pleased by your participation in the lawsuit we feel that your financial support of the environmental groups at this time would present a potential embarrassment which would not further our objectives of defending the environment. Some might charge that our motives were financial rather than that of concern for the environmental

degradation the Tennessee-Tombigbee Waterway will cause. We would more than welcome your financial aid at a later time and hope you understand our delicate position with regard to this matter.[23]

Lanier responded that he understood EDF's position and would follow its activities with regard to the Tenn-Tom lawsuit. He said that the L&N would consider EDF's invitation to join the litigation. "At the moment," he added, "I am not sure that we can contribute anything to this litigation although we have a very definite interest in the outcome of it."[24]

L&N executives ultimately decided against joining the lawsuit. Instead, they told EDF and CLEAN officials that they thought it better for the railroad to remain in the background and to oppose the waterway from behind the scenes.[25] EDF officials had also invited the Southern Railway System to join the suit against the Corps, but it, too, declined the offer.[26]

The L&N maintained a keen interest in the litigation. When the National Water Commission recommended the establishment of user charges on the nation's inland waterways in its draft report released in November 1972, the L&N, like the railroad industry in general, favored the fees.[27] In April 1973, Koester met with Williams, Kuehne, and Clemmer in Bowling Green, Kentucky, to discuss the waterway user charge proposal and other ways of opposing the Tenn-Tom outside the courts, largely in the form of economic arguments.[28] As Koester later told Williams, "I believe that imposition of reasonable user charges along with a modest rise in the cost of money figure used to compute cost-benefit ratios would send the Corps back to their drawing boards on a good many projects—including the Tenn Tom."[29]

Williams, Kuehne, and Clemmer had come to Bowling Green to attend the annual meeting of the Association of Southeastern Biologists, an organization listed as the nation's "largest regional biological group" with 1,500 members. At the association's business meeting, the leadership adopted a resolution on the Tenn-Tom. Describing the upper Tombigbee River as "the largest remaining free-flowing, uncanalized and relatively unpolluted river in the Mobile Basin," the biologists urged that "all construction on the Tennessee-Tombigbee Waterway be deferred until such time that all three phases of the Corps of Engineers' environmental studies have been completed so the full impact of the waterway can be evaluated."[30]

The biologists' Tenn-Tom resolution reflected the unabated environmental concerns with the waterway, but it carried little weight. And with the Tenn-Tom lawsuit appearing to face a likely defeat in the Appeals Court, members of CLEAN and other environmental critics were attracted to the L&N's emphasis on economic issues. This approach made tactical sense because of the failure of environmental arguments to persuade the White House, the Congress, or the Courts to reconsider the waterway. It was also timely because of the Corps's growing vulnerability in this area; construction costs were skyrocketing, the economic benefits of the project were being questioned, and fiscal conservatives were beginning to suggest waterway user charges. Moreover, most environmental critics were convinced that the waterway was extremely questionable on economic grounds. This belief was especially strong among the environmental experts associated with the litigation as they gained detailed knowledge of the project's economics.[31]

The environmental and railroad interests therefore renewed their working relations in early 1973. Following the meeting with Williams, Kuehne, and Clemmer, Koester sent 200 copies of the Southeastern Biologists' resolution to Clemmer, along with suggestions on effective distribution. In addition, he sent a draft resolution for CLEAN, which he suggested they revise and adopt. The proposed resolution focused on the economics of the Tenn-Tom. Koester also began advising CLEAN members on publicity for the trial at the Fifth Circuit Court in New Orleans. Moreover, he passed on newspaper articles dealing with the Tenn-Tom which the L&N received from a clipping service, and he offered to prepare anti-Tenn-Tom fliers for CLEAN's distribution.[32]

The L&N flew Williams and Clemmer to Washington in May 1973 to lobby Congressional members and staff against additional appropriations for the Tenn-Tom and to testify before the House and Senate public works subcommittees.[33] Their Congressional testimony echoed the arguments presented by the plaintiffs in federal court, although they complemented their environmental arguments with a strong attack of the project's economic justification.[34] Williams and Clemmer found that skepticism of

Corps projects was growing in Congress. Senator Gaylord Nelson was particularly supportive of the two environmentalists, and he told them that he again planned to oppose Tenn-Tom appropriations when the public works bill reached the Senate floor.[35]

In August 1973, Koester asked the railroad's top executives to reconsider their low-profile stance on the waterway. "In the belief that railroad opposition to the Tenn Tom would only help accelerate the building of the big ditch, we have never really sought publicity concerning our contrary position," he told L&N vice president R. E. Bisha. "In addition, we have held to the dogma that since the Tenn Tom was considered the rainbow's end in Dixie we would only be asking for ostracism if we tried to swim against the tide. I think it is time we re-evaluate our posture in this regard."[36]

Koester argued that several recent events had coalesced to change public opinion relating to water resources development. He listed the recent National Water Commission report that called for a waterway user fee and generally brought into question the extensive federal subsidies to the barge industry. He added that environmental organizations "have paved the way for communication of our opposition" and "have demonstrated there are a good many heretics in the South who do not accept all the Corps of Engineers baloney." Congress had appropriated less than $35 million for the construction of the waterway to date, he stated, leaving at least $465 million more for future appropriations. Thus, the railroad industry had plenty of time and opportunity "to communicate the logic of our position." Finally, Koester cited the U.S. Water Resources Council's recommendation for more stringent economic criteria for evaluating water projects, including a substantial boost in the discount rate used to calculate benefit/cost ratios. The time seemed ripe for the railroad industry to oppose openly the Tenn-Tom and other Corps navigation projects.[37]

Bisha met with L&N executives to discuss Koester's recommendations. The L&N leadership adopted a policy of no public opposition to the Tenn-Tom (or any other federal navigation project) if waterway user charges were imposed. It became clear in the fall of 1973, however, that user charge legislation was making little progress in Congress, while proponents of the Tenn-Tom were preparing to request a substantial increase in appropriations. In November, Koester asked L&N officials whether the company's policy intended that the L&N's office of public relations expend all of its "limited resources to achieve waterway user charge legislation and neglect the big target (the Tenn Tom)."[38]

The improbability of Congress passing user fee legislation and the growing political momentum enjoyed by Tenn-Tom supporters led executives of the L&N to go public with their opposition to the waterway. In July 1974, Koester wrote Brent Blackwelder of the Environmental Policy Center (EPC) about the economic aspects of the project. He emphasized that the estimated cost of the waterway had doubled in four years, yet despite this enormous jump in costs, the Corps was resisting a recalculation of the benefit/cost ratio. Koester said that 1974 estimates pegged the Tenn-Tom's cost at approximately $2 million per mile. Because interstate highways cost about $1 million per mile to construct and railroad track about half that, he told Blackwelder—who was one of Washington's premier environmental lobbyists—that "you could build both an expressway and a railroad from Mobile to Corinth, Miss., for less than the cost of building Tenn Tom." He added: "I believe the President's [Nixon's] anti-inflation program is inadequate but it does contain an invitation to cut all the fat out of Federal spending. I can't think of a better place to start than the Rivers and Harbors Bill and tax reform should begin by imposing waterway user charges."[39] Blackwelder was only too eager to tackle the Tenn-Tom. The EPC had tried unsuccessfully to reduce the Corps's fiscal year 1974 budget for public works by supporting amendments to cut either the agency's appropriation by 15 percent across the board or to eliminate completely 13 water projects, including the Tenn-Tom, from the appropriations bill.[40] With the L&N's expression of interest to cooperate in stopping the Tenn-Tom, Blackwelder and the EPC intensified their efforts to curtail Congressional funding for the project.

In July 1974, the Corps announced that the mountains of earth to be excavated from the Tenn-Tom's divide cut would be deposited in 51 adjacent valleys.[41] Blackwelder and other environmental critics responded quickly to this disclosure, especially because the Corps's plan did not provide specific solutions for the inevitable erosion, siltation, and revegetation problems. Blackwelder wrote to Office of Management and Budget

Director Roy Ash: "It is essential we feel for the Office of Management and Budget to do all it can to prevent work on the divide cut from going forward until the Corps spells out more specifically what the impacts of its dirt disposal will be." He urged Ash and the OMB "to do all you can to see that any construction is confined to the lower reaches of the waterway south of Columbus, Mississippi."[42] To the chairman of the Council on Environmental Quality, Russell Peterson, Blackwelder stated EPC's belief that the Corps should be required to "file an environmental impact statement before beginning work on the divide cut."[43]

Blackwelder sent blind copies of these letters to Philip Lanier, Senior Vice President-Law for the L&N Railroad. He told Lanier that the Corps's recently announced plan of filling 51 valleys with earth excavated from the divide cut "should provide us with a whole new way of getting at the project." He added, "I think there is a possibility of new legal action since the Corps' plans differ so dramatically from its previously stated intentions for dirt disposal."[44] Lanier, as it turned out, had already written to President Gerald Ford, responding to the request of the White House for "suggestions relative to inflation and to the Federal budget." Lanier recommended halting construction of all new waterway projects "pending a complete review of each and every one of them." His prime example of wasteful federal expenditures on navigation projects was the Tenn-Tom, which he claimed was "justified through false figures and phony accounting."[45]

Blackwelder continued to pressure various offices of the federal government. In February 1975, he told Russell Peterson about developments on the Tenn-Tom since the litigation to stop the project had failed. Blackwelder outlined a litany of problems: the Corps's revised plans for spreading excavated material on 51 valleys near the divide cut rather than filling "only a couple" of valleys as originally proposed; recent findings that indicated important archaeological sites located in the project area; the Endangered Species Act of 1973 which extended protection to those imperiled species in the Tombigbee watershed likely to be affected by the project; the increased economic costs to be shared by the poor 12-county area in northeast Mississippi; the "dramatic escalation of project costs" that were approaching $1 billion and which, Blackwelder warned, might reach "as much as $2 billion" before the project was complete; and the low job creation payoff per federal dollar spent on water projects. Calling the Tenn-Tom "one of the most environmentally destructive projects ever conceived," Blackwelder said that "in light of these current developments, the Environmental Policy Center believes that CEQ should call for a moratorium on the construction of the Tennessee-Tombigbee Waterway and should do everything in its power to bring about a reevaluation of the project."[46]

Although lacking the political sophistication and budget of national environmental organizations like EPC, CLEAN continued its regionally based opposition to the Tenn-Tom. During that spring, CLEAN printed 2,000 copies of its fact sheet—"Tenn-Tom: Review '75"—giving 1,200 copies to the Alabama Conservancy for inclusion with its monthly newsletter, and mailing the remaining copies to newspapers across the nation.[47] This two-page flier addressed such issues as the spiraling costs of construction, including the increasingly heavy financial burden on Alabama and Mississippi taxpayers for the non-federal share of the cost, the absence of flood control and hydroelectric power production, the effects on groundwater, environmental and archaeological impacts, and "unfair subsidies" to the barge industry.[48] Influenced by the arguments, the Mobile Bay Audubon Society issued a resolution that summer calling for the economic feasibility of the Tenn-Tom to be recalculated in light of the vast cost increases.[49]

As plans for the spring round of Congressional appropriations hearings began, Koester organized a meeting with Blackwelder, union officials from the Brotherhood of Locomotive Engineers, and executives from the L&N Railroad to refine a strategy of opposition to the Tenn-Tom which would more closely integrate environmental and economic criticisms. The labor leaders sought to demonstrate how the competitive navigational project would cut back on the number of regional railroad jobs.[50]

Around this time, Randall Grace, a young assistant professor of geography and environmental sciences at Tuskegee Institute, became heavily involved in the effort to halt the Tenn-Tom. He sympathized with the strengthening coalition of environmental and railroad interests,[51] and in May he testified before the House and Senate appropriations subcommittees for public works against further funding of the Tenn-Tom.[52] Following his testimony, he spent two days lobbying the offices of key Senators. Although Grace reported being well received by the legislative aides of the twelve offices he visited, he found that several Senators supported his stance yet feared confronting the powerful Senator John Stennis over the issue. A solid core of opposition against the waterway existed within the Senate, but political momentum promoting the project appeared overwhelming to many observers.[53]

Despite concerns over the strong political backing of the Tenn-Tom, project opponents were nevertheless heartened by the growing skepticism in Congress. The problem for the critics was how to mobilize this political base. Railroad officials, representatives of the railroad unions, and environmentalists continued to meet during the summer and fall of 1975 to discuss this issue. In September, the group decided to focus its efforts on telling Mississippi taxpayers about the financial burden local cost-sharing arrangements with the Corps would create. In 1975, the estimated cost of replacing Mississippi's highway bridges that spanned the Tombigbee was $100 million. They considered a campaign to publicize this fact to be an effective way to generate local opposition.[54]

While railroad and environmental opponents spoke with journalists and elected state officials in Mississippi about the large cost-sharing burden, Blackwelder raised this issue with the Senate Public Works Committee. He told the Senate committee that it was questionable whether the counties in northeast Mississippi, which were "among the nation's poorest," could meet these obligations. Moreover, he argued, "it is also questionable whether the entire state of Mississippi will be willing to take on these additional costs when there is a great problem statewide over inadequate funding of teachers and educational facilities." He suggested that Congress consider the prospect "that the U.S. taxpayer will be forced to cover the local share of this mammoth project."[55]

In addition to expanding its anti-Tenn-Tom efforts in Congress, L&N executives were interested in attacking the projects from other angles. In December 1974, L&N's Philip Lanier called Jon Brown, an attorney for the plaintiffs in the initial Tenn-Tom lawsuit, to discuss the possibility of reopening the litigation. Lanier said the L&N might be willing to join EDF as co-plaintiffs in any future court action, with the expectation that the railroad would underwrite all expenses. "I believe the interest of the railroad springs from their realization that the waterway will have very severe financial consequences for the company," Brown told EDF's Dennis Puleston, adding: "I believe a second factor involves the success which a coalition of railroads and environmental groups have had in halting the Upper Mississippi River Navigation Project, particularly Locks & Dam 26 near Alton, Illinois."[56]

The EDF leadership decided to pursue the opportunity offered by the railroad. In March 1975, EDF staff attorney James Tripp met with seventeen other Tenn-Tom opponents representing environmental and railroad interests to discuss future courses of action. The participants formalized their concerns by establishing a dedicated anti-Tenn-Tom group. They elected a three-person executive committee with representatives from the railroad Brotherhoods, railroad management, and environmental organizations, and established two "exploratory committees." The legal committee, chaired by Tripp, began "an investigation to determine the possibility of court action against the Tenn Tom." John Marlin of the Coalition of American Rivers headed the public affairs committee, which was to "prepare a report on how opposition to the Tenn Tom can best be communicated in the media."[57]

The group sought to expose any weaknesses in the waterway's economic justification. Toward this end, EDF retained transportation economist Joseph L. Carroll, who had worked with Tripp in the EDF lawsuits challenging the Tennessee Valley Authority's Tellico Dam and the Corps's Cross-Florida Barge Canal and Locks and Dam 26, to examine the economics of the Tenn-Tom.[58] Based upon Carroll's work, EDF appealed directly to President Gerald Ford in November 1975 to terminate appropriations for the waterway and thereby save the federal government over $1 billion. "Tennessee-Tombigbee is without question the most expensive navigation project ever undertaken by the Corps of Engineers," wrote EDF Executive Director Arlie Schardt, "and represents an outlay of money equalling one-fourth the total of all expenditures on navigation projects expended during the entire history of the Corps of Engineers." Schardt stressed the final burden of the project on federal and state taxpayers, the questionable economic justification of the waterway, and the subsidy of waterway users which "will threaten directly the financial viability of at least three railroads presently operating within the same service area."[59]

Buttressed by Carroll's findings and by the argument that there existed "a reasonable likelihood of obtaining an injunction to prevent the Corps from continuing construction,"[60] the L&N and EDF leadership decided to initiate another lawsuit aimed at stopping the Tenn-Tom. On 30 November 1976, the two organizations filed separate but parallel complaints with the U.S. District Court for the District of Columbia.

Tombigbee River Conservation Council

Despite the earlier loss in court, grass-roots environmental opponents of the Tenn-Tom had remained strident. In June 1975, CLEAN's Glenn Clemmer submitted a proposal to the L&N Railroad for financial sup-port to establish "a full time position for a person to fight the Tenn-Tom Waterway under the heading of an independent organization." He stressed that such an organization would provide the coordination necessary "for an effective political attack on the project." Because only $70 million out of the then estimated $870 million had been appropriated, the "greatest part of the project is still subject to the yearly mood of Congress," he wrote, and "design changes, advanced plannings, and economic re-evaluations make the merits of the project even more questionable." He mentioned that additional litigation was currently being considered, and that environmental lobbyists had detected "a general displeasure with the Tenn-Tom among legislators outside the project area." These skeptics in Congress shied away from openly opposing the project, Clemmer claimed, because the local population appeared to be in full support of the waterway. "While there is a growing sentiment against Corps projects in the area," he charged, "that opposition has not been organized and made visible in the case of Tenn-Tom." He called for a concerted effort to coordinate local opposition, work with the media in publicizing dubious aspects of the waterway, and work with members of Congress to request "a General Accounting Office investigation of Tenn-Tom economics, and ultimately a deletion in funding for the project."[61]

The L&N's leadership evaluated this request with caution, despite Leo

Koester's optimism,[62] but Clemmer's effort was rewarded a year later. By that time, L&N executives had become convinced of the need to challenge the project simultaneously from all angles, and that those efforts should be coordinated to ensure maximum effect. As a result, the L&N Railroad gave a $6,000 grant to establish the Tombigbee River Conservation Council (TRCC) in August 1976. In December, the TRCC was incorporated as a civic improvement society in Mississippi. Randall Grace, the biologist who had been working closely with the L&N in organizing opposition to the Tenn-Tom, headed the Starkville-based organization, whose main objective was to stop the Tennessee-Tombigbee Waterway. (Tangentially, the TRCC also opposed related projects, such as the expansion of the Black Warrior-Tombigbee Waterway and stream channelization in the Tombigbee drainage area.) The TRCC's strategy was to influence public opinion and government action through the systematic collection and dissemination of information about the Tenn-Tom. It solicited the support of other interest groups in the project area, such as landowners and hunting and fishing clubs, by informing them of the detrimental effects of the waterway, and it sought to coordinate the ongoing efforts of environmental groups opposed to the Tenn-Tom.[63]

Grace elaborated on the purposes of the TRCC in his initial membership drive letter. Referring to the Tennessee-Tombigbee Waterway Development Authority, Grace said that "the proponents of the waterway have pushed their one-sided version of the Tenn-Tom 'dream' for many years." He contended, however, that the Tenn-Tom dream was in reality "a nightmare." He promised to bring public attention to the waterway's economic, ecological, and human costs. "The TRCC will engage in lobbying, litigation, investigative reporting, development of grass-roots opposition to the Waterway, and whatever other means necessary to bring about the objective of stopping the Tenn-Tom Waterway." He urged interested parties to join the TRCC by pledging a modest $3.00 membership fee, for which they would receive periodic newsletters, special bulletins, "and requests for letter-writing and other action."[64]

Koester remained a close adviser to Grace and the TRCC. In November 1976, he sent Grace a detailed letter of suggestions in which he advised the TRCC director to list the specific goals he hoped to accomplish within the next ninety days. "The success of the Council will largely depend upon how well you can get volunteers to perform purposeful tasks," Koester said. He recommended that Grace and his associates "be somewhat circumspect about involvement in radio talk shows and TV presentations" because "stations want to use local programming to fill the difficult hours—when the audience is minuscule." So, too, did Koester suggest caution in relying on "various kinds of advertising gimmicks, viz. bumper stickers, badges, posters, etc.," because to be effective "you have to buy them by the tens of thousands and have to have an organization to distribute them."[65] To generate popular opposition in Mississippi, he advised telling residents that they "are getting the short end of the stick"—that is, that Mississippi taxpayers were being asked to contribute far more for the building of the waterway than were Alabama taxpayers, while Alabama was expected to reap the greatest share of the benefits.[66]

Grace mailed the first of many "TRCC Action Alert" letters in December 1976. Addressed to "Tenn-Tom Fighters," the alert attempted to initiate a letter-writing campaign to President-elect Jimmy Carter. Project opponents were given suggestions for the content of their letters, and told where and when to write.[67] In January 1977, the TRCC mass-mailed a pamphlet challenging, point-by-point, the assertions made in an earlier promotional pamphlet by the TTWDA.[68] In response to the Authority's occasional newsletter, *Tenn-Tom-Topics,* the TRCC began issuing its own occasional newsletter, the *Tenn-Tom Review.*[69] In its January 1977 issue, the TRCC challenged the waterway on the project's spiraling costs, the financial burdens placed on Mississippi and Alabama taxpayers, the "hidden cost" of expanding the Black Warrior-Tombigbee Waterway south of Demopolis, the "unfair subsidies" to the barge industry, and the variety of adverse environmental impacts. These hard-hitting accusations prompted the Corps's Mobile District to issue its own detailed "Fact Sheet" responding to each charge.[70]

The contest between the TRCC and the TTWDA to influence public opinion grew heated. "Those groups and individuals who are opposed to any further development of the Tennessee-Tombigbee Waterway are using every means available to them to stop the project," Glover Wilkins warned the Mobile District Engineer: "In addition to the pending litigation, the

Waterway opponents have launched an extensive public relations campaign, including speaking engagements and press releases, to help broaden their coalition against the project." Wilkins stated that opponents often employed information gained from the Corps, and that reporters from the news media were conducting telephone interviews with a variety of Corps employees. He therefore recommended that the agency "designate a person or persons who understand the policy and legal issues of the Waterway (the fewer the better) as spokesman for the Tenn-Tom with the news media." He also took the liberty to recommend that the Corps "establish a central control system for handling written and oral requests for information on the Tenn-Tom and the responses to those requests."[71]

The TRCC received an $8,000 grant from the L&N Railroad to help finance its operation during its second year, and it projected an additional income of $2,000 from membership receipts. Grace summarized the TRCC's successes and failures in his first annual report. The TRCC's activities included 47 presentations before civic, conservation, church, and labor organizations, 23 television interviews, 60 radio interviews, and extensive newspaper coverage. The TRCC also assisted in the litigation, most tangibly by convincing the Birmingham Audubon Society and the Alabama Conservancy to intervene. On the other side of the ledger, failure to stop the waterway headed the list. Grace then mentioned inadequate financing, lack of office staff and the corresponding weakness in Congressional lobbying, low membership (113 individual members in August 1977), the loss of support from landowners who sold their property to the government, the inability to gain the backing of the region's African-American community leaders, and the unsuccessful bid to vote down the Alabama bridge bond.[72]

The TRCC concentrated much of its effort in 1978 on assisting the plaintiffs with the litigation preparation. In February, the National Audubon Society matched the funds provided by the Audubon chapters in Mississippi and Alabama to award the TRCC $5,000.[73] Attempts continued to increase opposition to the waterway among the people living in the project area. Speaking before a public meeting in Jackson, Mississippi, sponsored by the TTWDA, Leo Koester stressed the enormous cost increases of the project and the special interest nature of the expenditures.

Despite the desperate need in Mississippi and Alabama for schools, hospitals, sewage treatment works, and waste disposal plants, Koester told the audience that Congress was instead providing the barge industry with "a $1.8 billion subsidy." "It is a sad irony," he said, "that while the Congress gives little more than sympathy to the grievous problems of the local communities, it goes about the task of providing millions upon millions to build a free right-of-way for the barges."[74]

In the Congress, critics continued to lobby against the annual public works appropriations. They also campaigned for the implementation of waterway user charges. "It has long been the belief of our people here that the most effective way to put a noose around the neck of the Tenn Tom is for Congress to impose waterway user charges which would recapture some of the construction cost," Koester told Richard Briggs, L&N's vice president for economics. "Even if there is only an outside chance we can achieve that kind of a bill, I think we should pull out all stops to accomplish it. It will be a long, long time before we have an equivalent opportunity."[75]

Just as Koester had hoped, the noose around the Tenn-Tom's neck did begin to tighten, but not because Congress imposed a waterway user tax. The project's troubles sprang instead from within the Corps of Engineers, which found itself being held accountable for an increasingly untenable economic justification. Outlandish cost increases, at first hidden from Congress, prompted the army engineers to recalculate (and significantly increase) the benefits claimed for the waterway, which in turn drew impassioned criticism from the environmentalist-railroad coalition and the news media, detailed audits from the investigatory arms of both the Army and the Congress, and a critical reassessment (and near abandonment) by the new Carter Administration.

CHAPTER VII **Economic Accountability**

DRIVEN BY THEIR CONVICTION that the Tennessee-Tombigbee Waterway would inescapably damage the Southern landscape, environmentalists were nevertheless forced to buttress their environmental arguments with economic criticisms. Having failed to kill the project in court, critics focused on the annual Congressional appropriations cycle, where the waterway remained vulnerable. It was at the appropriations hearings, in theory at least, that the Corps was held economically accountable.

Federal public works projects pass through at least two major Congressional processes: authorization and appropriation. Each process is handled by separate committees. Authorization is the necessary first step, but it does not guarantee appropriations will follow. Many authorized projects are never built. The Tenn-Tom, like numerous other projects, experienced years of delay between its authorization and the appropriation of construction funds. The provision of construction funds in 1970 did not assure the waterway's completion, however, for the Congressional appropriation process requires federal agencies to seek project funding on an annual basis. Appropriating money for only year-long periods enables Congress to monitor the progress of a project and makes the Corps and other federal agencies more closely accountable for their actions.

This annual appropriations process also makes it possible for detractors to work year after year to oppose continuing funding for a project. Changes in the political climate on Capitol Hill can therefore have a profound effect on a large and controversial project like the Tenn-Tom. This means that proponents can rarely rest—and, in fact, the Tenn-Tom remained a heated subject in the Congress throughout its entire period of construction.

Congressional oversight of water projects like the Tenn-Tom usually focuses on economic viability, measured by benefit/cost ratios. Because benefits are estimates of future occurrences, economic accountability rests primarily on reviews of project costs. Given the Tenn-Tom's enormous size and its long history of dubious economic feasibility, Corps officials were careful to maintain a positive benefit/cost ratio for the project as they came back to Congress each year for construction funds.

Early Appropriations

Evidence came quickly that the cyclical nature of the appropriations process would force Tenn-Tom proponents regularly to monitor and defend the huge public works project. In the year following the initial $1 million appropriation, Senator Gaylord Nelson of Wisconsin questioned the wisdom of following through with the waterway. It came as no surprise that Nelson orchestrated Congressional opposition to the Tenn-Tom, as the former Wisconsin governor had already established himself as one of Congress's leading advocates of environmental protection. Among his credits, Nelson had proposed the idea of Earth Day, which was first held on 22 April 1970, as a nationwide "environmental teach-in" that attracted over 20 million participants.[1] In attacking the Tenn-Tom, he charged the federal government with repeating past mistakes by "rushing headlong into a massive project whose environmental damages might outweigh any possible benefits." He warned his colleagues against appropriating the $6 million requested for the project's second year because the "waterway could well turn out to be an environmental disaster and an economic flop." Nelson announced his intent to introduce an amendment withholding construction funds until completion of "a thorough review of both the environmental and the economic justification for the waterway." Given the environmental awareness raised by Earth Day, he claimed the nation was headed on a new course, in which America would establish "new cost-benefit ratios more accurately reflecting the values and the problems of new technology and the building of massive public works."[2]

When the water resources appropriations bill reached the Senate floor in July 1971, Nelson offered his amendment, arguing that the spirit and "clear intent" of the National Environmental Policy Act (NEPA) would be violated by appropriating construction funds for projects, such as the Tenn-Tom, that had not fully complied with the new law's provisions. Moreover, there was an inherent conflict associated with federal agencies preparing environmental impact statements for their own projects. "I can just see some agency," he said sarcastically, "after spending millions of dollars coming in and saying, 'It is all a mistake. We have finally made a study in compliance with the intent of the law, and now we find that there is irreparable environmental damage of significant consequence; we are sorry we spent the money in the first place and [we] now recommend termination of the project.'" No agency, he suggested, would make such an admission.[3]

Nelson's amendment failed on procedural grounds, so he introduced a reworded amendment to withhold construction funds for a specified period of time—until 1 February 1972.[4] Mississippi Senator John Stennis led the opposition, charging that the amendment was not "practical," that it was contrary to the intent of NEPA, and that it would hold public works projects hostage to anyone willing to file a lawsuit. In fact, except for Nelson, everyone participating in the floor debate spoke against the amendment.[5] In the end, the amendment failed in the roll call vote, 56 to 17.[6]

Stennis sensed the political importance of proceeding rapidly with the Tenn-Tom, however, and he pushed the Corps's leadership to advance construction as quickly as possible. To maintain such a pace, Stennis told the Corps's Chief of Engineers that the agency must "avoid compromises which will reduce the funding below what is not only desirable but necessary." In the Senator's experience, "such compromises can only breed similar half measures in the future, and compound the delays." He warned the Corps that "the penalties of permitting a situation of that kind to develop are too serious to contemplate."[7]

Stennis asked Caspar Weinberger, then Deputy Director of the Office of Management and Budget, for a large Tenn-Tom appropriation in the President's fiscal year 1972 budget proposal. He outlined the need for a substantial budget to carry construction through a full year, and set $7 million as a recommended minimum. "Only with an amount of this magnitude will it be possible to get the kind of start that will provide the future funding flexibility that is so valuable in a large project of this nature," he said. Stennis stressed his was not "a mere pro forma endorsement" for the Tenn-Tom, but rather "a special plea for a substantial sum in the 1972 budget for the Waterway . . ."[8]

When Judge John Lewis Smith enjoined construction of the Tenn-Tom in September 1971 pending a hearing on the Corps's compliance with NEPA, Congressional supporters of the project publicly condemned the court action. In the House of Representatives, Tennessee Democrat Joe Evins said "the court in hearing the arguments of a few environmentalists apparently was swayed and influenced by their statements rather than the long considered and thorough studies and evaluations by the U.S. Corps of Engineers." The Congress, he asserted, "has heard much more testimony over the years concerning this project than any judge will ever hear," and the Congress had voted to proceed with the waterway.[9] Alabama's Jack Edwards claimed that "nothing, in my estimation, could create a more serene and beneficial effect on the environment" than the Tenn-Tom. He maintained that the vast benefits of the project were "being detoured from the road of progress by one judge and a handful of unbending ecologists."[10] Mississippian Jamie Whitten, the second highest ranking member of the House Committee on Appropriations in 1971, delivered the hardest attack: "Lawsuits are seriously crippling actions of the Corps of Engineers, the Soil Conservation Service, and other agencies where commitments have been made, people have levied taxes, and the projects are long overdue." Moreover, by "assuming the powers formerly claimed for kings under divine right," federal judges "are actually holding up construction of such projects as the Tennessee-Tombigbee navigation project, on letters filled with innuendoes and unproven charges."[11]

Gaylord Nelson, on the other hand, reminded his Congressional colleagues that Judge Smith's preliminary injunction "involved the same issue, the same debate, and the same problem" that he had previously raised before the Senate, leading him to conclude that "a Federal judge did what the Senate should have done."[12] With the Tenn-Tom now threatened in court, however, attorneys for the Environmental Defense Fund urged Nelson not to bring the waterway up for a vote again in Congress, fearing that such an action might adversely prejudice the litigation.[13]

Although Stennis remained confident that the courts would ultimately rule in favor of the Corps, he understood that public support for the Tenn-Tom must be maintained if he and the project's other Congressional champions were to succeed in garnering the necessary funding through the annual appropriations. He also realized that a favorable court ruling would not end public concern with the waterway's environmental impact. Project proponents, he told Glover Wilkins of the Tennessee-Tombigbee Waterway Development Authority, must "build up a justified conviction on the part of all who are interested in environmental matters—and that includes most of the American public—that this project is being carried out with firm resolve to protect the environment to the maximum degree." The extended construction period meant that Congress would have to make "repeated annual efforts" to get adequate funds budgeted for the project. These efforts, Stennis reminded Wilkins, would undoubtedly occur "in years of fiscal restraint, and the Authority must persevere if the Waterway completion is not to drag out for ten or fifteen or more years."[14]

With the battle over the Tenn-Tom's environmental impact confined largely to the courts, Congress quietly continued to appropriate construction funds during the early 1970s. This was made easier thanks to the presence of senior Southern Congressional members on the appropriations committees. Political correspondent Ward Sinclair has called the promotional campaign for the Tenn-Tom "one of the more remarkable public works lobby efforts of our time." He referred to the proponents' hearings testimony as "legislative wizardry," as "witness after witness and congressman after congressman extoll[ed] the wonder of the Tenn-Tom," cheered on by advocates on the committee. Sinclair added that, at these hearings, "no opponents, no critics, no environmentalists were heard," for voicing opposition at appropriations hearings headed by leading Tenn-Tom proponents took considerable courage.[15] As the editors of the *Chicago Sun-*

Times observed, "many lawmakers are afraid to oppose it [the Tenn-Tom] because it runs through the districts of powerful committee chairmen who control spending on pork projects nationwide."[16]

Cost Increases

The 1970s proved to be a volatile decade for the construction industry in the United States. The 1973 Arab oil embargo precipitated an energy crisis that drove up the price of petroleum products and, along with it, the cost of energy-intensive construction projects involving large-scale excavations. General inflation during the 1970s also edged construction costs higher. The Tenn-Tom was vulnerable to these trends, and the rising cost estimates for the project posed a major problem for the Corps and the waterway's supporters. The Corps found, in fact, that the spiraling costs substantially exceeded both the rate of inflation and the price of petroleum products. Tenn-Tom cost estimates began edging upward long before the first spade of soil was turned. In 1962, the Corps established the baseline cost at $263 million; by 1966, the estimate was $289 million; by 1969, $316 million. By the time the agency issued its first call for bids, in April 1971, projected total costs exceeded $346 million, plus another $40 million in state contributions from Mississippi and Alabama. Even these increases paled in comparison to the unbridled escalation unleashed once construction had started.[17]

As a matter of course, the Corps annually updates the estimated costs and benefits of all its civil works projects. The updates reflect design alterations initiated during construction, as well as changes in construction price levels. The latter adjustments are based upon published construction price indices. The Corps uses the *Engineering News-Record* (ENR) Construction Cost Index as its standard. The ENR index, which sets 1913 as its base year, considers the rate of inflation throughout the construction industry. This annually-adjusted index incorporates such unit costs as structural steel, cement, lumber, equipment, fuel, and common labor. Excavation, or earthwork, is not a specific item in the ENR indices, however, although it constituted a major item in the Tenn-Tom construction. The Corps used the ENR index until late 1974 when bids for the Tenn-Tom's Aliceville Lock and Dam revealed price increases that vastly exceeded the levels predicted by the indices. Earlier bidding on the Gainesville Lock and Dam had also exceeded the government estimate, but the bids on the Aliceville segment magnified the trend. Based upon these substantial increases in actual costs, agency officials launched a reanalysis of estimated project costs in January 1975. Instead of applying the ENR index, projections in this reanalysis began to use actual bid experience to formulate cost estimates.[18]

Concern among Corps programmers and executives about the rapidly escalating construction costs long predated the crucial January 1975 reanalysis. As early as March 1973, Major General Daniel A. Raymond, the South Atlantic Division Engineer, warned his superiors about the likelihood of substantial cost overruns. Raymond predicted cost increases well above "normal price level escalation" on all parts of the project, but especially on the divide cut.[19] A year and a half later, Raymond's successor, Brigadier General Carroll N. LeTellier, likewise expressed profound concern about justifying the cost increases to Congress.[20]

The question of what to tell Congress, and *when*, fell within the purview of Corps Headquarters. Major General John W. Morris, the agency's Director of Civil Works, conferred with his senior programmer, Bory Steinberg, about the technicalities of informing the Congress about the Tenn-Tom's mushrooming costs. They fully understood the political impact that such a public announcement would have, as the waterway's critics had for decades lambasted the project's dubious economic rationale. Without a substantial increase in the Tenn-Tom's projected benefits, the rising costs would erode the benefit/cost ratio to indefensible levels, handicapping the agency's ability to gain additional Congressional appropriations. The Corps's new economic analysis would have to address the problem by reexamining both costs and benefits. Morris decided, therefore, to bend the rules and forestall the presentation to Congress of the full implications of the massive cost increases until completion of the economic report, and instead give an incremental boost that would carry the agency through the next fiscal year: specifically, the Corps was to present an estimate of $815 million, up from the previously approved estimate of $732 million.[21]

As the South Atlantic Division Engineer, LeTellier had responsibility for presenting the House and Senate Appropriations Committees with the annual cost estimates for the Tenn-Tom. (It is common practice for the Corps's division engineers, rather than headquarters officials, to testify before the appropriations committees.) Because LeTellier was troubled that the estimated costs for the waterway far exceeded "normal escalation," he requested the Mobile and Nashville District Engineers to provide their "best possible estimate of current Federal and Non-Federal costs." To trim those estimates, he advised the District Engineers to "seek the most cost-effective solutions that will insure a prudent degree of safety and are most economical considering the intent of the project." He ordered this cost trimming to be applied across the board, including those aspects of the waterway addressing environmental concerns, just so long as "the minimum provisions" of the National Environmental Policy Act were met.[22]

Mobile District programmers sent the South Atlantic Division Engineer their honest appraisal within a month. Even with the cost-cutting prescribed by LeTellier, however, recent bids on the project and general economic trends within the construction industry as a whole led to an inescapable conclusion: the Tenn-Tom would cost at least $1.4 billion (or four times the estimate at the time of the first appropriation). LeTellier favored informing the Congress of this finding, but Morris and his senior advisors at headquarters disagreed, and they ordered LeTellier to delay reporting the higher figure to Congress until the Corps was ready to submit its newly calculated benefit/cost ratio sometime next year. Like a good soldier, LeTellier obeyed, rescinding the $1.4 billion figure his division had initially accepted.[23] At the Mobile District, the public affairs office was under strict orders not to announce the spiraling construction costs until officials at Corps Headquarters had completed "a presentation to appropriate people."[24]

With its economic reanalysis of the Tenn-Tom finished in January 1976, the Corps publicly announced—by means of a press release—both the cost increases and the results of its economic reevaluation. The agency revised the $815 million estimated project cost it had reported to Congress the previous year to $1.36 billion—an increase of $545 million. The Corps countered this news by declaring that the project benefits had also

Representative Jamie Whitten (D-Mississippi), chairman of the House Committee on Appropriations and a central political champion of the Tennessee-Tombigbee Waterway *(Courtesy of U.S. Army Corps of Engineers, Mobile District.)*

increased dramatically, from a projected 17 million tons of commodities expected to be shipped on the waterway during its initial year of operation, to 27 million tons. The agencies also extended the Tenn-Tom completion date by four years, from 1982 to 1986.[25]

Once made public, the withholding of increased cost estimates from Congress generated heated criticism of the Corps for conducting a major coverup of its affairs.[26] The army engineers realized that they would have to change the way they operated with regard to the Tenn-Tom. Major General Ernest Graves, the new Director of Civil Works, spoke of "the need for promptly incorporating changes in cost estimates into the system," given that delays in reporting such changes "can contribute to loss of our credibility in cost estimating with the Congress."[27]

The Corps tried to mute the public outcry over the Tenn-Tom announcements. But a reported cost increase of $545 million in one year and a fall in the benefit/cost ratio from 1.6/1.0 to 1.1/1.0 was startling news. In February 1976, members of the House Subcommittee on Public Works

Economic Accountability   157

called a meeting with senior Corps officials to address the issue. Subcommittee Chairman Joe Evins and Representatives Jamie Whitten and Tom Bevill—all staunch supporters of the waterway—wanted to discuss the controversy and forthcoming appropriations hearings.

The Corps's press release had especially angered Whitten. He thought it unnecessarily inflamed the debate and was tantamount to admitting to wrongdoing before being accused. What troubled him most, of course, was that the Corps had issued the news release without first advising him or the committee. Such forewarning, he stressed, would have allowed him and the other Congressional supporters to "grease the skids" for the Corps. Whitten told agency officials in no uncertain terms that they should inform the committee of any future problems with the Tenn-Tom, and the committee would help solve those problems. Evins and Bevill expressed concern about the additional authorization for expansion of the waterway south of Demopolis, which the Corps had stressed was important for maintaining the highest possible benefit/cost ratio. As far as the authorized project itself, Whitten said he expected the appropriations hearings to proceed smoothly and rapidly.[28] And they did. Whitten and Evins orchestrated a tempered hearing on the issue one week later.[29]

Kearney Report

In recalculating the Tenn-Tom's benefit/cost ratio, the Corps's leadership decided that the cost estimates could be safely projected by agency personnel, as could the estimates of all non-transportation benefits, such as recreation, area redevelopment, and fish and wildlife enhancement. Assessment of the project's primary benefits (transportation savings), however, presented another matter. Rather than determining those benefits internally, Corps officials hired an independent consulting firm, which they hoped would lend credibility and increase public confidence in the report. They awarded the contract to A. T. Kearney Management Consultants, Inc., a Chicago-based company that enjoyed an international reputation in transportation planning, economic analysis, and business management. The firm had already worked for the Corps by preparing an economic study of another controversial navigation project, the Cross-Florida Barge Canal.[30]

Kearney was to determine the potential waterborne commerce and the average annual transportation benefits that could be expected over a 50-year economic life (1986-2035) of the waterway and to make these estimates for three alternative plans. Between March and August 1975, Kearney representatives interviewed approximately 1,000 shippers and receivers as part of a field traffic survey to estimate the amounts and types of commodities that might be transported on the waterway. To determine the likely transportation savings for the project, they calculated waterway rates for movements on the Tenn-Tom and compared them to traffic rates for existing modes and routes.[31]

Of the three alternative plans—A, B, and C—Kearney considered, Plan A represented the project described in the 1967 supplement to the General Design Memorandum; that is, the plan Congress authorized. Plan B encompassed Plan A plus substantial improvements to the water course south of the Tenn-Tom between Demopolis and Mobile (the existing Black Warrior-Tombigbee Waterway), including numerous river bend widenings, 21 miles of channel cutoffs, and several bridge improvements. Plan C added the construction of duplicate locks at Coffeeville and Demopolis (again, both south of the Tenn-Tom) to the project described in Plan B.[32]

Kearney estimated that the waterway's first year of operation would generate traffic and total transportation savings of 28,071,000 tons and $55,523,000 for Plan A, and 28,148,000 tons and $64,476,000 for Plans B and C. In both cases, coal represented the most significant commodity movement at 18,400,000 tons, or 65 percent of revenue traffic. Kearney projected that about 2,000,000 tons of metallic ores and 1,859,000 tons of chemicals and allied products would be shipped via the waterway during its maiden year, while the volume of other individual commodities was significantly less.[33] The crucial aspect, of course, was long-term use.

In Kearney's estimates of the tonnage and savings that would accrue over the 50-year life of the project, the projections varied widely between the three plans. The waterway's carrying capacity under Plan A was said to be 44 million tons annually, a level that would be reached in 1991; commerce was then held constant at 44 million tons annually through the year 2035, producing an average annual equivalent benefit of $66.4 million, based on a 3.25 percent interest rate and a 50-year project life. Plan B's estimated maximum annual capacity was higher—55 million tons—but would be reached somewhat later, in 1997, and that led to average annual equivalent benefits of $85.6 million. Although the maximum annual capacity of Plan C was also estimated to be 55 million tons—a constraint imposed in this case by the limits of Bay Springs Lock—its estimated average annual equivalent benefit was higher: $96.8 million.[34]

In October 1975, the South Atlantic Division forwarded to Corps Headquarters Kearney's preliminary benefit and cost figures with information calculated for the three project alternatives: for the Tenn-Tom emptying into the unmodified Black Warrior-Tombigbee Waterway (BWTW) system; for the BWTW with new locks; and for the BWTW with a widened and straightened course.[35] These findings troubled Major General Ernest Graves, Director of Civil Works, and he called a meeting of his top civilian advisors to discuss the issue. From the preliminary data, he said, "it appears that the authorized alternatives have a BCR [benefit/cost ratio] less than 1 and the only alternatives with a BCR greater than 1 include downstream improvements which are not currently authorized. If final estimates confirm these preliminary results, it will be a very serious problem."[36]

When the thirty senior Corps officials gathered to discuss the economic restudy, Graves outlined the agency's position regarding the waterway south of Demopolis, warning that "at best, authorization for the duplicate locks and/or cutoffs is tenuous and we could become vulnerable in these areas in the future if we tried to claim authority exists." He stressed the need to resolve the authorization question on this part of the waterway as soon as possible, and Corps officials agreed on the two most promising solutions: either add the duplicate locks and additional channel work to the existing Black Warrior-Tombigbee Waterway authorization or seek a completely separate authorizing bill.[37]

When General LeTellier testified before the Senate appropriations subcommittee on public works the following February, however, the Corps pursued neither option. Instead, LeTellier simply called for the completion of all proposed navigational improvements between the Tennessee River and Mobile, listing their accompanying benefit/cost ratios as follows: 1.1 to 1 for improvements between Demopolis and the Tennessee River; 3.4 to 1 for improvements between Mobile and Demopolis; and 1.3 to 1 for the combined improvements.[38]

Economic Evaluations

Within the Department of Defense, the Corps's civilian head, Victor V. Veysey, Assistant Secretary of the Army for Civil Works, was exceedingly troubled by the Tenn-Tom cost increases and by the manner in which the Corps handled the issue. Veysey turned to the Army Audit Agency (AAA) in December 1975, asking the Inspector General to initiate a full-scale audit of the waterway. In a thumbnail sketch, Veysey traced the history of the Corps's Tenn-Tom cost estimates, which had ballooned from $815 million in 1974 to $1.360 billion the following year. He asked the Inspector General to determine the reasons behind these cost increases, whether the Corps's "cost estimating and contract award procedures have complied with applicable regulations, and whether there was an attempt to suppress cost increases."[39]

The AAA completed the audit in September 1976 after a thorough examination of project records at the Mobile and Nashville districts, the South Atlantic Division, and Corps Headquarters. AAA auditors had difficulty locating all the pertinent documentation supporting the Corps's economic calculations, as agency personnel repeatedly told the auditors that the requested documents could not be found or were destroyed. Nevertheless, AAA investigators found enough evidence to charge the Corps with misleading the Congress and the public by not taking into account all the cost factors when requesting Congressional appropriations. The economic justification for the project, the AAA concluded, was less than honest.[40]

AAA auditors contended that the Corps's budget requests were based on insufficiently detailed engineering and planning reports. These deficiencies led to general uncertainty in estimating the ultimate cost of the project. The AAA said the cost estimate for fiscal year 1967 had been "understated by about $28 million" because of using "unrealistic" unit prices for concrete and omitting the costs of clearing and grubbing.[41] The AAA found that the Corps's 1972 cost projection focused solely on parts of the waterway where design memoranda were already approved, but ignored parts of the projects not yet at that stage. As a result, "about $344 million of costs was not included in the fiscal year 1975 cost estimate for the fiscal year 1976 appropriation." Moreover, this underestimate of costs falsely inflated the project's benefit/cost ratio.[42] The AAA cited documentation implicating senior Corps officials in a decision to withhold news of the cost increase from Congress until it could undertake and submit a recalculation of the project's benefits. "Reporting an estimated project cost of $1.159 billion during fiscal year 1975," the AAA stated, "would have shown that the project might be only marginally justified from a benefit/cost standpoint."[43]

The audit concluded that the $1.360 billion cost estimate could not be considered "realistic" because it failed to provide for any future cost escalation, did not include probable federal subsidies to Mississippi and Alabama for highway relocations and bridge construction, and "did not include the added cost of work south of Demopolis to achieve the ultimate potential of the Waterway."[44]

The AAA was not the only government agency able or mandated to assess the economic underpinnings of federal water projects. Within the Legislative Branch, the U.S. General Accounting Office (GAO) had earned a reputation for probing audits and tough assessments of government projects and programs. Environmental activists had long recognized the usefulness of GAO investigations of questionable water projects, as a thorough audit gave Congressional opponents the added credibility of a professional, third-party assessment, while at the same time helping to generate increased public awareness of the economic shortcomings at issue. For members of Congress skeptical of a project, yet reluctant to challenge directly the powerful supporters of that project, a GAO study could open the subject for debate.[45]

Although the Tenn-Tom remained one of the most costly and debated water projects in the United States, it was but one of a whole host of controversial projects being built around the country during the 1970s. Many of these projects belonged to the Corps, but the Bureau of Reclamation (BuRec), the Tennessee Valley Authority (TVA), and the Soil Conservation Service (SCS) also were involved in projects then under fire. Questions about the economic justifications of these projects prompted Congress in 1975 to request the General Accounting Office to investigate the methodology and procedures used by these agencies in preparing their benefit/cost analyses for water resource development projects. The GAO planned to base its report on six case studies—four from the Corps (including the Tenn-Tom), and one each from BuRec and SCS. This study represented a follow-up to a similar report issued earlier the previous year.[46]

By February 1977, GAO field investigators had finished their preliminary evaluations of the six water projects. As customary, GAO submitted its draft report to Congress for review and comment prior to final publication. It was at this point that Senator John Stennis and his staff learned of the Tenn-Tom study, and they were reportedly enraged. The draft report questioned the Corps's economic analysis of the waterway and accused the agency of inaccurate and misleading methods in its calculation of the benefit/cost ratio.[47] Stennis called GAO Comptroller Elmer B. Staats and demanded that he remove the waterway from the report.[48] Although Staats subsequently denied responding to pressure from Stennis, he nevertheless eliminated the Tenn-Tom case study.[49] *Washington Post* staff writer Ward Sinclair later obtained a copy of an internal GAO memorandum addressing the decision to drop the Tenn-Tom from the agency's report. According to Sinclair, the memorandum contained a hand-written comment by the GAO auditor in charge of the study which read: "It seems this decision was made solely to keep from offending Stennis. Where is GAO's impartiality and moral obligation to the American public?"[50]

The final GAO report appeared in August 1978 under the title *An Overview of Benefit-Cost Analysis for Water Resources Projects—Improvements Still Needed*. The Tenn-Tom went unmentioned, but the Corps as well as the BuRec and SCS received poor ratings for their benefit/cost analyses. "Despite the importance of benefit-cost analysis to informed decisionmaking," the report read, "Federal water resources agencies have had a continuing problem preparing accurate, uniform, and logically developed benefit-cost ratios."[51] This problem stemmed, the GAO argued, from the agencies' internal regulations for economic evaluation of water projects, which were "frequently too general or incomplete and lack formal criteria to document the validity of some benefits."[52] Although Tenn-Tom supporters had deflected a direct attack by the Congressional auditors, the GAO report nevertheless increased skepticism of federal water projects.

Local Cost Sharing

During the early 1970s, Congressional supporters had had relatively little difficulty obtaining annual appropriations for the Tenn-Tom, in large part because the venue for the debate and settlement of environmental criticism of the waterway had shifted to the federal courts. As the debate expanded to include economic aspects of the project, the Congressional calm seemed ever more fragile. Problems in obtaining state contributions for the waterway, for example, rose to Capitol Hill for political resolution in 1975 and proved controversial as local cost sharing of federal projects emerged as a general issue in Congress.

Bridge replacements constituted the major cost-sharing responsibility for Mississippi and Alabama. This commitment seriously burdened the public treasuries of two of the nation's poorest states, especially as inflation drove up construction costs. Mississippi, which had the longest stretch of the waterway within its borders, faced the greatest challenge. In 1964, the estimated cost of replacing Mississippi's ten bridges over the Tenn-Tom ranged between $30–40 million. Ten years later, the estimated costs rose to nearly $100 million. Alabama's contribution in 1974 was estimated at $40 million.[53]

Even before construction began, concern developed within the Corps that the states were not responding with the speed and enthusiasm for which the agency had hoped. The states' role—significant in light of their budgets, but small compared to the overall work to be done on the waterway—was important to the completion of the project. In 1972, the Mississippi State Legislature voted down a bill to issue a $10 million bond to start replacing bridges along the Tenn-Tom. The state legislature rejected a similar bill the following year for $15 million.[54] And so, in 1974, the Corps presented the state of Mississippi with a March 1976 deadline for the completion of the first two bridges. Instead of generating overwhelming support within the state legislature, the Corps's deadline set off heated debate and controversy. Once again, attempts to pass state bond issues for the bridge replacements failed. Some members of the legislature, especially those representing regions of Mississippi outside the Tenn-Tom corridor, strongly opposed allocating state funds to the project, despite the previous commitment made by the state highway department.[55]

By early 1975, concern about the states' commitments had surfaced in Congress. M. Barry Meyer, Chief Counsel and Chief Clerk for the Senate Public Works Committee, explained to Senator James Eastland that although the state of Mississippi had agreed in 1964 to pay the full costs of relocating the highway bridges affected by the Tenn-Tom, by 1975 the cost estimate had risen to $138.1 million, an amount the state was "unable to bear." He suggested "a legislative remedy" to the problem, and said that Corps officials had expressed their willingness to assist "in drafting appropriate legislation to deal with this problem." Meyer maintained that state representatives could not possibly have foreseen the bridge replacement cost increases when they signed the original agreement, and therefore "legislation authorizing the Federal assumption of a large portion of the additional costs seems indicated."[56]

In a carefully crafted political maneuver, Mississippi's Congressional delegation devised a solution that would allow the U.S. Department of Transportation to assume much of the state's financial responsibility on the Tenn-Tom. After conferring with Representatives Whitten and Robert E. Jones (D-Alabama), Mississippi Republican Representative Thad Cochran proposed an amendment to the 1975 Federal Highway Act bill then being drafted in the House Public Works and Transportation Committee which would authorize $100 million to be spent to build highway bridges across the Tenn-Tom in Mississippi and Alabama. Although intended to relieve the states of their cost-sharing obligations, Cochran's amendment took care to mention neither the waterway nor any specific states by name. Rather, it authorized the Secretary of Transportation to spend up to $100 million in construction funds during the next three fiscal years for highway bridges "across any Federal public works project" when there was "a substantial change in the requirements and costs of such highway or bridge since the public works project was authorized, and where such increased costs would work an undue hardship upon any one State."[57]

Despite the amendment's deliberate low profile, it did not go unnoticed by the Tenn-Tom critics. The Environmental Policy Center (EPC) called the provision "a dangerous precedent of relieving local governments of their already small percentage of project costs and sets them free from their agreed upon responsibilities." The amendment represented "the quintessence of special interest legislation" and set a precedent that "would remove all incentive for responsible planning on the part of state governments."[58] Because the Senate version of the bill did not contain such a provision, EPC urged interested parties to ask members of the House and Senate Public Works Committees to reject the House provision when the two committees met in conference in January 1976.[59]

Brent Blackwelder of the EPC lobbied Office of Management and Budget Director James Lynn, telling him that if such provisions "are allowed to become law, then we have taken a step to allowing local governments to shirk responsibilities whenever they feel that there is some burden in meeting them." He emphasized the irony of this amendment being proposed "when concerned citizens have been urging that beneficiaries of projects pick up the complete tab for project costs." He characterized the amendment as the "latest covert attempt to eliminate the local cost sharing requirements" and as "yet another indication of how unsound the Tennessee-Tombigbee Waterway project really is."[60]

Leaders of the Sierra Club, American Rivers Conservation Council, and Friends of the Earth joined Blackwelder in writing to the chairman of the House Public Works and Transportation Committee, calling on him to "launch an oversight and investigation hearing into various aspects of the Tennessee-Tombigbee Waterway."[61] Thanks in part to such lobbying, Cochran's amendment failed. However, the waterway's opponents justly feared that a similar amendment might be inserted into the next Rivers and Harbors bill.[62]

Congressional supporters of the waterway could not give up on this issue, of course, for any delays in replacement of the highway bridges would translate into postponements in the completion of the waterway. In fact, the Tenn-Tom's accelerated schedule called for earlier construction of the bridges than had been originally planned. So Whitten and his allies angled to gain support for a clause within the Federal Aid Highway Act of 1976 which would provide federal assistance for the highway bridge replacements.[63] Section 156 of the Federal Aid Highway Act, signed by President Gerald Ford in May 1976, authorized the Secretary of Transportation to spend up to $100 million in federal funds during fiscal years 1977 and 1978 to assist states faced with constructing or reconstructing highways and/or bridges that cross federal public works projects if there had been substantial changes in design requirements or construction cost subsequent to the project's authorization. By so doing, the Highway Act relieved Mississippi of a large share of its financial obligation for bridge replacements over the Tenn-Tom.[64] The Corps did not include these additional federal (via the Department of Transportation) expenditures in calculating the project's overall cost, a fact later publicized by the waterway's detractors.[65] This criticism, along with election-year concerns about the accelerating national debt, led President Ford to rescind the highway bridge replacement funds in September 1976—a procedural move quickly overturned by the Congress.[66]

President Carter and Growing Criticism of Tenn-Tom

Controversy about federal funding of highway bridge replacements over the Tenn-Tom added to the growing number of issues that began undermining political support for the waterway in the late 1970s. The vast scale of the project and its rapidly escalating costs gave pause to Congress, which found itself grappling with ways to control the spiraling budget deficit. Sympathy for greater local cost sharing on federal water projects was growing in Congress, and the Tenn-Tom scored poorly on that account. This change in the political climate meant that for the first time during the 1970s members of Congress stood to be harmed politically for supporting the waterway.

Jimmy Carter won the Presidency in 1976 in part by appealing to this growing skepticism of water resources development policy. During the campaign, for example, he promised to get the Corps of Engineers out of the dam-building business. Following Carter's election, noted conservation writer George Reiger published an article in *Georgia Sportsman* detailing the Tenn-Tom's environmental and economic problems, and urging his readers to write their Congressional representatives in opposition to the project and to "send this article to President Jimmy Carter."[67] The Birmingham Audubon Society followed with its own "Resolution Opposing Uneconomic Water Projects, Including the Tennessee-Tombigbee Waterway," which it submitted to the new President.[68]

Waterway supporters naturally worried about Carter's intentions, and they wasted little time before launching a campaign to convince the incoming administration of the value of the Tenn-Tom. Soon after the election, Glover Wilkins and the governors of Tennessee, Alabama, Kentucky, and Mississippi congratulated Carter on his victory. Reminding him that the South had firmly supported him in the election, they turned to the Tenn-Tom, which they praised as a worthwhile federal project that provided needed employment and economic development to a depressed part of the Deep South, while adding substantially to the national transportation network. "We are hopeful," they stated, "that you will continue to give the Tenn-Tom a high priority in the allocation of funds."[69]

Tenn-Tom enthusiasts received little satisfaction when President Carter issued his water projects review list (or "hit list," as it was popularly called) in February 1977. The President recommended that Congress revoke funding from 19 federal water projects for fiscal year 1978. Although he acknowledged the importance of water projects in developing the nation's economy, he stated that "many of the 320 current projects approved in the past under different economic circumstances and at times of lower interest rates are of doubtful necessity now, in light of new economic conditions and environmental policies."[70] The 19 projects—11 falling under Corps jurisdiction and 8 under the Bureau of Reclamation—were to be thoroughly evaluated on the basis of current environmental, economic, and safety criteria by the two water agencies, the Office of Management and Budget, and the Council on Environmental Quality. Although Carter did not include the Tenn-Tom on his original list, he subsequently added it along with 13 other water projects.[71] To assist the administration in its review, public meetings were to be held on each project. The Tenn-Tom public hearing occurred in Columbus, Mississippi, at the Mississippi University for Women's Whitfield Auditorium on 29 March 1977.[72]

Tennessee-Tombigbee Waterway Development Authority (TTWDA) officials did everything they could to orchestrate events in Columbus. Although the Corps did not publicly announce the meeting until 23 March, the TTWDA received advance notice and began an underground marshalling of support on 17 March. Backers of the waterway thus had twice as long to prepare for the meeting as did the project's critics. In a widely-circulated letter to friends of the Tenn-Tom, Glover Wilkins outlined the threat of President Carter's water project review and the importance of the public meeting in ascertaining "whether 'the people' are for or against the waterway." He asked supporters to inform their close associates about the hearing, but stressed that they should not "publicize this meeting yet, since the hearing has not been officially announced by the Corps of Engineers." Wilkins asked that they ready their associates "to come en masse to Columbus prepared to make a five-minute statement in support of Tennessee-Tombigbee," and he offered assistance in preparing testimony and press releases, and dozens of participants availed themselves of that offer.[73]

Waterway opponents also prepared for the Columbus meeting, although their resources paled in comparison to the TTWDA's. The Tombigbee River Conservation Council (TRCC) spearheaded the opposition. TRCC director Randall Grace urged various groups to speak against the project, and he provided information and suggestions for presentations. On the eve of the meeting, Grace assured TRCC members that representatives from twenty-seven organizations in Mississippi, Alabama, Kentucky, and Tennessee would testify against the waterway.[74] TRCC's sister organization, CLEAN, stressed the importance of the Tenn-Tom public meeting to its membership: "A great deal of national as well as local attention will be focused on this event, and we simply must make a convincing appearance. . . . So dig out your 'Save the Tombigbee' shirts and be ready to go."[75]

Over five thousand people, the majority of them waterway enthusiasts, descended upon the small town of Columbus to attend the public meeting. They found the town wearing its most festive attire, thanks to the

OPPOSITE, BOTTOM: Project advocates pack the public hearing on the Tennessee-Tombigbee Waterway held in Columbus, Mississippi on March 29, 1977. President Jimmy Carter called for this review of the environmental, economic, and safety ramifications of the Tenn-Tom as part of his water projects "hit list." *(Courtesy of U.S. Army Corps of Engineers, Mobile District.)*
OPPOSITE, TOP: A union member demonstrates her support for continuation of the Tennessee-Tombigbee Waterway outside the March 29, 1977 public hearing in Columbus, Mississippi. *(Courtesy of Glover Wilkins Tennessee-Tombigbee Waterway Archives, Archives and Museums Department, Mississippi University for Women, Columbus, Mississippi.)*
ABOVE, TOP: Seated from *left to right* are Tennessee-Tombigbee Waterway Development Authority Administrator Glover Wilkins, Senator Howell Heflin (D-Alabama), Senator John Stennis (D-Mississippi), and Mississippi Governor Cliff Finch, all staunch supporters of the Tenn-Tom. *(Courtesy of U.S. Army Corps of Engineers, Mobile District.)*

efforts of the TTWDA, the Columbus Chamber of Commerce, and other influential backers of the project. Mississippi Legislators recessed their session to participate in the meeting. Pro-Tenn-Tom bumper stickers, posters, and hand-held signs were prominently displayed along the town's streets and sidewalks. Chartered buses brought in crowds; the Mobile Area Chamber of Commerce, for example, provided free, round-trip bus transportation to anyone willing to speak out for the waterway.[76] A gala parade preceded the official testimonies. Mississippi Governor Cliff Finch, driving a front-end loader, led the procession through the heart of town, his construction vehicle's up-raised bucket brimming with photojournalists from the local and national news media.[77] Following the thirteen-hour meeting, Mobile District officials hurriedly compiled the speeches and written statements into five, thick volumes of documents for submission to the White House.[78]

Senator Stennis, who then chaired the Appropriations Subcommittee on Public Works, devoted a full day to hearings on the Tenn-Tom two weeks later. The Mississippian took full advantage of the Congressional anger precipitated by Carter's water projects review. As Stennis told his colleagues: "At the first meeting in the White House after the President's public rejection of a list of ongoing water projects I told President Carter, 'If there is one thing you have done you have brought all of us in Congress together.'" He reminded those at the appropriations hearing that "the reanalysis of these projects was initiated unilaterally by President Carter, without prior consultation with or even prior notice to Congress." Moreover, he added, Carter unilaterally established the criteria used to select the projects under review and ignored the careful analysis of each project conducted by the Congress and the Executive Branch during the authorization process for each project.[79]

The political costs of Carter's water projects review were substantial to the new administration, and the President was forced to back down on nearly all his objections. On 18 April 1977, Carter announced his decision to continue funding the Tenn-Tom.[80] This brought negative responses from the waterway's critics. Randall Grace, on behalf of the Tombigbee River Conservation Council, sent Carter a point-by-point rebuttal of the President's "reasons" for approving the project. Grace charged that the Tenn-Tom was not examined from the "strict set of economic and environmental criteria" established in the President's review of water projects. He charged that Carter had made an error in continuing with the Tenn-Tom, but added that the "mistake can be rectified, even in the face of the blind political support that the project holds locally. The people of this area will stand behind you."[81] The TTWDA, on the other hand, took full advantage of Carter's decision to continue funding the waterway, issuing a slick brochure proclaiming in bold letters: "President Carter Approves Tenn-Tom."[82]

In spring 1977, a score of environmental organizations had formed the Coalition for Water Project Review to support Carter's hit list. When the reform effort collapsed in the hands of Congress, the Coalition continued to oppose the federal government's support of what it regarded as environmentally destructive and economically wasteful water projects. Among these projects, the Tenn-Tom stood out for intense criticism.[83]

Prior to the mid-1970s, conservative Republicans generally viewed environmentalists as obstructors of progress and rarely sided with their opposition to large federal projects. Environmental organizations and railroads opposed to the Tenn-Tom had formed the major coalition against the waterway throughout the 1970s. That coalition grew even more potent during the Carter era with the addition of a third major interest group—fiscal conservatives, represented especially by the public interest organization, the National Taxpayers Union (NTU). The nonpartisan, fiscally conservative NTU fought against tax increases and against large, wasteful federal expenditures. The Tenn-Tom exemplified the latter concern, and in 1978 the NTU joined forces with environmentalists and railroad representatives in opposing the waterway, and the group inserted an entirely new perspective into the debate. In successfully linking economic issues with environmental concerns, the Tenn-Tom played an important role as a training ground and test case.

In lobbying against the Tenn-Tom, members of the National Taxpayers Union coordinated their efforts with several other Washington-based opponents of the waterway. Most notably, they worked with Brent Blackwelder and Peter Carlson of the Environmental Policy Center. Both men had diligently opposed environmentally destructive water projects, and had successfully organized grass-roots opposition. They, too, had attempted to link fiscal conservatives with environmentalists, so their joint efforts with the National Taxpayers Union were mutually beneficial. The NTU also worked closely with Edward Osann of the National Wildlife Federation and with sympathetic Congressional staff—especially with the staffs of Representatives Robert Edgar and Joel Prichard.[84]

Project supporters grew extremely concerned with keeping the construction moving, if not accelerating it. Mounting opposition led proponents to fear that any delay in construction might enable the critics to win allies and strengthen their coalition. After the Tenn-Tom survived President Carter's hit list, for example, long-time waterway advocate Senator James O. Eastland wrote the Mobile District Engineer: "I am pleased that we were able to preserve the Tennessee-Tombigbee Waterway and I hope that progress can be made and the project completed before the environmentalists can bring this project to a halt."[85]

Sometimes, of course, unanticipated events tip the balance in political fights. In 1977, Democratic Representative Joe Evins of Tennessee retired from Congress, leaving vacant the chairmanships of the powerful House Appropriations Committee and its Public Works Subcommittee. For Tenn-Tom promoters, the new chairmen could not have been better. Jamie Whitten, the Committee's highest ranking Democrat, succeeded to the head of the full committee, while Tom Bevill, then the lowest ranking Democrat on the subcommittee, astonished Congress watchers by breaking tradition and gaining the subcommittee chairmanship. Bevill, like Whitten a staunch proponent of the waterway, now became the author of the annual House appropriations bill for federal water resources projects.[86]

Despite these well-placed Tenn-Tom allies, opposition in the Congress gained strength. In October 1978, for example, Senator William Proxmire of Wisconsin inserted James Nathan Miller's damning *Reader's Digest* article, "Trickery on the Tenn-Tom," into the *Congressional Record.* Proxmire, who was a member of the Senate Appropriations Committee, claimed that the Tenn-Tom "exemplifies all that is wrong with water resource development in areas where such development is not only unnecessary but positively destructive." In typical Proxmire rhetoric, he called the waterway "an outrageous waste of . . . tax dollars" and "the most expensive public works boondoggle in American history."[87]

At the White House, the Carter Administration continued to be besieged by Tenn-Tom critics to withdraw support of the project. In November 1978, Carter asked Secretary of the Army Clifford Alexander to comment privately and candidly about "allegations that the Corps of Engineers had used incorrect or misleading factors" in assessing the Tenn-Tom.[88] Alexander said the allegations were true; prior to the budget request for fiscal year 1977 funds, "the Corps reported benefit-cost information on the Tennessee-Tombigbee based on contemplated work which was beyond the scope of the authorized project." Alexander said that the decision to delay reporting the dramatic cost increases pending an updated benefit analysis did not technically violate procedures then in place. "Doing so," he said, "avoided possible embarrassment to both the Corps and the Congress, and thereby continued momentum toward project completion."[89]

Alexander reminded Carter that "the Tenn-Tom's weak economics were evident at the time of the 'hit list' review." He warned that reversal of the administration's decision to continue funding the waterway "would generate enormous opposition." He added that the current litigation would test the various contended issues associated with the project.[90] Carter's chief domestic policy adviser, Stuart E. Eizenstat, agreed that the Tenn-Tom suffered from "serious economic problems" and "perhaps in-adequate authorization," and that the Corps "misled the public and the Congress in 1976 about project costs," but he concurred that it would be "extremely unwise" to place the project under administrative review once again, especially considering the pending litigation. Eizenstat recommended waiting for the court's decision, and warned that should the Corps lose on the authorization question, the administration would find itself in "a difficult situation because project proponents will seek our support for additional authorizing legislation."[91] Carter followed that advice, and did not intervene in the controversy.

CHAPTER VIII

Social Justice: "A Cruel Hoax"

DESPITE THE WIDESPREAD environmental and economic objections to the Tenn-Tom, the waterway enjoyed a broad base of support among the citizens of Alabama and Mississippi. This enthusiasm was rooted in the faith that expenditures of hundreds of millions of federal dollars in their states could only help revive their sagging economies. Although Congress used national economic benefits (primarily in the form of transportation cost savings) to justify the allocation of public funds for the project, the popularity of the waterway in the Southeast came from the promise of local jobs and business opportunities. "In the years ahead, with the completion of the Tennessee-Tombigbee Waterway," proclaimed the *Alabama News Magazine* in typical fashion, "every forecast is that West Alabama will experience an industrial expansion without equal anywhere in the nation."[1]

The appeal of economic and developmental benefits was heightened by the fact that the Tenn-Tom crossed one of the most impoverished areas of the country. In 1970, for example, one-third of the families living in the counties adjoining the waterway fell below the poverty line. And African-Americans, who made up approximately 40 percent of the population, were twice as likely as whites to be impoverished.[2] Advocates portrayed the waterway as a panacea for regional ills in much the same manner as apologists of the Tennessee Valley Authority had defended their project during the 1930s. Noting the high percentage of chronically un- and under-employed African-Americans living in the Tenn-Tom corridor, waterway proponents sought national support from the civil rights movement by stressing the prosperity the project would generate for the region's minority population.

As the Tenn-Tom came under increasing attack in the federal courts, the Congress, and the national news media because of its harsh environmental consequences and its dubious economic justification, the project's promoters attempted time after time to shift the focus of attention to the economic benefits likely to accrue to this poor region of the United States with its large minority population. Such emphasis on social benefits invited close scrutiny of the Corps's accomplishments because regional economic development had never been part of its mission (unlike the TVA). African-American community leaders became enthusiastic supporters of the Tenn-Tom, taking at face value the promises of plentiful employment opportunities and job training, but their hopefulness gradually soured after construction got underway. All of the Corps's initial prime contracts went to large construction firms headquartered outside the Southeast, and, in the absence of employment goals for minorities and local workers, these contractors brought with them much of their own labor supply, especially in the high-skilled, high-paying crafts. In light of the rhetoric used to justify the Tenn-Tom, the local minority population felt betrayed, as they found themselves being excluded from the jobs, contracts, and decision making associated with the waterway.

Racial discrimination was not new to this region of the Southeast, whose African-American citizens had experienced a long history of social and economic injustice. The failure of the Tenn-Tom to provide meaningful employment and business opportunities to the local minority community gave every appearance in the early 1970s of being yet another example of a prolonged pattern of mistreatment. The legacy of the civil rights movement, however, was not to be underestimated, especially in the human terms of training and empowering a generation of community leaders. By the mid-1970s, a handful of civil rights movement veterans residing within the Tenn-Tom corridor drew from their earlier experiences to launch a community-based, adversarial campaign to force the government to live up to its promises of business and employment opportunities for the local minority population. They created the Minority Peoples Council, a nonprofit organization formed to ensure that the benefits of the Tenn-Tom were distributed equitably to all members of the local community. They understood that the environmental and economic controversies were steadily undercutting Congressional support of the waterway, and that the project could ill afford to lose further support on other grounds. The Minority Peoples Council consequently pressured the Corps to increase both the employment of African-American workers and the award of contracts to minority-owned firms, if the agency wanted to ensure the endorsement of the Congressional Black Caucus. The army engineers capitulated in 1976 by issuing minority personnel utilization goals, listed by craft, for all work done on the Tenn-Tom.

President Jimmy Carter instructed the Corps to strengthen its tepid efforts to use its projects to stimulate local economic development. In 1977, the Carter White House issued its Rural Initiative Program, based upon the premise that large-scale public works had profound impacts beyond their stated technical objectives and should therefore be directed to further the social goals of the nation. The Tenn-Tom became one of the demonstration projects of the Rural Initiative Program to increase the independence, skills, and the general quality of life of the people living near the waterway. Despite this high-level support, African-American community activists found it necessary to monitor and prod the Corps to force it to abide by its stated goals for minority participation.

Rising Expectations

It was not difficult to convince the people living in the Tenn-Tom corridor that their communities would benefit from the infusion of millions of dollars in federal public works expenditures. The promise of a better tomorrow can be irresistible, especially for a region whose own economic prosperity has lagged behind that of the rest of the country. The Tenn-Tom's apologists knew this, and took full advantage of the hope and expectations of the people in western Alabama and eastern Mississippi.

The promoters' constantly exaggerated claims led the Corps's Nashville District Engineer to remark: "It is no wonder that the people of the region considered [the Tenn-Tom] a cure for all their troubles."[3]

Support of the local population was not sufficient in itself to build the Tenn-Tom, of course, because federal public works projects must be justified on national, not regional grounds. Although transportation savings constituted the project's raison d'etre, the Tennessee-Tombigbee Waterway Development Authority stressed the Tenn-Tom's added contribution of easing the mounting social problems in American cities. Authority members, paraphrasing scholarly analysis of the urban crisis confronting the nation during the 1960s, claimed that the steady decline in rural job opportunities in the South had propelled millions of poor and ill-trained people to migrate to northern cities in search of employment, "causing grave and concentrated social and economic maladjustment." Sociologists, they stated, agreed that deprived people are better assisted by opportunity in their home environment. This, the Authority members said, could be provided through the "economic stimulus" of the Tennessee-Tombigbee Waterway, which would help stem the tide of Southern out-migration. "It is grimly ironic that the economic cost of the Detroit riots, alone, is far in excess of the entire cost of the Waterway," they asserted. "Its potential for constructive assistance in solving the crucial problem of metropolitan congestion is a compelling reason for its prompt initiation."[4]

Republican Representative Jack Edwards of Alabama concurred: "This is the sort of federal government expenditure that will really help in the fight against poverty and malnutrition in rural America." Using the codewords of the Republican establishment, he said that the Tenn-Tom "will help many good people move from the growing welfare rolls with well earned paychecks in their hands for honest work—work resulting from countless jobs created by the construction of the waterway and the commerce which it will generate."[5] Seeking to solidify Congressional support for the project in 1970, Edwards claimed that "the waterway brings with it jobs, both on a short-term and long-term basis. Immediately, it offers jobs on construction crews and the use of local resources. In the future, it offers new industry creating new jobs."[6] The waterway provided "new hope and a new way of life" to the local population, he argued. "For many, it represents their only hope for the future."[7]

The Nixon Administration was anxious to highlight the social benefits promised by the waterway, believing that that would help deflect the criticisms hurled at the President by environmentalists protesting his endorsement of the project. With construction slated to begin with the Gainesville Lock and Dam in Greene County, Alabama, one of the nation's poorest counties and the county with the highest percentage of African-Americans (73 percent) bordering the waterway, the President's special assistant, Robert J. Brown, told the Corps that the White House maintained "a very strong desire that job opportunities resulting from the Gainesville project be made available to residents of Greene County" to help alleviate this "desperate situation." He stressed, however, that employment of these people depended upon the agency exercising "extreme vigilance" to ensure "compliance with Federal regulations relating to non-discriminatory hiring practices."[8]

Under Secretary of the Army Thaddeus R. Beal assured the White House that the Corps would "encourage contractors to employ local people," and that equal opportunity requirements would be stressed.[9] African-American community leaders in Mississippi and Alabama were encouraged by the agency's rhetoric, which led them to look upon the Tenn-Tom as a long-term economic stimulant for the impoverished area. This hope caused most community organizers to oppose the environmental lawsuit brought against the Corps. In general, they shared a common perception among African-American activists that the mainstream, predominately white environmental organizations valued outdoor recreational amenities over basic issues of social justice and economic equity.[10]

Greene County Probate Judge William McKinley Branch became one of the most outspoken black leaders to question the environmentalists' motives. Branch asserted that the African-American population of Greene County and other communities to be affected by the waterway stood to benefit enormously from the construction effort, "since the Corps says 90 to 100 per cent of the unskilled construction labor would be drawn from the immediate area."[11] Referring to the thousands of jobs to be created, Branch contended that "the environmentalists know that, too. Let them prove their goal is not to kill these jobs."[12] He charged the national environmental groups with caring "more about wildlife than the desperate needs of suffering poor people," and added that "there's got to be a way the ecologists and engineers can get together and build this project safely."[13]

CLEAN chairman and fellow Alabamian James D. Williams spoke to this concern. "I think if people looked into the economic arguments," he said, "they would see that the money going for the dam [at Gainesville] could be channeled into something more directly beneficial to the people of the area. Why not put the money into a highway? Why not develop other types of programs? We're not against jobs for the people of that area, but other things should be taken into account."[14] Williams later recalled that he and other environmentalists tried to convey to the local minority leaders the lessons that were learned from the fights over the Cross-Florida Barge Canal and the Gillham Dam in Arkansas. That experience showed that large, out-of-state contractors would come into the area with their own work forces to build the project. Environmentalists therefore warned the community activists not to be misled by promises of abundant employment opportunities. In fact, Williams speculated that there might actually be a net *loss* of jobs if the waterway brought an end to such local employers as trucking firms and railroads.[15]

As Williams discovered, however, environmental organizations had a poor track record in the early 1970s of caring for the needs and concerns of poor and minority groups, and, as a result, leaders of the African-American communities in Alabama and Mississippi remained unswayed by CLEAN's arguments and clung to the hope of economic rejuvenation. Their optimism held fast in December 1972, when ground breaking ceremonies for the Gainesville Lock and Dam officially launched construction of the Tenn-Tom. The Corps had awarded the contract to the Guy H. James Construction Company of Tulsa, Oklahoma, one of the country's largest civil engineering firms. To the dismay of the local population—black and white alike—Guy James imported the majority of its workers from Oklahoma and housed them in trailers near the work site. Construction at Gainesville therefore resulted in the hiring of only a scant number of local workers—a small proportion of which were African-Americans—

and those few locals who did find employment filled the low-skill positions.[16]

This pattern continued throughout the early years of construction. In addition to the poor representation of minority workers on the Tenn-Tom, small, local contractors also faced a variety of disadvantages. Most of the prime contracts were in the range of several millions of dollars, which restricted the viable bidders to large, experienced companies with bonding and financing capability. Few firms in the Southeast met these basic requirements, and of those that did, none of them were minority-owned businesses. Several hurdles stood in the way of substantive im-provement of minority business participation: prime contractors often used the same subcontractors from job to job, only a small percentage of which were minority business enterprises; prime contractors, who usually insisted that their subcontractors be bonded, claimed that few minority subcontractors were so qualified; the large contracts generally exceeded the range of most minority business enterprises; and no formal regulations assured the participation of small and minority contractors.[17] So, despite efforts by the Corps to encourage the prime contractors to involve minority-owned firms as subcontractors and suppliers, there was no real incentive for them to do so. As late as August 1976, therefore, no minority firms were at work on the waterway's largest section, the divide cut.[18]

The Growth of Community Activism

This lack of progress attracted a wide array of public and private organizations to fight to improve minority participation in construction of the Tenn-Tom. The Mississippi Economic Development Corporation (MEDC) exemplified private sector efforts. Established in 1972 as a nonprofit corporation dedicated to promoting economic development in Mississippi, the MEDC took an early interest in the opportunities inherent in the building of the waterway. In 1975, the MEDC created a Tennessee-Tombigbee Division with the stated purpose of facilitating the involvement of minority-owned businesses.[19] To accomplish this goal, the Columbus-based Tenn-Tom Division held a series of public meetings and workshops to provide technical assistance in the preparation of bids, to help sharpen the management and marketing skills of minority and disadvantaged businesses, to advise on contract compliance, to explain the bonding and financial assistance available to minority contractors, and to disseminate general information about contract and subcontract opportunities.[20]

The MEDC's Tenn-Tom Division changed its name to the Tennessee-Tombigbee Construction Assistance Center in 1976 and hired Esther M. Harrison as its executive director. Among other things, the center attempted to link local minority-owned construction firms and supply companies with the major construction firms bidding on the prime contracts.[21]

While the MEDC focused on opportunities for disadvantaged businesses, the Federation of Southern Cooperatives, headquartered near the waterway in Epes, Alabama, concentrated on employment practices. Federation leaders viewed the Tenn-Tom as an engine to stimulate economic development in the rural South, helping especially the region's poor and minority populations. John Zippert, a civil rights movement veteran and director of the Federation's Research and New Program Development office, assumed a leading role in pressing the Corps to assure that the waterway directly benefitted the people of limited income living in the project's vicinity.[22]

Federation officials became increasingly dismayed by the failure of African-Americans to gain fair and proportional representation in either employment on waterway construction projects or in contracts (or subcontracts) for construction. In January 1974, the Federation hosted the "First Peoples Conference on the Tennessee-Tombigbee Waterway" to address the lack of minority involvement. Over 200 community residents attended the meeting, where they resolved to create the Minority Peoples Council on the Tennessee-Tombigbee Waterway as an ongoing grass-roots advocacy organization committed to full minority participation. The council's manifesto asserted: "For Black people to share equitably in the far reaching benefits of the project, we must be informed and organized. We will need to monitor closely every aspect of the Tennessee-Tombigbee Waterway Project to insure Black people share fairly in its benefits." The council proposed to serve as "a countervailing, locally-controlled force"—a need created by the fact that no minority representatives sat on the "existing public decision making bodies relative to the Tennessee-Tombigbee Waterway."[23]

The Minority Peoples Council quickly established a formal list of objectives, which included: a minimum employment goal of 40 percent minorities in all aspects of the construction and management of the waterway (including employment within the Corps); promotion of training programs for minority workers; 40 percent minority involvement in decision-making boards; involvement of minority people in the economic development of the region accruing from the waterway; and the promotion of minority land ownership and black institutions of higher learning.[24]

The council benefitted from its members' experiences in the civil rights movement, most notably by incorporating such programs and tactics as coalition building, public protest, close interactions with the news media, report writing, fact-finding, attacks on the project where it was most vulnerable, and elevation of the minority participation question as an issue before Congress.[25] In May 1975, for example, council chairman (and veteran civil rights activist) Wendell H. Paris told Representative Joseph Evins, Chairman of the Appropriations Subcommittee on Public Works, that the council remained "doubtful and suspicious" of the Corps's claim that the waterway "was established to help poor and unemployed in the impact area."[26]

Brent Blackwelder of the Environmental Policy Center probed this chink in the Corps's armor when he testified before the Senate Public Works Committee that "a basic question of social justice arises over the expenditure of more than a billion dollars to put a barge canal through a very poor area of the country, an area which certainly has much more compelling needs than a barge canal." He noted that the majority of residents in many of the counties traversed by the waterway were African-American, yet, he said, "there are no Blacks on the Tenn-Tom Waterway Development Authority to help insure that planning would contribute to the betterment of Blacks and poor people in the area."[27] Blackwelder was not alone. Senator Edward W. Brooke, a Massachusetts Republican and one of Congress's prominent African-American leaders, complained to Assistant Secretary of the Army for Civil Works Victor V. Veysey about the lack of minority board members on the Tennessee-Tombigbee Waterway Development Authority (TTWDA). Veysey responded that the federal government had stepped up its efforts to increase minority participation in the construction of the Tenn-Tom as well as in its planning and design. He explained, however, that the federal government had no say in the appointment of board members to the TTWDA—it being an interstate agency—and recommended the Senator raise the issue with the appropriate state governors.[28]

TTWDA officials, whose principal purpose was to sustain enthusiasm for the Tenn-Tom in their member states and in Congress, found themselves frustrated by the community activists who berated the Corps and its associates for their poor minority hiring practices. Convinced that publicity surrounding the minority issue might endanger the entire project, the TTWDA staff thought it better to address this question quietly than to air the Corps's dirty laundry. However, the leaders of the Minority Peoples Council were not about to pursue a soft-spoken approach. As Lamond Godwin explained to Glover Wilkins: "I am sure that a person of your proven political acumen can understand and appreciate why we must be persistent on the matter of equal employment opportunity."[29] Wendell Paris elaborated on this, telling reporters at Meridian, Mississippi's newspaper: "If our legitimate requests are not accepted we will be forced to oppose this project at every level; at the appropriation hearings in Congress, in state legislatures, in the courts and at the construction sites."[30]

Some of the Minority Peoples Council's criticism bore fruit. In 1977, the TTWDA added African-American members to its board of directors through Alabama Governor George Wallace's appointment of Tuskegee Institute President Luther Foster and Greene County Probate Judge William McKinley Branch to his state's delegation. The TTWDA's ten-person staff, on the other hand, remained all white, "except one part-time black college student," and Glover Wilkins explained that "with such a small staff there is little turnover and, consequently, little opportunity to increase minority employment." Nevertheless, Wilkins claimed that TTWDA was going out of its way to make use of minority-owned businesses. He stated, for example, that they had "conducted an exhaustive search for a minority printing firm in the area," and although the nearest company was in Birmingham and "posed some logistic problems and was inconvenient," they had asked this business to do some of their printing.[31]

Waterway advocates needed to assure all possible votes for the continued Congressional appropriations; the mere possibility that controversy over minority participation might result in a turnover of a block of votes against the project moved them to action. In October 1975, Mississippi Senator John Stennis hosted a meeting at the request of the Minority Peoples Council to address Tenn-Tom affirmative action plans. The Assistant Secretary of the Army for Civil Works and representatives from the Office of Federal Contracts Compliance, the Corps, labor unions, and contractors attended. The two-and-a-half-hour meeting ended with Stennis expressing his strong desire that a comprehensive affirmative action plan be implemented within six months—a plan that would establish a minimum minority hiring goal of 40 percent and work for increased minority participation in planning activities.[32]

Major General Carroll N. LeTellier, the South Atlantic Division Engineer, echoed Stennis's concern, telling his superiors on the eve of his departure from office that, despite the tumultuous environmental and economic controversy surrounding the waterway, no other consideration immediately threatened the political viability of the project more than minority participation. He recommended a high-level meeting on the issue, stating: "There is no question in my mind that it is normal and proper to expect that the largest public works project in the nation going through one of the most depressed areas in the country with an extremely high minority density should raise great expectations from the Corps."[33] When briefing his successor, Brigadier General Kenneth E. McIntyre, LeTellier dwelled on the minorities issue, which he believed was "about to blow sky high."[34] McIntyre received assurances from Corps headquarters that the issue would be elevated in priority. "I agree," wrote Director of Civil Works Major General Ernest Graves, "that some positive means of ensuring that minorities receive their share of the economic benefits of the Tennessee-Tombigbee Waterway construction program is necessary."[35]

Senator Howard H. Baker, Jr., Republican from Tennessee, added his voice in calling for greater minority involvement. As early as March 1976, he expressed his concern that no contract had yet been let to a minority-owned business. "The Tennessee-Tombigbee Waterway is a showcase project for the Corps," the Senator remarked. "I believe that the Corps must do all it can to assure fair consideration of black-owned firms." Baker asked the army engineers to look into the various problems associated with minority participation on the Tenn-Tom and to report back to him.[36]

California Democrat Augustus F. Hawkins, chairman of the Subcommittee on Equal Opportunities of the House Committee on Education and Labor and a prominent member of the Congressional Black Caucus, pressed the Secretary of the Army on this issue. When he presented his subcommittee's report, "Equal Opportunity Employment on the Tennessee-Tombigbee Waterway Project," just prior to consideration of the House fiscal year 1977 public works appropriations bill, Hawkins emphasized the failure of the Corps to assure employment opportunities for minority citizens. The report charged the agency with a "lack of action and cooperation" with the Minority Peoples Council's proposed affirmative action program, which it recommended the Corps adopt.[37]

Victor Veysey gave the standard comeback, assuring Hawkins that the Corps "has cooperated fully with the efforts of the local minority organizations."[38] Veysey later reminded the Chief of Engineers that "contractor employment practices have aroused considerable Congressional and public interest from the beginning." He instructed the Corps to convince contractors to improve their hiring practices. "While escalating costs are a major concern on this project," he said, "even a higher priority must be our national policy of equal employment opportunity."[39]

Responding to the intensified Congressional and administrative pressure, the Corps finally issued an affirmative action plan in July 1976. The plan established four zones along the Tenn-Tom, each with its own minority hiring goals based upon the percentage of minorities in the local work force. The hiring goals, from south to north, were 26 percent, 40 percent, 30 percent, and 12 percent.[40]

The Minority Peoples Council's leadership was livid at this response and lobbied vigorously against the plan. It argued that zones with the highest minority hiring goals covered the areas where a substantial amount of construction had already been completed, while the lowest percentage zone in northeast Mississippi encompassed the divide cut, the project's largest construction site. The Corps responded to this complaint, submitting an Areawide Affirmative Action Plan that established a timetable and employment goals applicable to the entire waterway.[41] The compromise plan, which was approved by the U.S. Department of Labor, set a goal of 19 to 20 percent of minority workers to be employed immediately in all trades. The goals thereafter increased to 21 to 22 percent in 1977, 23 to 24 percent in 1978, 25 to 26 percent in 1979, and 27 to 30 percent in 1980 and beyond.[42]

From this moment forward, special bid conditions on construction contracts established the affirmative action requirements. The minority hiring percentage *goals* (the government was careful not to call them quotas) and timetables were listed, and contractors had to demonstrate "good faith efforts" in meeting these requirements. Good faith efforts meant assigning a company official to oversee the affirmative action plan; notifying minority recruitment sources, community organizations, and labor unions of the company's intent to hire minority workers; maintaining a file on each minority worker referred to the company and documentation of why that person—if not retained—was not hired; participating in local training programs; encouraging minority employees to recruit their family and friends; assuring that the company's hiring and promotional practices did not discriminate on the basis of race; and continual monitoring of its equal employment program. Failure to uphold these good faith efforts could lead to suspension, termination, or cancellation of the contract.[43]

The Minority Peoples Council reluctantly supported the revised plan. Council members were disappointed because they believed the employment goals were still too low (an ultimate goal of 30 percent by 1980, as opposed to the council's 40 percent target) and that the "good faith effort" requirement of contractors was primarily a record-keeping and public relations exercise. Within a year, the council demonstrated that the good faith clause actually constituted "a blueprint of loopholes for contractors to follow to subvert vigorous compliance with the Plan."[44]

Implementation of the new plan was flawed. Although several contractors had made significant progress toward meeting the overall percentage goals of minority employment, they failed to recruit workers for skilled occupations. The Office of the Secretary of the Army ordered the Corps to strengthen the training programs, but protests had already begun.[45] On 1 June 1977, nearly a hundred people, most of them African-Americans, demonstrated in front of the Corps's divide section field office in Holcut, Mississippi. The protesters marched for nearly two hours carrying such signs as "Wake Up Corps of Engineers" and "The U.S. Corps of Engineers Is Not Enforcing the Minority and Local Hiring Requirements of the Contracts." The Tupelo, Mississippi, television station covered the event on its evening news broadcast. During the demonstration, a ten-member delegation—headed by Wendell Paris of the Minority Peoples Council and Aaron Henry of the National Association for the Advancement of Colored People's (NAACP) Mississippi chapter—met with agency officials to discuss the group's complaints that the contractors had not hired the required percentages of minorities and locals, nor had they provided them with adequate training opportunities.[46]

In addition to criticism about the low numbers of African-American construction workers, the Corps's Mobile District also came under attack for racial discrimination within its own ranks. Veysey noted that "only 124 of the 1534 Corps employees were minority group members, and none of the 184 GS supervisory personnel was a minority group member." He chided the Mobile District for having "no goals or timetables established for increasing minority group hiring in the District," and for allowing segregated facilities to exist in the district's Tuscaloosa Area Office until 1976. The Assistant Secretary demanded a full accounting of the reasons for the agency's poor performance in minority hiring. He also requested the names of those individuals responsible for the segregated rest rooms, lockers, and water fountains at the Tuscaloosa office and a description of the contemplated disciplinary action.[47]

President Carter's concern with environmentally and economically questionable water projects, with civil rights, and with uplifting economically depressed areas of rural America came into sharper focus under his Secretary of the Army, Clifford L. Alexander, Jr., an African-

American whom the President mandated to improve opportunities for minority members throughout the Army, including the Corps of Engineers.[48] When Alexander addressed the Corps's annual meeting of Division Engineers in Lake Lanier, Georgia, in spring 1977, he devoted 25 minutes of a 30-minute speech to a discussion of the minority problems within the agency, and most of that discussion revolved around the Tenn-Tom. Alexander reminded Corps executives that they worked for him and that they had better correct this problem. Brigadier General McIntyre paraphrased Alexander as saying: "There are two ways that I, as Secretary of the Army, can approach this. I can use the carrot approach or the stick approach. And I am here to tell you I just put my carrot in my pocket."[49] Although Alexander never actually used his "stick," his tough rhetoric contributed to the climate of change that enveloped the army during the late 1970s and which resulted in modest improvements in the military's minority recruitment and promotion policies and practices.

Within this context, John Zippert, director of research for the Minority Peoples Council, and Robert Valder, director of the Southeast Regional Office of the NAACP Legal Defense and Education Fund, prepared a primer on minority participation in the Tenn-Tom construction. The "White Paper," as it was known, summarized the environmental and economic criticism of the waterway, the gap between the Corps pronouncements and practices regarding minority participation, and the initiatives needed to rectify the situation. "A cruel hoax is being perpetrated against the minority and poor people of the Tennessee Tombigbee Waterway area," Zippert and Valder argued. "Their economic plight is being manipulated to justify the construction of a major development project which will serve local, regional and multinational economic interests, but whose benefits the people themselves may never fully enjoy."[50]

The minority advocates presented several recommendations: that the Corps submit a comprehensive affirmative action plan by October 1977 to bring the waterway project into full compliance "with the requirements of the Constitution and all relevant civil rights statutes"; that the Tenn-Tom affirmative action plan be designed to serve as a model for all large federal public works projects; that the Corps improve its own internal equal employment opportunity program; that government agencies, private contractors, and labor unions provide better training programs for minority workers; and that Congress legislate a "Minority Resource and Oversight Center" to coordinate affirmative action programs on the waterway.[51]

The Minority Peoples Council and the NAACP Legal Defense and Education Fund submitted this blistering exposé (which ran to nearly 150 pages) to Alexander and the Chief of Engineers in August 1977, along with a request for a formal meeting to resolve the issues. They then distributed the document widely, giving copies to every member of Congress, to the President and his staff, to relevant federal, state, and local officials, and to environmental organizations and other interested parties.

Carter's Rural Initiative

Having underestimated the political clout of the Tenn-Tom's Congressional supporters when he challenged the project with his hit list, President Carter was determined to wring as much good out of the waterway as he possibly could. As the Minority Peoples Council and the White Paper argued, the Tenn-Tom presented the administration with broader social opportunities. These opportunities fitted into Carter's rural development policy to link social goals to large public works projects and to emphasize the socio-economic advancement of disadvantaged peoples living in the vicinity of such projects. When Carter withdrew his opposition to the waterway after the hit list review, he recommended that the Corps develop a program to assure full employment opportunities for minorities and local residents.[52]

In August 1978, Carter created the Interagency Coordinating Council to facilitate the implementation of the federal government's urban and regional policy. Jack H. Watson, Jr., Assistant to the President for Intergovernmental Affairs, chaired the council. Watson looked for promising cases of interagency coordination to serve as models for the type of efforts the council sought to promote. When Sara Craig, chairwoman of the Southeastern Federal Regional Council, brought the Tennessee-Tombigbee Waterway to his attention, Watson immediately recognized its potential. By bringing representatives of local, state, and federal agencies together with members of local minority and labor organizations, Watson believed regional employment and economic development opportunities would be realized.[53]

Watson made the development of federal guidelines for the expansion of job opportunities for local residents and minorities on federal construction activities a high priority for the Interagency Coordinating Council. "Because of its size and job creating potential," he told the council members in January 1979, "the Tennessee-Tombigbee project will serve as a 'proving ground' for this effort."[54] Two months later, the Carter White House convened a two-day meeting to consider area development associated with large-scale federal construction projects. Seventy people from across the country joined in the deliberations, including representatives from the Minority Peoples Council and other regional minority interest groups.[55]

The Interagency Coordinating Council incorporated the ideas formulated at this meeting into the Small Community and Rural Development Policy announced by President Carter in December 1979. As Jack Watson described the policy, "it provides a clear purpose and program of action for addressing small community and rural needs." By February 1980, "Area Development from Large-Scale Construction" became a major policy initiative and featured two Corps of Engineers projects as demonstrations: the Tennessee-Tombigbee Waterway and the Red River Waterway in Louisiana.[56]

To further its policy objectives, the White House Rural Initiatives Program spawned the Tenn-Tom Project Area Council (PAC) in October 1980. The PAC comprised representatives from federal and state agencies, from local and private organizations, and from interested citizen groups. Although it was an intergovernmental group, funding came solely from the Corps and the Tennessee Valley Authority. PAC encouraged employment of local minorities, women, and economically disadvantaged persons on the waterway construction; participation of small and minority businesses; and maximum economic development of the region (notably by attracting new industries to the area).[57] Despite its promise and strong community appeal, the PAC was dismantled the following year—along with the entire White House rural initiative—by the new Reagan Administration. The PAC's short life span precluded its accomplishment of any substantive results and left President Reagan free to abolish the program (along with its associated concerns) quietly and without stated justification.[58]

Congressional Vigilance

As challenges to continued Congressional appropriations for the Tenn-Tom intensified during the early 1980s, the minority participation issue assumed ever greater importance. In their attempt to persuade members of the Congressional Black Caucus to vote against the project, critics resurrected and distributed excerpts from the Minority Peoples Council's 1977 "White Paper." Waterway supporters viewed this action as unwarranted, as they believed the Corps had made great strides toward improving minority participation. In July 1980, Donald G. Waldon, Deputy Administrator of the Tennessee-Tombigbee Waterway Development Authority, asked Minority Peoples Council Chairman Wendell Paris to provide Representative Tom Bevill with an assessment of the minority participation on the Tenn-Tom during the past three years. Waldon acknowledged that Paris and his colleagues were "not completely satisfied with all the efforts required for full minority involvement in the waterway," but he was confident Paris would agree that "considerable progress" had nevertheless been achieved.[59]

Paris submitted a report, but it was not the document Waldon had envisioned. Paris told Bevill that the Minority Peoples Council had always supported "full funding and construction of the Tennessee-Tombigbee Waterway provided that minority and poor people receive an equitable share of the benefits in the construction and post-construction phases of the project's development." As requested, Paris compiled a list of minority participation accomplishments since 1977. Although this list documented improvements, it also emphasized the failure to reach any of the stated hiring goals, especially in the higher-paying, higher-skilled crafts. Paris included a detailed set of recommendations and bluntly expressed his conviction "that your concern to investigate minority participation in the Tennessee-Tombigbee Waterway was prompted by opposition to the funding of the project." Paris did not mince words: "Your concern seems more related to an uninterrupted flow of money to build the project than a genuine concern for human and civil rights."[60]

Members of the Congressional Black Caucus also remained suspicious of talk without action, and in 1980—with the waterway facing live-or-die votes on continued appropriations—Caucus members became increasingly influential. As the senior African-American Congressional member on the House Appropriations Committee, Representative Louis Stokes of Ohio became a natural target for lobbyists on both sides of the issue. Proponents such as Bevill and Jamie Whitten realized the waterway could ill afford additional controversy. Because boosting the government's affirmative action efforts seemed to offer an important course of action, Bevill took the lead in trying to convince Stokes that the Corps was making a strong effort toward minority hiring and that most African-American workers and businesses in the Tenn-Tom area favored the project. Bevill invited the Ohio Democrat to visit the waterway and talk to members of the local community. Stokes accepted the invitation. He became convinced that the project was, indeed, helping the local minority community, and he thereafter spoke out in favor of the waterway—both officially at appropriations hearings, and privately when soliciting the support of his Black Caucus colleagues.[61]

Stokes's endorsement proved vital, for when the House of Representatives called its vote for Tenn-Tom appropriations in 1980, the waterway won by a thin margin of ten votes, nine of them cast by members of the Congressional Black Caucus. As South Atlantic Division Engineer Major General John F. Wall told a meeting of Tenn-Tom contractors, that vote "illustrates that what happens to the disadvantaged minorities in the [Tenn-Tom] area is definitely of interest to Congress and the Black Caucus in particular." He continued, stating that "what is at stake is more than a conception of social justice that you may or may not subscribe to. It's more than good faith efforts toward meeting a contractual commitment. We're talking about the life of the project itself."[62]

The deal that was struck in 1980 did not go unnoticed, however. The *Cleveland Plain Dealer* questioned Stokes's role in "persuading some of his black House colleagues to vote for" the Tenn-Tom.[63] Stokes explained to the newspaper editors that "it's a political situation we're in here"; federal funding for the Cleveland harbor improvement and Euclid Creek flood control projects demanded his support of the Tenn-Tom, he claimed, because Alabama Representative Tom Bevill chaired the Appropriations Subcommittee on Energy and Water. As Stokes said matter-of-factly: "I couldn't say thanks for the $20 million for my project, but I can't vote for your bill. This is what politics is about."[64]

Although not challenging Stokes directly, environmental activist Brent Blackwelder continued his reference to minority participation and contracting on the Tenn-Tom in August 1982 by writing all members of the Congressional Black Caucus (CBC). Blackwelder accused the Corps of racial discrimination and contended that completion of the waterway would provide few significant benefits to African-Americans living in the project area. "Even though blacks comprise 40% of the population of the area," he wrote, "black contractors will get less than 3% of the outlays on the project." Blackwelder contended that "a more blatantly discriminatory project would be hard to conceive. Almost any other type of program could not get away with such flagrant bias. Why should the Army Corps of Engineers and the Southern congressional delegation get away with such activity?" He urged CBC members to vote against further appropriations for the project.[65]

Stokes took issue with Blackwelder in a letter to CBC member Katie Hall, a Democrat from Indiana. Stokes detailed his personal investigations into minority involvement in the waterway construction, stating: "My findings were that while the Corps of Engineers had not yet achieved its goal of 30% participation for minorities in this $1.7 billion project, a great deal of effort was being expended by the Corps of Engineers and Black leaders to achieve this goal." He then attacked Blackwelder's organization, the Environmental Policy Center (EPC), because it had "no affirmative action plans in Washington, D.C., a city more than 70% Black." Stokes noted the irony that Blackwelder, "while excoriating the Corps of Engineers for racial discrimination on this project is Executive Director of an organization whose record on employment for minorities is far worse." He concluded that Blackwelder and the EPC thus "have no credible basis to offer the Congressional Black Caucus their unsolicited advice on this project," and recommended that the CBC continue to promote the interests of minorities in Mississippi and Alabama by continuing to support the Tenn-Tom.[66]

The Limits of Reform

With less than three years of construction remaining for the waterway, South Atlantic Division Engineer John Wall attempted to place the Tenn-Tom affirmative action program in perspective for the project's principal contractors in 1982. Explaining that the Corps's mission has been in engineering, design, and construction, he said that "until the Tenn-Tom, the Corps has never had a project in which such a conscious effort has been made to accommodate and insure the needs and desires of the local residents in the project area." He called the waterway a "grand experiment" that pushed the agency "to link project policy decisions with socio-economic consequences of those decisions."[67]

Looking back at the Tenn-Tom experience shortly after the waterway's completion, the Minority Peoples Council gave the army engineers mixed reviews. "All objectives set were not obtained," Council officials said of the Tenn-Tom affirmative action plan. Not only did the Corps fall short of the stated goals, but "blacks and women tended to perform a disproportionate share of labor and other lower wage jobs." Still, they wrote, without the affirmative action plan the numbers would have been lower.[68]

Along with transportation savings, one of the principal rationales offered by Congressional backers of the Tenn-Tom was the economic benefits to be generated (through regional growth and construction-related employment) by the waterway for low income and minority people living in the project area. The fact was that these promises, which had raised the hopes of the area's poor and African-American communities, went largely unrealized (and really unpursued) until the late-1970s. By the early 1980s, after prolonged adversarial confrontations spearheaded by local minority activists and the Congressional Black Caucus, some progress was finally achieved on this front, although those benefits were largely temporary—increased employment and contracting opportunities during the construction of the waterway, as opposed to long-term, stable economic development.[69]

The success of the grass roots organization was enhanced by the national visibility of the project and its increasing political vulnerability. The waterway had great difficulty standing on its engineering, economic, and environmental merits; any additional controversy could be enough to tip the balance. Therefore, more out of a cynical sense of self-preservation than with any commitment to social justice, the Corps addressed the issue with far greater attention than it would have otherwise. The Minority Peoples Council understood this, and took advantage of it.

The particulars of the Tenn-Tom case suggest the limits of reform. Although the Corps proved insensitive to the needs and desires of the minority population of the Southeast, it more than demonstrated its political adaptability by co-opting its critics with comparatively minor concessions. These concessions, which ultimately had little or no effect on the cost or quality of the waterway, may well have saved the entire project. Thus, for a very small price, the Corps and its political backers were able to achieve their goal.

CHAPTER IX   **In Court–Act II**

HAVING HAD THE START of construction postponed for over a year as a result of litigation initiated by the Environmental Defense Fund (EDF) and the Committee for Leaving the Environment of America Natural (CLEAN), Corps executives recognized the necessity of following closely the letter, if not the spirit, of the laws pertaining to the controversial waterway. Yet, by the mid-1970s, agency officials felt confident that they had weathered the worst of that storm, at least as far as meeting their strict legal obligations. The lawsuits's dismissal in 1974 lulled them into lowering their guard and into treating the chorus of environmental complaints as largely a matter of public relations.

Unknown to the Corps executives, however, attorneys for the Louisville and Nashville (L&N) Railroad and the EDF had been meeting periodically throughout 1975 and 1976 to discuss a joint suit to stop the Tenn-Tom. Both the railroad executives and the environmentalists were convinced that it would serve their interests to oppose the waterway in court, as well as in Congress, and so they agreed that the L&N would finance the litigation. The railroad retained Jon T. (Rick) Brown and Stephen E. Roady, the Washington, D.C., attorneys who had served as counsel on the previous Tenn-Tom litigation for EDF, and Joseph V. Karaganis, a Chicago-based attorney affiliated with Brown and Roady's Washington law firm who had successfully litigated the Corps on other projects. Staff attorney James T. B. Tripp represented the EDF.[1]

On 30 November 1976, the L&N and EDF simultaneously filed separate but parallel complaints with the U.S. District Court for the District of Columbia. The L&N filed its complaint as a sole plaintiff, while EDF joined with four co-plaintiffs: the organization CLEAN and three individuals—Glenn H. Clemmer, G. Randall Grace, and F. Glenn Liming—who were members of both EDF and CLEAN.[2]

Phrasing the complaints posed a critical early step for the plaintiffs because issues litigated in an earlier lawsuit can be barred from subsequent trial. To distinguish this suit from the first, the plaintiffs argued that the project now under construction differed significantly from the project originally authorized by Congress. The plaintiffs, following the nomenclature used in the Corps's 1976 economic reanalysis of the Tenn-Tom, labeled the waterway described in the 1971 litigation as "Original Plan A." They contended that the Corps had abandoned this design "to undertake a new and significantly different and larger project in its stead." This "new project" involved three major alterations: "(1) significant design changes in Original Plan A; (2) major new navigation works south of Demopolis, Alabama—the southern terminus of Original Plan A—and (3) the designing and planned construction of duplicate locks at Demopolis and Coffeeville, Alabama, all in an effort to produce estimated benefits from the waterway sufficient to exceed estimated costs." The plaintiffs labeled the alternatives under points 2 and 3 as "Plan B" and "Plan C" (again following the categories in the 1976 reanalysis). Contending neither Plan B nor Plan C was authorized by Congress, the plaintiffs argued the Corps was acting in an illegal manner by building the revised waterway. Moreover, the environmental impact statement had addressed none of these major changes.[3]

The plaintiffs raised other issues that were not litigated in the original lawsuit. For example, they addressed the waterway's economic justification, especially the federal subsidy of a transportation mode directly in competition with the nation's railroads. In its complaint, the L&N explained that its main line roughly paralleled the projected course of the

Tenn-Tom, and because of the federal subsidies enjoyed by the barge industry, the railroad stood to lose substantial revenue from the diversion of bulk commodities from rail to barges. The 1976 economic reanalysis had raised the projected initial project year traffic from 5.7 million tons to nearly 28 million tons. The restudy also projected a major shift in the types of commodities to be shipped on the waterway; coal, for example, was negligible in the earlier study, but it constituted 60 percent of the predicted base year tonnage in the restudy. This increased level of traffic and shift to coal heightened the concern of L&N managers about the potential adverse impact on its own traffic. The railroad had had to purchase its right of way, finance its line construction, pay taxes on that property, and underwrite the costs of operating and maintaining its lines. Such was not true for barge operators, who bore none of the costs of right of way acquisition, construction, property taxes, operation, or maintenance. The L&N thus argued that the Tenn-Tom would place the company "at a decided competitive disadvantage in the economic competition for coal and other commodities now moving over plaintiff's rail lines."[4]

The L&N and EDF accused the Corps with changing the overall design of the Tenn-Tom in order to bolster a declining benefit/cost ratio. As the estimated price of the waterway rose from $323 million in fiscal year 1971 to $1.428 billion in fiscal year 1976, the benefit/cost ratio fell from 1.6 to 1.08. To strengthen the project's economic justification, the plaintiffs stated, "the defendants have continually increased their projections for barge traffic via a completed Original Plan A." The limited capacity of the watercourse immediately south of Demopolis, which could accommodate no more than six barges in an individual tow, further constrained the traffic that could be legitimately projected for the Tenn-Tom. The plaintiffs argued, therefore, that "to continue the artificial escalation of project benefits," the army engineers had proceeded "to a new and unauthorized project."[5]

Plan B represented the Corps's proposal to widen, deepen, and straighten the watercourse south of the Tenn-Tom from Demopolis to Mobile; that is, to enlarge the existing Black Warrior-Tombigbee Waterway. Plan C consisted of providing duplicate locks at Demopolis and Coffeeville, Alabama. The plaintiffs also described the Corps's "New Plan A," which involved "radically" altering the design of the canal section of the Tenn-Tom from a "perched canal" to a "chain of lakes." They said that the chain-of-lakes involved new dams that would flood more agricultural and timber land than the original design. The construction of 36 miles of roads in the divide cut and the dredging of wetlands where the waterway joined Pickwick Lake, as described in New Plan A, were likewise unaddressed in the initial design document and the environmental impact statement. In addition, the New Plan A added a major design feature calling for the disposal of excavated material in 51 valleys adjacent to the divide cut. The plaintiffs asked that construction (which was approximately ten percent complete at this time) be enjoined "until such time as Congress may authorize such work and until such time as defendants are in compliance with all applicable statutes and regulations."[6]

The lawsuit was far more sophisticated than the litigation initiated five years earlier. Environmentalists and water project opponents had learned a lot since the enactment of the National Environmental Policy Act, and much of that knowledge stemmed from the numerous court suits brought against the Corps.[7] The Tenn-Tom complaint drew heavily from this experience, most notably from the 1974 lawsuit filed to stop construction of Locks and Dam 26 on the Mississippi River near Alton, Illinois. The Locks and Dam 26 litigation, which succeeded in gaining a court injunction, had incorporated a three-pronged attack on the army engineers—environmental, economic, and authority. Perhaps the most novel aspect was the questioning of the Corps's discretionary authority; the plaintiffs argued that Congress had authorized the existing structures at Alton in the 1930s, but had not re-authorized the substantially larger replacement locks and dam that the Corps was constructing.[8] Strategically, the Locks and Dam 26 litigation forged a coalition of environmental organizations, railroads, and fiscal conservatives interested in imposing waterway user fees.[9]

The imbroglio surrounding the Locks and Dam 26 project led L&N executives to believe that they might well win the Tenn-Tom contest. It also brought them additional support from people like J. A. Austin, a vice president for the Missouri Pacific Railroad Company (which was a plaintiff in the Locks and Dam 26 suit), who offered his assistance to those railroads likely to be affected by the Tenn-Tom.[10]

Aside from stiffening the resolve of the nation's railroads, the Locks and Dam 26 litigation also set an important legal precedent. The Court's decision established that discretionary authority could be judicially challenged. This ruling had broad implications for the Corps's civil works program, and the agency's legal counsel began urging caution in proceeding with "post-authorization changes"—that is, unilaterally altering the scope of projects as specifically authorized by Congress.[11]

As the plaintiffs' lead attorney in the Locks and Dam 26 case, Joseph Karaganis gained valuable insights into how the Corps operated, how it was organized, where and how it filed its records, what its regulations stated, and how it interacted with Congress. He also learned the conceptual underpinnings of the agency's benefit/cost analysis. Armed with this highly pertinent knowledge, Karaganis proved extremely effective in orchestrating discovery for the Tenn-Tom litigation, in uncovering inconsistencies within the Corps, and in preparing the legal argument for the case.[12]

The new litigation outraged the Tenn-Tom's supporters. They had already gone through a prolonged trial and appeal, and had won. The challenge by the environmentalists therefore seemed unfounded to them. The L&N Railroad's complaint was another matter, but no less disagreeable. The Mobile *Register* had long defended the Tenn-Tom, and its December 1976 editorial exemplified the frustration felt by many advocates: "We cannot help but be amazed at the audacity and greed demonstrated by the Louisville and Nashville Railroad in its suit which seeks to block further construction of the Tennessee-Tombigbee Waterway."[13] Alabama State Senator Sid McDonald accused L&N President Prime F. Osborn of "using court action to hold the Tenn-Tom project as hostage for the acquisition of . . . waterway user charges on America's inland waterway system." McDonald warned Osborn: "I can assure you that your company will harvest a tremendous amount of ill feelings from the business and political community" that "will undoubtedly translate into financial harm for the L&N Railroad."[14]

Osborn denied McDonald's charges, claiming that the L&N brought suit because it thought the Corps failed to comply with the law as it affected the railroads. In justifying the litigation, Osborn wrote, "the Corps of Engineers simply did not consider whether that transportation could better be accomplished by existing modes of transportation, namely, railroads, motor carriers and existing waterways."[15]

Organizing the Defense

A month after the suit was filed, Manning Seltzer, chief counsel for the Corps, organized a meeting of the agency's top officials to hear two staff attorneys present both sides of the case in a mock hearing. The exercise succeeded in highlighting what the agency might face. Beyond the strong possibility of losing the litigation, the Corps risked considerable political damage. It could be perceived that the agency had acted illegally, had squandered public funds, and had violated its public trust. The ramifications thus extended far beyond the fate of this particular waterway. This revelation led the Corps leadership to assign top priority to the litigation.[16]

Because it was responsible for the actual planning, design, and construction of the waterway, Mobile District became the Corps's most active office in preparing for the lawsuit. Assistant district counsel David H. Webb took the lead in assisting Department of Justice attorneys with the Corps's defense and with coordinating the agency's responses to the plaintiffs' interrogatories.[17]

For its part, the Tennessee-Tombigbee Waterway Development Authority (TTWDA) continued to monitor the political vitality of the Tenn-Tom, doing whatever it could to ensure that the project was built in its entirety. After the new lawsuit was filed, John Gullett, the Washington, D.C.–based attorney who had represented the TTWDA in the first litigation, told the Authority's local attorney, Hunter Gholson, that the EDF and L&N complaints "finally bring out into the open and identify the fact that the environmentalists and the railroads are in bed together against Tenn-Tom." He warned that the L&N brought "unlimited resources" to the lawsuit and that "in these days of the new Carter administration, EDF will wear a white hat." He worried too about the plaintiffs' "very capable lawyers," and that TTWDA's interests would not necessarily be protected by the Justice Department and the Corps of Engineers. After all, he said, "it is our meat that is in the fire." Gullett recommended that the TTWDA intervene early and then "lie back and wait; but at all times keeping an eye on the ball."[18]

The TTWDA did exactly that. And as it had done in the 1971 litigation, it worked closely with the Corps and the Justice Department in crafting the overall litigation strategy, providing background research, and lining up expert witnesses. The TTWDA acted in many ways that the Corps, as a federal agency, could not. Perhaps most importantly, the Authority was free to lobby openly in Congress. As the TTWDA contract economist Joseph R. Hartley told the Authority's board members, "if we had left everything up to the Corps the last 15 years, Tenn-Tom would be extinct."[19]

The TTWDA played the critical role of filling the gap between the way the Justice Department represented the Corps and the way the Corps wanted to be represented. As in the first court case, the Authority helped the army engineers retain the service of attorneys in the Civil Branch of the Justice Department rather than those attorneys in the Land and Natural Resources Branch who would normally hold jurisdiction. Gholson met periodically with the Chief of Engineers, Lieutenant General John Morris, to discuss those questions the Corps thought should be raised in the litigation but which the Justice Department wanted to avoid. As an intervenor, the TTWDA could be more bold, aggressive, and antagonistic in probing the plaintiffs' motives than could the government attorneys.[20]

The Justice Department has traditionally suffered a high turnover rate among its lawyers. Attorneys working for it are typically bright and hardworking, but it is a common career strategy for a young lawyer to work a few years at the department, gain experience and personal connections, and then move into private practice where the salaries are far more lucrative. For a complex and protracted litigation such as that with the Tenn-Tom, this situation presented a problem of continuity.[21] The September 1978 departure of Nicholas Diacou, Justice's lead counsel in the Tenn-Tom litigation, exemplified this turnover problem. Diacou was a seasoned and highly competent attorney whom both the Corps and the TTWDA hoped to see replaced with someone of equal effectiveness. When it appeared that one of the two junior attorneys working on the Tenn-Tom litigation would be elevated to chief trial attorney, Hunter Gholson, knowing that the Corps could not interfere with internal Justice Department decisions, suggested using the TTWDA's political clout to appeal directly to Senator James Eastland, who chaired the Judiciary Committee. As Gholson later told the Authority's board of directors: "The Corps is trying to get more attention from the Attorney General, and we have asked Senator Eastland to help in getting this case put on the list of serious cases. This means the Attorney General would review this case each week."[22] After conferring with Bill Simpson of Eastland's office, Gholson met with Attorney General Griffin Bell's top aide, Mike Eagan. He told Eagan that the Tenn-Tom litigation held top priority within the Corps, that Eastland likewise considered it of great importance, and that the lawsuit required the best possible lead attorney. Eagan concurred, and he convinced Bell to place the suit on the list of serious cases and to assign Edward S. Christenbury, a senior attorney with proven trial experience, to manage the case.[23]

Court Actions

Within days of the filing, United States District Judge Joseph C. Waddy ordered the L&N and EDF complaints consolidated into a single action.[24] Justice Department attorneys quickly filed a motion for transfer of venue to the U.S. District Court for the Northern District of Mississippi, the same District Court that had ruled on the earlier Tenn-Tom suit. After considering all arguments pro and con, Waddy granted the transfer of venue in March 1977 and Judge William Keady again had the responsibility of hearing the case against Tenn-Tom.[25]

Discovery consisted of three main parts—requests for production of documents, interrogatories (written questions to which written answers are required), and depositions (oral questioning of individuals)—and the plaintiffs were allowed to pursue their inquiry far more thoroughly than they had been during the previous litigation, where Judge Keady's discovery restrictions had hamstrung the plaintiffs' ability to develop their best possible case. Keady ruled far more generously in 1977. The plaintiffs submitted their first request for production of documents and their interrogatories in July 1977.[26] The interrogatories totalled 51 pages and 114 separate questions; the defendants' response, submitted in October 1977, totalled 173 pages.[27] Following the plaintiffs' second request for the production of documents in 1978, counsel from both parties reached an informal agreement allowing the plaintiffs to microfilm Corps documents using their own equipment and providing the agency with duplicate copies of each microfilm reel. This procedure allowed the Corps to monitor the plaintiffs' discovery by pinpointing the documents they thought were important enough to copy. Hundreds of thousands of pages were made available to the plaintiffs at Corps offices in Atlanta, Mobile, Nashville, Vicksburg, and Washington, D.C.[28]

In January 1978, enlightened by the information gained through discovery, the plaintiffs—now joined by the National Audubon Society, the Birmingham Audubon Society, and the Alabama Conservancy as intervening plaintiffs—filed a 15-count, 69-page amended complaint. The extensive discovery process had enabled the plaintiffs to grasp fully the complexities of the Corps's decision-making process and to gain a detailed understanding of the waterway and its legislative history. The plaintiffs asked the court to enjoin "further planning, design, and construction" of the Tenn-Tom until the Corps "obtained Congressional authorization for any and all project work not presently authorized."[29]

The plaintiffs argued that the Corps had learned in 1974 that the overall costs were going to exceed greatly its earlier estimates. "In a desperate and secret attempt to avoid termination of the project," the agency then withheld information on the escalating costs from the President and the Congress in order to buy the time necessary to conduct a new economic study. In the words of the plaintiffs, "the Corps intentionally submitted false and misleading project cost data to lessen the 'emotional impact' and to avoid disclosure that the project's economic justification had disappeared."[30] The plaintiffs also noted that while the agency was conducting its economic restudy in 1975, "high Corps officials learned that the construction of the Tennessee-Tombigbee had been undertaken and justified to Congress on the secret and unstated assumption of a 300-foot-wide channel all the way from Pickwick Pool to Mobile . . ." Agency officials soon learned that they lacked the authority to build a 300-foot-wide channel for any portion of the waterway; nevertheless, the plaintiffs charged, "the Corps has hidden this lack of authority from the President, the Congress, and the public and has illegally continued to construct the 300-foot channel without necessary Congressional authorization."[31]

The amended complaint stated that the Corps's leadership, having discovered this lack of authority in 1975, decided not to stop construction while they sought the required Congressional authority, but instead "decided to illegally segment the 300-foot Pickwick to Mobile channel at Demopolis and to continue construction of the unauthorized segment north of Demopolis." They then argued publicly that the two projects were entirely independent of one another, even though internally agency planners understood that "such piecemeal construction prejudices the decision to build a 300-foot channel south of Demopolis."[32]

The plaintiffs challenged the economic restudy, claiming that the Corps "manipulated data, utilized false and unsubstantiated factual as-sumptions, and inserted erroneous data to bloat its benefit calculations." They also charged the agency with suppressing a 1976 Army Audit Agency report of the Tenn-Tom—requested by the Assistant Secretary of the Army for Civil Works in 1975 when he learned of the enormous cost increases—which had found the Corps to be misleading Congress on the project costs. In attacking the economic justification for the project, the plaintiffs also claimed that the army engineers misled Congress by failing to report several significant costs, including: the cost of expanding the navigation system south of Demopolis, which was required to justify the project; the cost of mitigation lands required to offset the destruction of wildlife habitat; the economic losses to other transportation modes; and the decline of revenues to local governments caused by the removal of lands from the local tax base.[33]

In questioning the Corps's authority to build the redesigned waterway, the plaintiffs stressed that the project was authorized in 1946 with a channel width of 170 feet (150 feet in the divide cut). In the 1966 economic restudy, however, the Corps "assumed a 300-foot-wide channel (280 feet in the Divide Cut) all the way from Pickwick Pool to Mobile." Moreover, the plaintiffs argued, "Corps personnel blithely and arbitrarily assumed a 300-foot width was available all the way to Mobile in order to claim the economic benefits of wide, eight-barge tows." No Congressional authorization was requested for this major change in design, thus the plaintiffs concluded that the agency lacked the authority to build a waterway of those dimensions.[34]

The plaintiffs further contended that Corps executives asked their lawyers "to review the authority for the entire channel," both north and south of Demopolis. Counsel had reportedly advised that authorization for the 300-foot channel from Demopolis to Pickwick was "cloudy," "doubtful," or "non-existent." Plaintiffs said that senior agency staff met "on or about November 14, 1975" and decided "to proceed illegally on the 300-foot channel north of Demopolis," and thus to disregard the advice of counsel.[35]

The case now brought against the Corps focused on the agency's internal decision-making process. It probed deeply the manner in which the Corps developed its economic justifications for marginal projects. Should the plaintiffs win the suit, the Corps could be publicly embarrassed. Agency officials outside the Office of Counsel now finally realized how serious and potentially damning the charges were.[36]

The First Trial

Judge Keady divided the plaintiffs' fifteen causes of action into three areas of judicial concern. The first area dealt with the Corps's alleged lack of legal authority to construct the Tenn-Tom. This issue centered on the expansion of the waterway to 300 feet without Congressional authorization, but included other added structures and changes in design made without specific approval from Congress. The second area of judicial concern involved the Corps's method of calculating benefits and costs, and whether it had the legal justification for continuing the project based on economic benefits from major changes to the Black Warrior-Tombigbee Waterway—a completely separate project that had yet to receive Congressional authorization. Finally, the third area focused on the Corps's compliance with the National Environmental Policy Act and other environmental statutes.[37]

The first trial addressed the agency's authority to construct the waterway as designed in the mid-1970s. Once a project was authorized, the Secretary of the Army was granted limited discretionary authority to make minor design changes, but major alterations in the overall project dimensions were to be approved through specific request and vote of the full Congress. The appropriations committees allocated annual construction funds, but these committees could not grant authority for project changes. Of particular importance for the Tenn-Tom case was a February 1967 request by Major General Frederick J. Clarke, then Acting Chief of Engineers, "that the Secretary of the Army approve a 300-foot wide waterway as the basis for further planning."[38] Secretary of the Army Stanley R. Resor approved the change from 170 to 300 feet the following month.[39] The plaintiffs charged that this change was illegal, as it exceeded the Secretary's discretionary authority.[40]

The plaintiffs submitted the affidavit of Richard A. Hertzler, who, as a senior civilian staff member within the Department of the Army, had monitored the Tennessee-Tombigbee Waterway. Hertzler maintained that Resor's 30 March 1967 memorandum, which approved the widening of the waterway, "did not increase the authorized width of the Tennessee-Tombigbee Waterway to 300 feet," and that it "was, at most, intended to permit the Chief of Engineers to conduct further planning and design of the project . . ."[41] The plaintiffs added that, although the Corps cited Resor's memorandum as the authorizing document for the 300-foot channel, this document was never mentioned in the legislative history of the Tenn-Tom from 1967–1976.[42]

While developing their legal challenge to the Corps, the Tenn-Tom opponents always recognized that litigation was but one part of their overall strategy to kill the waterway. Six years of litigation under the National Environmental Policy Act (NEPA) had taught the environmental community that lawsuits could not permanently halt federal projects; only the U.S. Congress could. As CLEAN newsletter editor Sherrie Clemmer remarked in December 1976: "A lawsuit is only a stopgap measure."[43] The decision to terminate the Tenn-Tom had to be made by the national legislature, and elected officials were sensitive to public opinion. Opponents therefore approached the new lawsuit as a way of delaying the project while public opinion—or so they hoped—changed, and as a vehicle to educate and foster such change in public opinion.

As one of their prime tactics, the opponents worked closely with the news media. The discovery process provided them with several opportunities to disclose the inner workings of the Corps, and they made good use of such occasions. A key disclosure was the Corps documents on project costs and benefit/cost calculations, particularly when these records showed cost increases of over $500 million. Such announcements made good copy. Typical of the newspaper responses was the *New York Times* editorial on 4 December 1978 entitled "The Engineering of Deceit." "The project appears to be pork barrel politics at its worst," the editorial asserted. "Thanks to an alliance of the Corps and powerful members of Congress, internal audits were ignored, legislative review was frustrated, one unfavorable report (from the General Accounting Office) was suppressed, and President Carter was forced to retreat when he tried to stop the project last year."[44]

Representative Tom Bevill responded to the *Times* editorial in a speech before the House. The Alabama Democrat claimed the editorial represented "part of a trial by press being carefully orchestrated by the plaintiffs." He said there was "evidence that the plaintiffs brought reporters to their attorney's office in Chicago for briefings before the start of the trial for the express purpose of promoting a trial in the news media."[45]

Bevill was correct. During the hearing, the plaintiffs met frequently with media representatives, held press conferences, issued news releases, and granted interviews. The considerable media attention outraged the waterway supporters, but the Department of Justice prohibited the Corps from talking to the media about the litigation. This ban worried senior Corps officials, who perceived the public relations beating of the Tenn-Tom as having long-lasting repercussions for the agency as a whole.[46]

As it did during the previous litigation, the TTWDA worked with the media where and when the Corps could not. In response to James Nathan Miller's scathing *Reader's Digest* article, "Trickery on the Tenn-Tom," the TTWDA issued a rebuttal that it distributed to newspaper editors throughout the Southeast. The Authority countered Miller's arguments, and added that his and other similar stories resulted from the plaintiffs "leaking distorted stories to media trying to arouse the public and inflame the press. Their purpose is to destroy public confidence in Tenn-Tom."[47] As media coverage of the trial intensified, the TTWDA published a 27-page fact sheet addressing the plaintiffs' allegations point-by-point. The widely disseminated document covered such issues as the status of construction, the economic costs of terminating the waterway, the transportation and regional benefits of the project, environmental planning, and minority participation.[48] Although the TTWDA had strong influence within the Southeast, media outside that region tended to be less receptive to its public relations campaign.

By all accounts, Judge Keady paid no heed to the media coverage. In March 1979, he handed the Corps a clear-cut victory by dismissing all the plaintiffs' authority claims.[49] However, Keady did not rule on the merits of the allegation that the Corps had illegally widened the waterway to 300 feet. Instead, the judge used the equitable doctrine of laches to bar the claim, writing that "a delay of more than nine years, in our opinion, before seeking a remedy against allegedly unauthorized widening of the channel, is clearly inexcusable." Keady found that the defendants would suffer undue prejudice from the belated assertion of the claim because nearly $600 million had already been allocated to the project.[50] He also ruled that the other, smaller design changes unrelated to the project's increased channel width were within the Corps's discretionary authority.[51]

Keady's decision on the first part of the case—dealing with the issue of authorization—did not absolve the Corps of its actions, at least on the question of expanding the channel width from 170 to 300 feet. Although Keady did not decide whether the Secretary of the Army had authority to increase the channel width, the implications of his decision were that the Corps should have obtained Congressional authorization for such a substantial project modification. Because Keady concluded that the equitable doctrine of laches barred the challenge (that is, that the plaintiffs had waited too long to file their suit on those grounds), he never formally ruled on the authority question. He did write, however, that the "law authorizing construction of Tennessee-Tombigbee Waterway contained no extraordinary or unusual language which would vary authority of Corps of Engineers to make other than minor modifications in design without seeking additional authorization of Congress."[52] He also wrote that "if laches were not an effective defense, and the question of width authorization remained open for determination upon the merits, plaintiffs' challenge would deserve careful scrutiny."[53]

The L&N and EDF immediately appealed the District Court ruling.[54] Keady, in turn, suspended the trial on the remaining issues pending the outcome of the appeal. The plaintiffs raised three main issues before the Circuit Court: whether the District Court erred by invoking the doctrine of laches; whether the Corps should be enjoined from building the 300-foot channel for the Tenn-Tom because Congress had only authorized a 170-foot channel; and whether the Corps should be enjoined from making major changes in the project design without Congressional authorization or consent. Having failed to gain a temporary injunction, the plaintiffs sought an expedited appeal, noting that "the project is currently under an accelerated construction schedule which might prejudice the ability of the plaintiffs to obtain a meaningful trial."[55]

In March 1980, the U.S. Court of Appeals for the Fifth Circuit affirmed Keady's judgment that the plaintiffs' action be barred by laches.[56] As the plaintiffs and defendants prepared their cases for the continuing trial at the District Court, EDF and the plaintiff-intervenors—National Audubon Society, Birmingham Audubon Society, and Alabama Conservancy—petitioned the Fifth Circuit Court of Appeals' decision in August 1980 for a writ of certiorari. The L&N Railroad decided not to join this attempt to take the case to the U.S. Supreme Court. When the Supreme Court denied the plaintiffs' request to review the lower court decision in October 1980, the legal challenges to the Corps's authority to modify the project ended.[57]

The Second Trial

With the authorization charges laid to rest, the litigation moved ahead within the District Court to the deferred counts dealing with economic and environmental issues. The plaintiffs contended that the Corps had failed to comply with statutory and regulatory requirements pertaining to the computation of benefits and costs, and that the agency had likewise failed to comply with several environmental statutes and regulations, including the National Environmental Policy Act, the Fish and Wildlife Coordination Act, the Federal Water Pollution Control Act, and the 1977 Clean Water Act. The major charge was the Corps's failure to file an environmental impact statement (EIS) or supplemental EIS for the major project changes made after the completion of the original EIS. These design alterations included the replacement of the "perched canal" with a "chain-of-lakes" design in the canal section; the changes in the river section, which initially was designed to follow the natural course of the Tombigbee River, but which was realigned to straighten the navigation channel by constructing artificial cutoffs, thereby isolating 21 miles of the river's channel; and the disposal of an additional nine million cubic yards of excavated material in the divide section.[58]

The plaintiffs claimed the Corps illegally divided the waterway into two sections—one comprising the original project between Demopolis and the Tennessee River, and the other south of that project between Demopolis and Mobile—and that it failed to prepare an EIS for both parts. They said the agency's 1976 economic reanalysis summary failed "to analyze and disclose project alternatives." Finally, they cited the Corps's failure to address major project design changes in a supplemental or revised EIS.[59]

The Corps held that the benefit/cost calculations were not subject to judicial review and, moveover, that Congress had tacitly approved those figures by continuing to allocate annual appropriations. It also claimed that Congress approved the 3.25 percent interest rate used to justify the project. The agency denied the charge that it must prepare an EIS for the waterway south of Demopolis before being allowed to continue construction of the Tenn-Tom, stating that such an EIS would be prepared for that (separate) project when it sought Congressional authorization. In response to the claim that it should prepare a revised or supplemental EIS for all substantial changes made subsequent to the original EIS, the Corps argued that it had appended the original EIS with supplemental environmental reports and that these reports failed to uncover "any significant environmental impacts not previously disclosed in the original EIS . . . Ironically, the principal changes . . . constitute design modifications implemented by the Corps specifically to minimize adverse environmental impacts."[60]

Keady ruled in favor of the defendants in October 1980, ordering that the Corps did not have to prepare a supplemental EIS. The other counts, addressing issues of local assurances, economic justification of the project, fish and wildlife mitigation, discharge of dredged materials, and failure to publish agency procedures, were likewise dismissed.[61]

The plaintiffs appealed the decision to the U.S. Court of Appeals for the Fifth Circuit. They also filed a motion for injunction pending appeal, arguing that the several months delay before the Circuit Court ruled on the appeal and the Corps's accelerated construction schedule would seriously prejudice their economic and environmental charges.[62] The Fifth Circuit Court denied the motion and heard oral argument in April 1981.[63] Three months later, it reversed in part the District Court's final judgment and ruled that the Corps discontinue the use of the 3.25 percent interest rate in calculating the project's benefit/cost ratio and that it prepare a supplement to the EIS for all project changes made since the filing of the original statement. Moreover, the Appeals Court directed the District Court to enjoin immediately "(1) the construction or letting of contracts on the chain-of-lakes segment and the cutoffs of the Tombigbee River channel, and (2) any activity significantly affecting the environment of the land acquisitions planned after 1971." The District Court was allowed to grant specific exemptions if the Corps could demonstrate, after an evidentiary hearing, that "the public interest will suffer irreparable harm" because of the injunction, although that exemption could not be based solely on future construction cost increases.[64] In conclusion, the Court stated:

> We are extremely reluctant to interfere with the construction of a project that Congress has authorized for the last ten years and that is now 55% complete. But the plaintiffs have established that the Corps has blatantly violated NEPA and its own regulations by refusing to prepare a supplemental EIS on the major changes in the Tennessee-Tombigbee Waterway that have occurred since 1971. . . . If the limited injunctive relief we have imposed seems harsh or late, the fault lies with the decision of the Corps to ignore its responsibility and to resist this lawsuit while continuing its default.[65]

This decision could hardly have arrived at a worse time for the Corps. The politics of the waterway had changed substantially to the point where appropriations votes in both houses of Congress barely achieved the needed majority. The anti-Tenn-Tom coalition had grown to include environmentalists, the railroad industry, railway unions, and fiscal conservatives.

The Fifth Circuit Court's ruling that the Corps had acted illegally by continuing to build the project without providing a supplement to the EIS stiffened the opponents' resolve, and gave further credence to their main arguments. Moreover, the injunction threatened to do two things: delay the completion of the project, thereby allowing opponents time to strengthen their coalition and possibly kill the project in Congress; and increase the overall cost of the waterway and thereby undercut its already tenuous economic justification. The court decision also heightened the media coverage of the controversy, which served to erode public confidence in the Corps and in the waterway.[66]

The Circuit Court's ruling seemed an ominous turning point for the Tenn-Tom, but during this crisis the TTWDA made its most significant contribution to the legal proceedings. TTWDA attorney Hunter Gholson suggested that the federal defendants file a petition for rehearing with the Fifth Circuit and argue that the project area would suffer severe economic adversity, especially lost jobs, if construction was abruptly curtailed, and that such curtailment would cost the federal government substantial shutdown and start-up costs. Gholson proposed asking the Court to allow the Corps to postpone only those sections of the waterway which related specifically to the plaintiffs' objections regarding the EIS and which would have a minimum economic impact on the region. The Justice Department refused to file such a petition, citing its policy not to question the opinion of an Appeals Court. So the TTWDA filed alone as a defendant-intervenor.[67]

Although the Circuit Court denied the TTWDA's petition for a panel rehearing, it did rephrase its earlier decision. The revised wording gave the District Court greater discretionary authority, allowing it—without a hearing—to forgo the imposition of an immediate injunction. Moreover, the Circuit Court stated that "if the current activity is so extensive that an immediate cessation would necessarily be greatly expensive and harmful, the District Court may in its discretion choose to delay the imposition of an injunction until after a reasonable period is allowed for the Army Corps of Engineers to complete and submit its supplemental EIS."[68] With this restatement, the intervenors achieved precisely what they wanted: Judge Keady had been given the opportunity to minimize, postpone, and perhaps totally avoid enjoining construction on the waterway.

The Circuit Court's restatement demoralized the plaintiffs, who had but one month earlier celebrated their first significant victory in the Tenn-Tom litigation. The ability of the District Court to delay an injunction if it believed an immediate cessation would cause great expense and harm, they contended, "has the potential for utterly dismantling the July 13 opinion of the Court." They urged the Circuit Court to adhere to that opinion and "issue an immediate *but limited* injunction covering carefully defined areas of construction and activity on the project" that would "minimize or preclude the layoffs of *any* employees on the project." They proposed that layoffs could be avoided by transferring employees from enjoined sections of the waterway to construction activities unaffected by the court order.[69]

The Circuit Court stood firm, sending defendants and plaintiffs back to the District Court to seek a compromise on what sections would be stopped pending completion of the supplemental EIS. Keady's discretion did not altogether solve the Corps's problems, but it did change the situation from one of despair to one of hope. With Keady ruling on what parts of the project would or would not be shut down, waterway supporters were justifiably confident. In addition to having a judge favorably disposed to their position, the Corps also had another advantage: unlike the plaintiffs, the army engineers had the privileged technical knowledge and information necessary to determine what parts of the waterway could be postponed without delaying the completion date for the overall project.

In September, Keady reversed part of his own final judgment. He ordered the Corps to cease using the lower interest rate of 3.25 percent in calculating the benefits and costs of the Tenn-Tom, and to substitute the higher interest rate of 6.875 percent which was being applied to current federal water projects. He also required the agency to prepare a supplemental EIS addressing the project changes made since March 1971. Keady set June 1982 as the deadline for submission. Finally, counsel for all parties agreed to negotiate among themselves the extent of work stoppages pending completion of the supplemental EIS with attention to avoiding widespread unemployment and public harm. Construction would be allowed to continue on all current contracts while the agreement was being worked out.⁷⁰

The Mobile District had established a task force in July to determine what portions of the project could be delayed without seriously disrupting the overall completion date of the project.⁷¹ The Corps then presented the plaintiffs with a array of items, none of them essential elements of construction, although some of them *appeared* to be crucial components. Although there was some heated give and take, the finesse worked; in December 1981 the plaintiffs agreed to the injunctions packaged by the Corps.⁷² The army engineers promised not to start new contracts, such as clearing trees and filling wetlands, if allowed to continue with work already under way—work that happened to be on the waterway's "critical path."⁷³

The Corps completed its Draft Supplemental EIS (DSEIS) in December 1981. The L&N Railroad followed with a stinging, 59-page critique of the DSEIS, which asked the Corps to revise the report because of the high numbers of alleged deficiencies in the document.⁷⁴ The Corps did not, and filed its two-volume Final Supplemental EIS (FSEIS) in April 1982.⁷⁵ Under the December 1981 work suspension order, the injunction remained in effect for sixty days following the Court's receipt of the FSEIS. In June, the plaintiffs filed their objections to the FSEIS and a motion for continuation of work suspension.⁷⁶ A 69-page objection to the FSEIS made the plaintiffs' case that the document should be declared inadequate.⁷⁷

Despite the fact that the L&N and EDF had submitted their detailed objections to the FSEIS only two days earlier, Major General E. R. Heiberg, Director of Civil Works, signed the Record of Decision approving the completion of the waterway on 30 June 1982.⁷⁸ The plaintiffs objected to Heiberg's action, telling the District Court that "the Corps had a maximum of 48 hours to review the plaintiffs' sixty-odd pages of objections to the SEIS."⁷⁹ Keady denied the plaintiffs' motion, freeing the Corps to continue construction on all remaining aspects of the project.⁸⁰

Return to Congress

The judge's decision marked the end of nine years of litigation for the Tennessee-Tombigbee Waterway. Few public works projects have endured such extensive and protracted environmental scrutiny in the federal courts. As with the close of the first litigation, the end of the second lawsuit underscored the limits of using the judiciary system to kill environmentally harmful federal projects. That was something best done in the Congress.

During the initial lawsuit, the anti-Tenn-Tom coalition had put all its eggs in the litigation basket, holding the optimistic view that the new National Environmental Policy Act would bring about substantial change in the programs and activities of the federal government. The coalition was less sanguine during the second litigation, although it was still hopeful that it could prevail through its challenge of the agency's discretionary authority. When that strategy failed, the Tenn-Tom opponents realized that their only real hope of stopping the project was to turn to the Congress, and that was where the final battle was staged.

CHAPTER X **The Last Congressional Challenge**

PRIOR TO THE APPROPRIATION of construction funds in 1970, the principal battles over the Tenn-Tom took place in the Congressional arena, where the waterway's advocates at first fought to gain authorization for the project and then to secure the needed appropriations. After 1970, despite some flare-ups within the House and Senate, the controversial project found itself being contested in other venues, most notably the federal courts. By the late 1970s, however, Tenn-Tom opponents realized that the litigation was likely to continue for years, during which time the waterway would be built at an accelerated pace. It became clear that the greater the amount of money poured into the project and the greater the extent of its completion, the harder it would be to stop. This led the critics to reconsider their previous reluctance to oppose the project simultaneously in the courts and in the Congress.

Tenn-Tom promoters, on the other hand, realized that the project was still not out of the political storm. The mood of Congress, for example, had changed dramatically across the decade, reflecting voters' growing anxiety about the escalating national deficit. The Tenn-Tom, which had ballooned from an estimated cost of about $350 million at the beginning of the decade to nearly $2 billion by 1979, just seemed one more shameful episode of rampant public spending. For the first time since the initial appropriation of construction funds, it became politically expedient for members of Congress to oppose the Tenn-Tom as one of the country's most conspicuously wasteful projects.

Gaylord Nelson

Gaylord Nelson had long been a champion of the environmental movement, and in 1971 he spearheaded an ill-fated attempt to withhold construction funds from the Tennessee-Tombigbee Waterway. When CLEAN and the Environmental Defense Fund (EDF) sued the Corps to kill the project, they urged Nelson and other Congressional critics to cool their attacks on the waterway and allow the judicial system to resolve the issue. The Wisconsin Senator obliged, and turned his attention to other issues.

Eight years later, however, reports of cost escalation, dubious economic benefits, and questionable legal authorization made public by the second lawsuit rekindled Nelson's concern and led him to introduce a bill in March 1979 to revoke authorization for the waterway.[1] Although the Louisville and Nashville (L&N) Railroad and EDF lawyers shared Nelson's goal of stopping the project, they worried that the Senator's deauthorization legislation might prove counterproductive to their legal argument that the Corps *already* lacked Congressional authorization to construct the redesigned waterway. If an amendment to deauthorize or withhold appropriations lost in the Congress, the attorneys reasoned, the judge might use that vote to throw the whole Tenn-Tom case out of court.[2] To avoid this potential outcome, Nelson's legislative assistant, Rebecca Wodder, worked closely with the plaintiffs in drafting an amendment that would specifically address the authority question pending in federal court. Wodder also attended the anti-Tenn-Tom strategy sessions organized by the National Wildlife Federation, American Rivers Conservation Council, National Audubon Society, Friends of the Earth, National Resources Defense Council, Association of American Railroads, and the L&N Railroad.[3]

Nelson announced his impending legislation to deauthorize "the biggest pork barrel boondoggle of them all" during his acceptance speech

Senator Gaylord Nelson (D-Wisconsin), one of the most vocal and persistent critics of the Tenn-Tom in the U.S. Congress *(Courtesy of Historical Office, U.S. Senate.)*

for the Great Lakes Sport Fishermen Club's Clean Water Award. He chastised the Corps of Engineers for deliberately misleading the Congress on the Tenn-Tom's costs, traffic projections, and economic analyses. "The Corps plans the project, the Corps does the economic projections, the Corps does the economic evaluations, and the Corps builds the project," Nelson thundered. "Should anyone be surprised that this built-in conflict of interest produces boondoggles rather than sound transportation policy?"[4]

Democratic Senators William Proxmire of Wisconsin, Donald W. Reigle of Michigan, and Howard Metzenbaum of Ohio joined Nelson in co-sponsoring the "Tennessee-Tombigbee Waterway Deauthorization Act of 1979." Speaking from the Senate floor, Nelson claimed that the project was an "economic bad joke," that "its benefits have been created from thin air," and that "its costs have been consistently understated." If the "severe and irretrievable environmental impacts" associated with the project were not bad enough, Nelson said that "the corps deliberately deceived the Congress as to the costs of the project." He carefully addressed the ongoing litigation by proclaiming that "neither the raising of these issues nor the appropriation of further funds for the project can be construed by the court as approval of the corps' activities or a limitation on the duty of the court to decide each of the issues in the case." Despite the funds already expended on construction, he urged his colleagues "not go along with the concept of throwing good money after bad."[5]

Senator John C. Stennis led the opposition to the deauthorization bill. In a speech before the full Senate, Stennis insisted that the Tenn-Tom "has been subjected to far more extensive analysis and reanalysis than the average water resources project." He defended the Corps's projected benefits and warned that "the costs of delaying the Tennessee-Tombigbee for any reason would be tremendous."[6] Republican Senator Thad Cochran, Stennis's junior colleague from Mississippi, also spoke against Nelson's bill, calling it "the most absurd thing I have ever heard. That's about like me introducing legislation to kill the Saint Lawrence Seaway."[7]

At Nelson's request, Alaska Senator Mike Gravel, who chaired the Water Resources Subcommittee of the Committee on Environment and Public Works, scheduled an investigative hearing on the Tenn-Tom for 14 June 1979.[8] This one-day hearing was to be part of a planned six days of public hearings intended to address a variety of "new and continued project authorizations" and "[n]ational water policy proposals." According to Gravel's press release, "evidence developed in these sessions will be used as the foundation for this year's omnibus water resources bill."[9]

Stennis, whose power in the Senate was nearly unmatched by virtue of his chairmanship of the Appropriations Committee, did not take kindly to the proposed hearing, and he quickly pressured Gravel to drop the matter, arguing that a Congressional inquiry would be injurious to the waterway, which was in litigation. Gravel complied, and in May, the Environment and Public Works Committee canceled the Tenn-Tom hearing, maintaining that no hearings would be held on matters currently in litigation.[10] Nelson learned of the cancellation indirectly through a staff assistant, which led him to complain bitterly to Environment and Public Works Committee Chairman Jennings Randolph about Gravel's decision. Nelson claimed that, to his knowledge, "this is the first time that your Committee has refused to hold hearings on a bill." The allegations of deliberate deception demanded Congressional consideration, he told Randolph, adding: "I believe the evidence is overwhelming that this project should be stopped and I would hope that the Congress at least be given a chance to hear the arguments."[11]

Nelson wrote Gravel that he was "not concerned with addressing the question of authorization," an issue to be decided by the judiciary. Instead, Nelson emphasized that his interest concerned the Tenn-Tom's economic justification, which was "a question legitimately in the purview of Congress." With regard to the Committee's "policy" of not considering issues in litigation, Nelson reminded Gravel that his own Water Resources Subcommittee had recently held hearings on the Tellico Dam, which was in litigation, and that "for four years in a row, hearings on Locks and Dam 26 were held by your committee while that matter was actively before the courts." Nelson went so far as to suggest that there had never been such a policy in the Senate, for, "if this policy were to be adhered to, all any one would have to do to prevent Congressional consideration of an issue would be to file suit." Having placed his objections on the record, Nelson asked Gravel to reconsider his decision and go forward with the Tenn-Tom hearings.[12]

Gravel refused to back down. In fact, he would not even talk to Nelson about the issue. Nelson therefore announced that he would conduct his own hearings on the "Economic Impact of the Tennessee-Tombigbee Waterway Project" before the Senate Small Business Committee, which he chaired.[13] He slated the two-day hearing for mid-August, and his staff began preparations in July. The first day of hearings would feature witnesses critical of the project: independent economists who would evaluate the Corps's benefit/cost analysis and argue that, even with the sunk costs, it made economic sense to stop the project now; railroad and railway labor union representatives who would address the adverse financial impact of the Tenn-Tom on Southeastern railroads and their employees; small business and farming interests opposed to the project because of the waterway's damage to forestry and the economic hardship it would impose on the states' taxpayers to replace highway bridges; and officials from the Army Audit Agency and General Accounting Office who would discuss their respective audits of the waterway.

The second day would begin with witnesses from the National Taxpayers Union, who would contend the project was a colossal waste of tax dollars and who would support Nelson's deauthorization legislation, and a panel of environmentalists who would address the extensive ecological damage caused by the project. The remainder of the second day would be turned over to witnesses from the Corps, the Tennessee-Tombigbee Waterway Development Authority (TTWDA), and the governors of Alabama and Mississippi.[14] In their notes for the hearing, Nelson's legislative assistants outlined their objective: "The Small Business Committee will be able to demonstrate that repeated fraud by the Corps has been perpetrated on the Congress, the American people and the people of the area." Moreover, they wrote, "it will be evident that the Corps will be unable to respond to the very serious charges made, thus confirming that it has been involved in a cover-up and that the project is without financial merit."[15]

Attorneys for the L&N Railroad and EDF cooperated fully with Nelson's staff in preparing for the hearing. By coordinating their efforts in the ongoing litigation with Congressional attacks on the project, L&N and EDF officials reasoned that the two approaches would be mutually strengthened. Lead attorney Jon Brown expressed this philosophy to Nelson's Administrative Assistant, Jeff Nedelman, adding that he was "convinced that Congressional hearings are the most feasible first step in bringing the project to a halt, and that the hearings, if conducted properly, will be of great benefit to the interests of your office as well."[16]

Despite his staff's efforts and the public announcement of the August hearings, Nelson continued to hope that Gravel would allow the hearings before the Water Resources Subcommittee, where their potential impact on the omnibus water resources bill would be greatest. Nelson thought Gravel might worry about jurisdictional infringement by the Small Business Committee, and in July Nelson offered to cancel his hearings if Gravel would hold his own hearings on Tenn-Tom before the public works subcommittee.[17] Gravel stood by his decision to postpone the hearings until after the Fifth Circuit ruled on the plaintiffs' appeal. Jennings Randolph backed Gravel, while reassuring Nelson that his bill to deauthorize the Tenn-Tom would be given "every consideration . . . when it is timely to do so." He added diplomatically that he regretted "that the hearings were postponed in such an abrupt manner," and promised Nelson that he would "be given adequate notice when hearings are rescheduled."[18] Nelson canceled the Small Business Committee hearings, only to wait in vain for the ones to be held by the public works subcommittee.[19]

Nelson did not sit back quietly, however, and in 1980 he shifted his criticism of the Tenn-Tom. Turning to the fiscally conservative Senator Alan Simpson of Wyoming, Nelson argued that the waterway's economic justification was "dramatically worse than the very grim picture last year." He recommended that the $209 million fiscal year 1981 request could be eliminated "with no loss whatever to the public interest," and he urged Simpson to support deauthorization, explaining: "I understand the support this project enjoys. At some point, however, we must realize that we simply cannot afford the 'get along, go along' politics of the past any longer."[20]

Maine Senator William S. Cohen supported Nelson's attempt to deauthorize the Tenn-Tom. In a handwritten letter, he thanked Nelson for his "heroic effort to take some of the pork out of the water barrel that seems to be running over the rim." Cohen was pleasantly surprised that forty Senators voted in favor of deauthorization and suggested that "the winds of change may be heading in our direction." He asked Nelson to "be sure to alert me to when the 'Mississippi Clone' comes roaring through for an appropriation—I'd like to help on that one."[21]

Requesting Supplemental Funds

Although the Tenn-Tom had been routinely criticized in Congress, the waterway's annual appropriations had always been shielded by its inclusion within omnibus legislation for all federal water and energy projects; meaning that the votes were cast either for or against the entire slate of public works projects, never specifically for or against the Tenn-Tom. This changed in spring 1980 when Corps officials notified the appropriations committees that they were running short of funds for the waterway for fiscal year 1980 (which ended on 30 September 1980) and would need a $58 million supplemental appropriation to meet the year's contract obligations.[22] The ability to offer amendments to the supplemental request from the floor of Congress allowed opponents to single out the Tenn-

Tom for an "up-or-down" vote, and, as Representative Tom Bevill later recalled, this "was the beginning of our real controversy."[23] The problem facing the Tenn-Tom enthusiasts was not simply that the waterway could now be singled out from the floor of the Congress to withhold funding, but that there now existed incentives to do just that; inflation and the mounting federal deficit combined to make challenging pork-barrel expenditures a means to impress the electorate.

As early as May, New York Senator Daniel Patrick Moynihan joined Gaylord Nelson in notifying their colleagues that they would "offer an amendment deleting all supplemental funds for the Tennessee-Tombigbee Waterway" when the fiscal year 1980 supplemental appropriations bill reached the Senate floor. "This unnecessary and obsolete project has already received $165 million in Fiscal Year 1980," they said, "the largest yearly appropriation for an individual public works project in this nation's history." They stated that the Corps had already transferred an additional $23.5 million from its other projects to the Tenn-Tom in fiscal year 1980, and that the agency was now requesting "yet another $58 million in supplemental funds." In soliciting support for their amendment, they asserted: "At a time when we are struggling to eliminate wasteful federal spending, the continued funding of the Tennessee-Tombigbee Waterway is an affront to the American taxpayer of incredible proportions."[24]

The Supplemental Appropriations and Rescission Bill of 1980 came before the House of Representatives in June. Two Representatives who were bucking the system in their attempt to develop reputations as Congressional reformers and pork barrel critics, Republican Joel Pritchard of Washington and Democrat Robert Edgar of Pennsylvania, offered an amendment to delete $58 million slated for the Tenn-Tom. "We are not charging today that this project should be abandoned or that the funds spent thus far have been wasted," Pritchard explained. "We are saying 'Hold on,' what is the great rush?" In light of the Tenn-Tom's initial $165 million appropriation for fiscal year 1980 and the additional $23.5 million that was "administratively transferred to it with approval of the Appropriations Committee," Pritchard claimed that "the project is not being delayed for lack of funding; quite the contrary. There are reports that people are working around the clock, I assume many at premium pay, so current funding does not seem to be any problem." Supplemental appropriations bills were for emergencies, Pritchard said, and the Tenn-Tom faced no such crisis. By denying the supplemental request, Congress and the American people would gain the opportunity to explore the "many serious questions being raised" about the project "by the press, by environmental groups, by taxpayers' organizations and by many people connected with the project."[25]

The Pritchard-Edgar amendment aroused the longest and sharpest floor debate on the Tenn-Tom within the House since construction funds were first appropriated.[26] The project's environmental impact and questionable economic justification attracted the greatest attention. The mushrooming costs proved extremely controversial, especially among the swelling ranks of fiscal conservatives. Representative Bill Grenn of New York articulated this view when he said that the vote on the Pritchard-Edgar amendment "will be a real test of whether the Members of the House are serious about holding down Federal spending."[27] Outside Congress, the fiscally conservative National Taxpayers Union (NTU) worked in concert with Pritchard and Edgar in lobbying for the amendment.[28]

Tom Bevill and Jamie Whitten continued their roles as the project's most ardent defenders in the House. Bevill suggested regional bias on the part of Pritchard and Edgar, and reiterated the national benefits to be derived from the waterway. He claimed the only railroad opposed to the Tenn-Tom was the L&N. Bevill and Whitten argued repeatedly that all the criticisms had been fully aired in the annual appropriations hearings and had been resolved to the satisfaction of the committee.[29]

Dissatisfaction with the Tenn-Tom continued to spread, but not enough to carry the Pritchard-Edgar amendment, which lost by a vote of 230 to 185.[30] "Although we are delighted in winning this up and down vote on Tenn-Tom," Glover Wilkins told Representative Jack Edwards of Alabama, "we feel that it was too close for comfort." Wilkins worried that the momentum generated by the opponents would be carried forward to fight the fiscal year 1981 appropriations bill scheduled for a floor vote later in June. He therefore provided Edwards and other Congressional allies a list of "facts" to help them refute criticisms of the waterway.[31] The TTWDA later expanded this fact sheet into a 28-page booklet—*The Truth about Tenn-Tom*—as part of its public relations campaign to answer the opponents' numerous allegations.[32]

Following the House vote, the Senate then considered the supplemental appropriations bill. The $58 million in additional funds for the Tenn-Tom was part of an overall $16.6 billion supplemental appropriations bill that included disaster relief to victims of the Mount St. Helens volcanic eruption, aid to Nicaragua, and additional funds for Conrail, food stamps, child nutrition programs, and retired military cost-of-living ad-justments. Again, the Tenn-Tom drew critical attack. Republican Senator John H. Chafee of Rhode Island offered an amendment to delete all supplemental appropriations for the waterway.[33] "I just think it is wrong," Chafee said, "for the Corps of Engineers and the contractors to speed up, to work 7 days a week, 24 hours a day, pour their concrete, get more done so they could get us further down the road and then say, 'We are out of money. Don't make us stop, that will be wasteful. Give us $58 million more to finish this year.'"[34]

Daniel Moynihan agreed, calling the waterway "a project to clone the Mississippi River."[35] This was Moynihan's first occasion to address the Senate as the new chairman of the Subcommittee on Water Resources of the Committee on Environment and Public Works. He announced that the subcommittee's first hearings under his chairmanship would address the Tenn-Tom and the transportation of coal. He said he hoped to provide the Senate with his subcommittee's findings in time for consideration of the 1981 appropriations.[36]

Stennis managed the opposition to the amendment, which included impassioned speeches by Democratic Senator Howell Heflin of Alabama and Republican Senator Howard H. Baker, Jr., of Tennessee. According to the *New York Times*, "debate on the Tennessee Tombigbee project pitted a bipartisan coalition of Northerners against a similar group of Southerners, in an old-fashioned regional wrangle."[37] The rollcall vote taken at the close of the prolonged exchanges defeated the amendment 47 to 36.[38]

Sunk Costs and Deferred Maintenance

By the late 1970s, Tenn-Tom advocates began using the "sunk costs" argument in defending the waterway: stop the project now, they warned, and the nation would receive no benefits, while it would lose the substantial investment *already* expended; complete the project, and society would reap the full benefits, while only having to expend a relatively modest additional amount from this time forward. The sunk costs argument was a well-worn tool of public works project supporters, and skeptics had anticipated its use on the Tenn-Tom even before construction started. Shortly after the Environmental Defense Fund filed suit in July 1971 to halt the waterway, the Louisville *Courier-Journal* editorialized that "once the project is begun and millions are spent, the pressures will grow to have it completed rather than waste the money already invested. The same argument kept the SST [supersonic transport] going for three years after recognition of its environmental dangers and carried the controversial Cross-Florida canal halfway across the state."[39]

Every year when Corps officials testified before the Congressional appropriations committees, they gave financial reports on each of their major projects: how much was spent, how much remained to be spent, what problems remained unsolved, and so on. They discussed and recalculated the projects' benefit/cost ratios annually. It was here, in the area of *updated* benefit/cost ratios, that the sunk costs issue typically aided project proponents. By definition, sunk costs could not be recovered. They were already invested and therefore, proponents argued, no longer comprised part of the ongoing project costs. The question for decision makers was: Should more money be committed to this project? To assist in this decision, Corps representatives provided a benefit/cost ratio based solely on the remaining benefits and costs. Because the benefits of navigational projects like the Tenn-Tom would accrue primarily in the future, such economic reanalysis showed ever higher benefit/cost ratios following completion of greater proportions of the project—less money required to finish the project, and nearly all benefits to be reaped only if construction seen through to the end.[40]

The Corps termed this sunk-costs, readjusted benefit/cost ratio the "remaining benefit-remaining cost ratio," and it was a major incentive to speed up construction. The logic was simple: the more money plowed into the project, the harder it would be to stop. In 1981, for example, the Corps justified its annual budget request by claiming that the Tenn-Tom's remaining benefit-remaining cost ratio was 3 to 1.[41] Robert K. Dawson, Deputy Assistant Secretary of the Army for Civil Works, defended this position in explaining away the earlier, critical assessment of the waterway. "The economic justification in favor of project completion no longer can fairly be challenged, when justification is considered on a remaining benefits-remaining costs basis," he argued.[42] The Senatorial delegations from Mississippi, Alabama, Tennessee, and Kentucky put it even more bluntly. Stopping the project, they said, would cost over $300 million in contract termination and project restoration, while nearly $1 billion already invested "would be almost totally lost." "It really comes down to this," they contended, "there can be a completed Tennessee-Tombigbee Waterway by spending $353 million more than it will cost to stop construction, terminate the contracts, and restore the area."[43]

While Congressional allies made frequent use of the sunk costs concept in defending the Tenn-Tom, the Corps employed other techniques in a quiet and shielded manner to bolster the project's political stability. Agency planners were particularly concerned with the rapid increase in construction costs, which threatened to undermine an already weak benefit/cost ratio. Beginning in 1975, the army engineers initiated various efforts to reduce costs without jeopardizing the project itself.[44] Some of these efforts focused on modest design changes that would lower construction costs (and perhaps quicken the completion date) but which would raise expenditures for long-term maintenance. In a sense, it was a form of juggling the books—postponing some fine-tuning or finishing aspects of the project until *after* it was opened for traffic, thus lowering the overall construction costs for more "hidden" (and often greater) expenditures at the maintenance and operations stages. The quiet deferring of costs illustrated the lengths to which the Corps would go to steal the wind from their opponent's sails.

By October 1975, the Corps reported $46,856,300 in savings from cost reductions already adopted. These changes included: reducing the amount of clearing in the dredged material disposal areas and eliminating the revegetation of those sites ($7,490,300 in savings); cutting back on slope protection (riprap) in the Rattlesnake Bend Cutoff ($1,533,300); altering the technique for dewatering construction sites ($4,379,500); and designing for flood waters to overtop structures, thus allowing smaller spillways ($13,888,800).[45] Several other changes under consideration, the Corps noted, also produced initial cost savings, although they substantially increased operations and maintenance expenditures in subsequent years. For example, the installation of timber guide walls at the lock structures would save the government an estimated $485,000 over the construction of concrete guide walls. Maintenance costs, however, due to the frequent replacement of damaged timber sections, were estimated to be $525,000. The Corps also considered postponing construction of mooring facilities near the locks to save approximately $3,340,000 in initial construction costs. Their provision, which remained a necessary part of the waterway, would simply be financed with operation and maintenance funds *after* completion of the project.[46]

The protection of certain embankments from erosion also exemplified "deferred maintenance." Where the channel was widened or straightened by cutting through a river bend, the Corps's hydraulic engineers estimated the potential erosion to those cuts based upon the action of the water and the makeup of the soil. Where high potential for erosion existed, engineers typically recommended the placement of riprap to protect the banks. In the divide cut, for example, Nashville District officials believed safety and practicality demanded that stone riprap be placed along both banks of this entire section. Engineers at the South Atlantic Division, however, preferred to lower the initial investment by leaving the banks exposed. Nashville's design staff disagreed, and fashioned a technical argument persuasive enough to change the minds of their superiors in Atlanta.[47]

Although some engineers at the Mobile District argued for heavy bank protection within the river and canal sections, top district officials proposed and followed another course of action: channels would be excavated and banks allowed to erode naturally. Corps personnel would then repair the damaged sections when and where needed after opening the channel. As they told the Army Audit Agency in 1976, "if adverse conditions occur after the waterway is placed in operation, corrective measures will be handled by operation and maintenance."[48]

This approach to embankment protection had two strong advantages.

Aesthetically, many people find long stretches of stone-lined embankment unsightly. Mobile District's approach meant that only those sections of the channel actually requiring protection would have it, leaving the rest of the waterway with a more natural appearance. Economically, deferred maintenance allowed accountants to lower the costs of construction by charging the necessary work to the subsequent operation and maintenance budget. That is: stone protection laid down before completion of the waterway would be counted as construction costs; stone protection laid down after completion of the waterway would be counted as maintenance costs. Clearly, it was impossible to foresee every stretch of the channel that would eventually require protection, but, on the other hand, there were some areas where the engineers knew with great certainty that erosion would occur. In these latter, vulnerable areas it was not a matter of "let's wait and see," but a deliberate attempt by the Corps and its political allies to mask the full costs of the waterway.[49]

The Eroding Tenn-Tom Coalition

Although the behind-the-scenes use of deferred maintenance to make the project appear less costly and the on-stage use of the sunk costs argument helped secure the passage of the supplemental appropriations bill in June 1980, Tenn-Tom supporters could not rest on their laurels. Like the tides, the Corps was part of a cycle that brought it back again and again to the Congressional shores in search of annual appropriations. Recognizing this, Representatives Joel Pritchard and Robert Edgar began rallying their colleagues to delete $200 million earmarked for the Tenn-Tom in the fiscal year 1981 budget. In a letter delivered to all members of the House the day before the floor debate, Pritchard and Edgar asserted that "we cannot afford to spend on a project beset with so many problems."[50] Two days later, they introduced an amendment to reduce the Tenn-Tom appropriation from $225 million to $25 million, enough to complete the southern portion of the waterway from Demopolis to Columbus, while Congress reassessed the remainder of the project.[51]

Iowa Representative Berkley Bedell delivered the last speech of the subsequent debate. "In the salad days of unending tax moneys," he said, "the Tenn-Tom project was merely a humorous anachronism. Now, however, it is no longer a laughing matter."[52] Many members agreed with him, but not enough to stem the flow of money: the Pritchard-Edgar amendment went down in defeat, 216 to 196.[53] Although victorious, the Tenn-Tom advocates worried about the continual erosion of their base of support. As Jamie Whitten told TTWDA attorney Hunter Gholson about the appropriations bill: "I surprised them [the project opponents] by getting the bill called up on Tuesday instead of Thursday as they expected. This, perhaps, made the difference."[54]

Whitten's sly orchestration did not go unnoticed. The *Washington Post*, whose writers had long criticized the Tenn-Tom and other dubious Corps of Engineers projects, ran an editorial entitled "The $3,000,000,000 Ditch," which discussed the Tenn-Tom's Congressional troubles and noted that "Whitten had to pull out all the stops to keep Tenn-Tom alive. There were veiled threats, and some not so veiled, about the fate of other water projects." The *Post* editors proclaimed that support for the waterway "is wobbling in the House and a full-scale assault is under way in the Senate." They added that "the effort to kill the Tenn-Tom and start reclaiming about half the federal investment in it by selling the land the Corps has acquired now shifts to the Senate. We wish it well."[55] The *Post*'s was but one voice in an editorial chorus that rose against continued funding of the Tenn-Tom during summer 1980.[56]

By this time, the National Waterways Conference (NWC), which routinely lobbied Congress on behalf of inland navigation projects, had a vested interest in the Tenn-Tom. The escalating controversy over this project threatened to taint the nation's entire waterways development program. "I hate to admit it," NWC president Harry N. Cook wrote Glover Wilkins, "but I believe that the Tennessee-Tombigbee Waterway is in a very tenuous position." Cook reasoned that support should be increasing because the project was half-way through construction, but instead, he thought that "continued funding for Tennessee-Tombigbee (if a vote were held today) would fail to pass the U.S. Senate by a margin of 10 to 15 votes." He offered Wilkins a number of suggestions to help him "counter this backslide in project support." Cook noted the tendency of first and second term lawmakers to sympathize with "the environmentalists' argument that now is the time to put a stop to 'water politics,'" and that these same members of Congress were suspicious of Corps of Engineers projects and justifications. Tenn-Tom opponents, he said, "have succeeded in convincing numerous lawmakers that project benefits are illusory, that potential traffic volumes have been fabricated, that the above-unity benefit/cost ratio is a sham." He urged the TTWDA to counter this with a vigorous campaign, including widely-distributed fact sheets and promotional booklets, letters of support to all Senators from the White House and Senator Howard Baker, and the establishment of "a close alliance with supporters of other major water projects." Cook emphasized the threat facing the very survival of the waterway. "In essence," he argued, "the central issue now confronting Tennessee-Tombigbee in the Congress is this: Is it a creditable project with the realistic prospect of substantive benefits to the region and Nation if it is completed, or is it just a boondoggle conceived by the Corps of Engineers and railroaded through the Congress by a few powerful Southern lawmakers?"[57]

Wilkins, of course, did not need to be reminded of the Tenn-Tom's troubles. He knew the fiscal year 1981 appropriations vote was critical to the waterway's survival. Working with Senator Stennis and his staff, the TTWDA coordinated the intensified lobbying effort. Stennis gave the Authority use of his hideaway Capitol office for this purpose.[58] Wilkins assured Stennis in a confidential letter that the TTWDA was working to get the support of "national, regional and single-project, water-oriented organizations" and that it would be "contacting Senatorial staffs and Senators, personally where possible, providing them with written and oral information on the Tenn-Tom."[59]

The TTWDA did not limit its lobbying efforts to the Senate. The Authority also organized meetings for various members of the House, providing informal forums for project supporters to invite their undecided colleagues to hear project status reports from senior Corps officers. Typically, these meetings took the form of steak luncheons underwritten by the TTWDA. Tom Bevill hosted many of these affairs in his committee hearing room. In concert with the TTWDA, Bevill and his staff organized a "water group" consisting of Congressional members interested in various water resources development projects (and who, presumably, adhered to the "all for one, one for all" philosophy when it came to appropriations votes). Members of the water group lobbied Representatives from their own state and perhaps an additional state who were neutral or opposed to the Tenn-Tom. On the day of the appropriations vote, Bevill's staff called every member's office to solicit support.[60]

The Senate did not meet to consider the 1981 Energy and Water Development Appropriations bill until September 1980. Unlike the House bill that included $225 million for the Tenn-Tom, the Senate bill reduced the amount to $208 million. Senator John H. Chafee of the Environment and Public Works Committee again led the opposition. He introduced an amendment to delete $200 million from the appropriations request and to prohibit construction north of Columbus, Mississippi.[61] Critics rallied behind Chafee and his amendment. Environmental Defense Fund attorneys working on the Tenn-Tom litigation turned their attention to the Senate votes.[62] Prime Osborn, chairman and chief executive officer of Seaboard Coast Line Industries (the parent company of the L&N Railroad), wrote all Florida shareholders, asking them "to write, wire or phone your Senators and urge them to vote for the Chafee amendment to delete funds for the Tennessee-Tombigbee Waterway."[63]

Despite the eloquent speeches and exchanges on the Senate floor, most votes were decided beforehand, many of them influenced by the prolonged and extensive lobbying on both sides. Opposition to the project had gained momentum, but the Chafee amendment still fell short, 52 to 37.[64]

The appropriations bill was not the sole vehicle for Congressional scrutiny of the waterway. In July 1980, the Water Resources Subcommittee of the Senate Committee on Environment and Public Works, under Daniel Patrick Moynihan's chairmanship, had held a three-day hearing on the Tenn-Tom to discuss the transportation needs for the nation's coal producers and the wisdom of completing the waterway. Project backers had been concerned that Moynihan's hearings would further inflame the controversy. The TTWDA, which had concentrated its lobbying efforts on appropriations, had attempted to have the hearings postponed until the Senate had approved the Tenn-Tom appropriations for 1981. Although that effort failed, the waterway's proponents came well prepared with witnesses and factual material. The TTWDA also urged Senator Jennings

Randolph, the full committee chairman, to attend the hearings to help keep Moynihan in check.⁶⁵

The New York Senator had told his colleagues in June that he hoped to provide them with the findings of his subcommittee's hearing when the Senate convened to consider the 1981 energy and water development appropriations bill. After the hearings ended in late July, his subcommittee staff sent the Corps a copy of the transcript for correction and a list of follow-up questions. However, the agency failed to meet the September deadline. Angered but left with few alternatives in the absence of the transcript, Moynihan inserted a summary abstract into the *Congressional Record* which suggested that the Corps's projections of coal shipments on the Tenn-Tom were gross exaggerations.⁶⁶

When the Senate and House energy and water development bills went to conference in September, the House's $225 million and the Senate's $208 million line item for the Tenn-Tom were compromised to $212 million.⁶⁷ Even at this amount, the waterway remained by far the single largest expense request in the appropriations bill—no other water project received over $100 million.⁶⁸

Avoiding a Showdown

Tenn-Tom opponents had mixed reactions to the national election in November 1980. Riding the coattails of Ronald Reagan, Republicans had gained the majority in the Senate.⁶⁹ This shift strengthened the influence of fiscal conservatives, but weakened the appeal of environmental criticism. Opponents therefore brought the economic deficiencies of the waterway to the attention of Reagan's transition team, hoping that if the new administration did not reduce funding for the Tenn-Tom or demand a greater percentage of cost sharing, then perhaps the Republican Senate might itself cut appropriations.⁷⁰

Disappointment with the Reagan administration came early, as the February 1981 Presidential budget requests indicated the Tenn-Tom would escape the broad-based budget cutting seen elsewhere in domestic spending. Reagan's appointment of an outspoken Tenn-Tom critic, David Stockman, as head of the Office of Management and Budget had raised false hopes. The *Los Angeles Times* captured this frustration when it editorialized: "Only a year ago, when he was representing a Michigan district in the House of Representatives, David A. Stockman was a vehement opponent of the Tennessee-Tombigbee Waterway. Now, as director of the Office of Management and Budget, he has the opportunity to halt the further waste of taxpayers' dollars on one of the costliest and least defensible boondoggles in the nation's history."[71] The editors of the *St. Petersburg Times,* who called the Tenn-Tom "a classic example of waste and fraud that cries out to be stopped," also lamented Stockman's failure to cut funding of the project. "President Reagan would have the government take food from children but continue digging a worthless, $3-billion ditch," the editors stated. "He wants to cut food stamps, school lunches and Aid to Families with Dependent Children. Yet he would finish the Tennessee-Tombigbee Waterway."[72]

On 13 March 1981, the Senate Environment and Public Works Committee recommended, in a 9-4 vote, stopping construction of the Tenn-Tom. Republican Senators Alan Simpson of Wyoming and Pete Domenici of New Mexico led the effort to restrict expenditures for the waterway to $25 million—that is, enough to complete only the existing portion of the waterway south of Columbus. Simpson, who initially sought to eliminate all funding for the waterway, had accepted Domenici's $25 million compromise.[73]

The political momentum against the Tenn-Tom was propelled in part by an expanding coalition of interest groups willing to battle the project in Congress. Since authorization of the waterway in 1946, groups and individuals had lobbied against it, and their voices grew in numbers and volume during the 1970s when construction began; but a broad-based and closely coordinated coalition of opponents did not emerge until the project was nearly half completed. By that time, the list of seemingly unlikely collaborators included: National Taxpayers Union, Common Cause, League of Women Voters, Sierra Club, National Wildlife Federation, Environmental Policy Center, Coalition of Water Project Review, Association of American Railroads, and Railway Labor Executives Association.[74]

In the summer of 1981, Tenn-Tom advocates faced their most serious challenge in Congress since construction began. The debate on the floor of each house was intense. Michigan Representative John D. Dingell voiced the concerns of many skeptics when he stated that Americans were being asked "to sacrifice large sums of money for legitimate human resource programs while the Tenn-Tom awaits even more taxpayer money. But for whom, how many, and for what reason?"[75] After spirited debate, the waterway's supporters managed a narrow 208 to 198 victory. The House of Representatives appropriated construction funds at the rate of $20 million a month for another twelve months, prompting a *Wall Street Journal* editorial that attacked the allocation as "economic nonsense."[76]

Polling Congressional members became crucial for proponents and opponents alike.[77] When the Tenn-Tom critics gathered in September, for example, they discussed writing and calling interested citizens, visiting Congressional offices, placing negative reports and articles in the *Congressional Record,* lobbying the Congressional Black Caucus, encouraging anti-Tenn-Tom editorials, and generating group letters.[78] They stressed the importance of identifying key Senators and Representatives who were most susceptible to pressure on the Tenn-Tom issue. Usually, these swing votes included Congressional members with high ratings on overall environmental issues (as measured by the League of Conservation Voters) or with records of voting against wasteful spending, but who had been voting in favor of the Tenn-Tom (presumably in exchange for support of their own district-based projects by the influential backers of Tenn-Tom, which included the chairs of both the House and Senate Appropriations committees). Opponents also targeted first-term Congressional members.[79]

To focus specifically on Congressional efforts to kill the waterway, opponents formed the "National Campaign to Stop the Tenn-Tom," with headquarters at the Capitol Hill office of the Environmental Policy Center. The National Campaign sought to rouse grass-roots organizations throughout the country which had fought other federal water projects. They provided citizens with information packets and encouraged them to call, write, and/or visit their Congressional delegates to urge them to vote against additional appropriations for the waterway. Understanding the power of humor, the National Campaign distributed hundreds of three-billion-dollar bills for the Tenn-Tom, labeled "legal pork tender," to publicize their cause.[80]

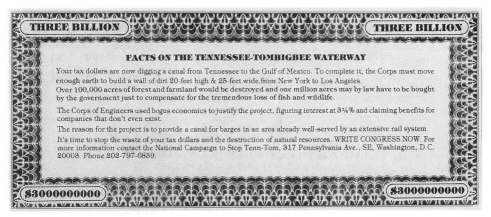

Handbill distributed in 1981 by the National Campaign to Stop the Tenn-Tom *(Courtesy of Brent Blackwelder.)*

Charles Percy, Daniel Moynihan, and six other Senators circulated a letter to their colleagues in October announcing an amendment to the energy and water appropriation bill which would eliminate $189 million for the Tenn-Tom. "At a time when we are struggling to balance the budget," they wrote, "it is unconscionable for us to continue full Federal funding for a project which will divert 70 percent of its future traffic from the Mississippi River, a perfectly adequate waterway in its own right." The Senators criticized the claims made for massive coal shipments, stating: "Studies conducted by both the GAO and the Congressional Research Service suggest that the project would be a net drain on the Treasury for its entire 50 year life." They questioned the wisdom of ever starting the Tenn-Tom, and added that "throwing good money after bad will not magically produce benefits that are simply not there."[81]

When the Tenn-Tom appropriation came up for consideration in the Senate on 4 November 1981, the project passed by a mere two votes, 48 to 46.[82] Len Rippa, who had headed the National Taxpayers Union's efforts against the waterway as NTU's Director of Congressional Affairs, later reflected on the loss of the "Tenn-Tom battle." "Undoubtedly," he wrote, "the coalition of organizations that joined in the fray was unique. Perhaps, more than any other factor, the wide range of interests that came together for the common purpose of stopping this boondoggle led to very close votes."[83]

Their narrow victory shook the confidence of Tenn-Tom's promoters. Senator Howell Heflin quickly gathered Alabama's Congressional delegation together to refine their strategy for the next appropriations vote. Heflin believed 1982 would be "the key year" for Congressional support, because thereafter the costs of shutting down the project would exceed those of completing it. Heflin's strategists discussed ways of raising private funds to underwrite the intense lobbying effort they deemed necessary. Part of that campaign would include field trips for Congressional members to view the waterway. Heflin told reporters that fund raising required "an all-out effort" to win the involvement of chambers of commerce, industries, and other people affected by the waterway.[84]

Members of the House and Senate worked closely together in early 1982 to expand grass-roots support for the waterway. Heflin devoted his March constituent newsletter to the political clash over Tenn-Tom. In an attempt to shock his constituents into action, Heflin declared that "if the vote were taken today in the U.S. Senate to decide the future of the Tennessee-Tombigbee Waterway, the Tenn-Tom would probably lose." He outlined his colleagues' efforts to bring Congressional and media skeptics for personal tours of the project, claiming that "a first-hand view of the waterway's potential and a generous dose of southern hospitality could change an opponent's position."[85]

The opponents likewise wasted little time in preparing for the next appropriations vote. The project's slim margin of victory encouraged a belief that events and political sympathy were moving in their direction. EDF attorneys were optimistic about the political climate in 1982. "In view of the Administration decision not to raise taxes to decrease the deficit," they claimed, "there will be a redoubled effort to cut Federal spending." As a first step, they recommended targeting key swing votes in the House and Senate for immediate lobbying.[86] The Sierra Club's Washington, D.C., office followed this advice in developing its political campaign agenda. Through an emphasis on budget deficit issues, the Club focused its efforts on "those members [who] should be with us, but switched their votes in 1981. This group of Representatives will be targeted for mailings in their districts, meetings with concerned constituents and as much press coverage [as] we are able to generate."[87]

The Sierra Club's effort was thorough and sophisticated. The members selected for lobbying fitted into one of two categories: first-term Senators and Representatives who had yet to vote on Tenn-Tom appropriations; and "swing votes"—members who had voted both for and against the Tenn-Tom in the past and whose overall voting record suggested they might turn against it in the future. The Club analyzed seven previous appropriations votes and tabulated members' voting records according to three separate ratings. A high rating by the League of Conservation Voters suggested receptivity to environmental arguments. A high rating by the National Taxpayers Union indicated a fiscal conservative who might be troubled by the project's exorbitant cost. A high rating from Americans for Democratic Action indicated a liberal philosophy and support for social programs that might compete for funds with "unnecessary" projects like the Tenn-Tom.[88]

Congressional supporters of the waterway became extremely cautious. They decided never again to ask for supplemental appropriations or any other legislation related exclusively to the Tenn-Tom. They had learned their lesson earlier; every bill presented opponents with an opportunity to introduce amendments to deauthorize the project or withhold funding.[89]

Throughout the summer of 1982, opponents remained optimistic that the waterway could still be stopped, and they stepped up their efforts to attract newspaper coverage. Len Rippa spoke with several reporters at the *New York Times*, praising the newspaper's earlier coverage of the project.[90]

Sierra Club president Denny Shaffer sent editorial page editors of the nation's major newspapers packets containing fact sheets on the Tenn-Tom, voting records of House members on the waterway appropriations, and sample editorials critical of the project. As he explained, "the Sierra Club, along with a coalition of other environmental groups, the National Taxpayers Union, Common Cause, League of Women Voters, and the Conservation Caucus, has spotlighted this Army Corps of Engineers canal project in Alabama and Mississippi as the most wasteful in the entire water development program."[91]

By August, the anti-Tenn-Tom coalition had added the National Gray Panthers and the Brotherhood of Railway and Airline Clerks to its ranks. The coalition urged members of the House to support the new Pritchard and Edgar amendment to delete funding for the Tenn-Tom. They emphasized the large percentage of the total Corps of Engineers's 1983 construction budget which went into the waterway. The $189-million Tenn-Tom request represented "more money than the individual allocation for each of 47 other states, nearly three times as much as for any other Corps construction project and more than 15% of the *total* FY '83 Corps construction budget."[92]

Typically, energy and water development appropriations bills clear both the House Energy and Water Subcommittee and the House Appropriations Committee before the August recess. That did not happen in 1982. Opponents realized that a delayed floor debate would seriously hamper efforts to delete funding for the Tenn-Tom. The Senate—where skepticism of the waterway ran deepest—would have little time to debate the bill, which would not arrive until after it passed the House. Len Rippa brought this issue up in a widely-circulated letter to newspaper editors. "Because of the shortage of time at the end of the session," he declared, "there is a distinct possibility that Energy and Water Appropriations will be included in a continuing resolution." Such a move would preclude a specific amendment on the Tenn-Tom.[93]

That is exactly what happened. When Congress proposed a continuing resolution, Edgar and Pritchard pragmatically altered their tactics. Rather than attempt to eliminate all funding for the waterway, they proposed an amendment to cut the $186 million request for fiscal year 1983 by $100 million. This shift acknowledged that the project would be built, but sought to restrict additional expansion of the waterway and to set a precedent for greater local cost sharing. As Edgar pointed out, "the amendment would leave the Tenn-Tom the highest-funded [at $86 million] Army Corps of Engineers construction project in the country, but we believe a substantial funding reduction would place the project more in line with other national priorities."[94]

Tom Bevill countered with a news conference in which he said that with the waterway "approximately 82% complete," it would cost more to stop the project and "put it in 'mothball' status" ($431 million) than it would cost to complete it ($387 million). Moreover, he added, "stopping work on the Tenn-Tom also would be putting 5,000 people out of work in one of the most economically depressed areas in the nation."[95] These were Bevill's public statements; privately, he connived to prohibit *any* amendments against the Tenn-Tom in the continuing resolution. The Rules Committee, however, denied Bevill's request for a prohibition against Tenn-Tom amendments, and ruled that the waterway, along with three other highly controversial public works projects (North Dakota's Garrison Diversion project, Nebraska's O'Neal Dam, and Tennessee's Clinch River Breeder Reactor) would all be subject to amendments. According to Bevill's later account, "before the rule was distributed, someone in authority had drawn a line through the language authorizing an amendment to Tenn-Tom." As Bevill boasted, "without an amendment, the funds were secure." Noting that funding for all three of the other projects was later removed from the continuing resolution by way of amendments, the Alabama Representative took further credit for this political savvy and foresight. "If the rule had stood as originally written," his official statement asserted, "the Tenn-Tom would have been stopped at that point."[96]

Although outraged at the procedural sleight of hand executed by Bevill and his collaborators, members of the anti-Tenn-Tom coalition recognized that the Corps still faced nearly three more years of construction. The chances of killing the project outright no longer existed, but with the waterway requiring over $200 million to complete, it remained subject to amendments in the next two appropriations bills. Bevill knew this too, of course, and to prevent even the opportunity for Congress to vote on the

A dredge cuts through "the plug," the final barrier for mixing the waters of the Tennessee and Tombigbee Rivers. The unknown consequences of joining two entirely separate river systems worried many ecologists. *(Courtesy of U.S. Army Corps of Engineers, Mobile District.)*

project, the shrewd subcommittee chairman removed all requests for additional funding for the Tenn-Tom from the energy and water development appropriation bill for fiscal year 1984. Instead, the bill directed the Corps to complete the waterway "within funds available" in the construction general account (that is, to use excess money appropriated for *other* Corps projects which sat unspent within the agency's fiscal year 1983 budget). To placate skeptics who feared the bill would give the Corps a blank check, Tenn-Tom proponents agreed to set a cap on spending: $202 million ($90 million for 1984, $90 million for 1985, and $22 million for 1986), the estimated amount needed to complete the waterway.[97]

The maneuver raised the question of the constitutionality of multiple-year appropriations (House rules required annual appropriations). Bevill, who managed the energy and water development appropriations bills, solicited and gained the approval of House Speaker Thomas P. O'Neill for a special exception.[98] The passage of this bill guaranteed enough funds to finish the waterway and ended the Congressional debate over appropriations. The Corps's Chief of Engineers was relieved that the three-year allocation would eliminate "another big battle" in Congress.[99] Senator Stennis was equally impressed, and complimented Bevill: "I actually believe that, everything considered, your work on the Tennessee-Tombigbee, from the time you started your special duties as Chairman of the Subcommittee until now, is the best legislative job I have seen done here."[100]

On the other side of the ledger, long-time Tenn-Tom opponent and environmental activist Brent Blackwelder reflected on the symbolic value of the fight. "If you could eliminate the biggest project and beat the Chairman of the Appropriations Committee . . . you could win a devastating blow to pork barrel," he said. "You know, it would show that nothing was sacred. And as it was, we forced them to play every chip and bargain everything. It is amazing what people must have gotten out of these guys for agreeing to go along with this dumb thing."[101]

Blackwelder's rationalization could hardly mask the frustration and sense of profound defeat felt by members of the anti-Tenn-Tom coalition. Bevill's procedural trick may have assured the funds necessary to complete the waterway, but it also typified the extremes to which the project's proponents had reached. The Corps, the Congressional advocates, and the Tennessee-Tombigbee Waterway Development Authority had all, at one time or another, played fast and loose with the rules and, in so doing, had failed in their responsibility to maintain the public trust. The environmentalists, good government activists, fiscal conservatives, and a few maverick Congressional members had dared speak truth to power because they perceived the Tenn-Tom to be a colossal waste. Although opponents had not been able to stop the waterway, they did succeed in a larger sense—by increasing public involvement in federal water resources development planning and review, tightening the rules in the economic justification of Corps of Engineers projects, and giving greater voice to environmental and social justice aspects of public works programs.

Conclusion

THE STORM OF POLITICAL and legal confrontation over the Tenn-Tom subsided in 1983, when the acquisition of the three-year Congressional appropriation and the dismissal of the litigation virtually assured completion of the waterway. Basking in the glow of victory, the Corps and the Tennessee-Tombigbee Waterway Development Authority (TTWDA) began preparing a grand dedication ceremony to mark the project's opening.

Confidence did not flow through the veins of all Corps officials, however, especially those engaged in public relations. Samuel R. Green, the chief of Mobile District's public affairs office, for example, told his commanding officer that the decades-long controversy would not end when the waterway was finished, but that "many opponents will continue to question both the justification for the project and the methods by which it was brought to completion and, perhaps worse, relate it to future projects." Green said that such situations raise red flags for public relations planners because they involve "seriously divided publics and the event being planned could trigger serious, unpredictable backlashes for the Corps." As he explained, "an overdone dedication for Tenn-Tom could easily result in the Corps being accused of being insensitive to environmental interests and those who want reduced Federal spending. Spokesmen for these interests have the savvy, wherewithal and communications expertise to use such an event for their own purposes." He worried that negative press coverage of the dedication could harm the agency's plans for future water resources projects. He therefore urged the Corps's leadership "to prudently limit" its exposure.[1]

Green's caution was no match for the understandable enthusiasm of the TTWDA, whose long, hard fight to build the waterway had been won. The TTWDA's public relations staff and fund-raisers went all out to assist the army engineers in staging a celebration worthy of the nation's largest navigation project. Although the Tenn-Tom opened to commercial traffic with the passage of the *Eddie Waxler* on 10 January 1985, the official dedication did not take place until warmer weather settled in the South. On 1 June 1985 in Columbus, Mississippi, an extravagant celebration complete with flotillas, fireworks, and plenty of politicians capped a week-long festival. It also, as Green had feared, drew unfavorable media attention. *Miami Herald* reporter Fred Grimm, for example, wrote that "as a politi-

The towboat *Eddie Waxler,* making the first commercial passage through the Tenn-Tom in January, 1985 *(Courtesy of U.S. Army Corps of Engineers, Mobile District.)*

Conclusion 247

cal accomplishment, the Tenn-Tom ranks right up there with the Seven Wonders." He claimed the waterway "became one of the most controversial pieces of public works of all times, the very epitome of pork barrel, the point of reference against which all other backyard projects were measured."[2] *Christian Science Monitor* reporter Robert Press predicted that "the project will be closely watched for years to see if enough people use it to justify its expense and environmental damage."[3] David Rogers of the *Wall Street Journal* focused on the activities of Jamie Whitten, titling his newspaper account "Rivaling Cleopatra, A Pork-Barrel King Sails the Tenn-Tom." Rogers began his story: "Move aside, Huck Finn. Eat your heart out, Cleopatra. Rep. Jamie Whitten is coming down the Tennessee-Tombigbee." In describing "the granddaddy of pork-barrel projects," Rogers wrote: "The 234-mile-long Tenn-Tom is the last hurrah for the Southern pork-barrel policies that have been Rep. Whitten's way of life in 44 years of Congress. The 75-year-old dean of House Democrats is the public man floating on public waters and under a bridge—the Jamie L. Whitten Natchez Trace Parkway Span—built with public money." Rogers noted how the waterway, by then opened to commercial traffic for six months, was strikingly empty, its greatest appeal proving to be recreation. "Cynics even suggest the government may have created a nearly $2 billion fishing hole."[4]

The lack of commercial traffic on the Tenn-Tom continued to generate criticism. "During its first year of operation the waterway accommodated only 500,000 tons of coal," the *St. Petersburg* (FL) *Times* editorialized in February 1986. "That's not a misprint. Actual coal tonnage came to only 2.4 percent of [Sen. Paula] Hawkins' 'conservative' estimate of 21-million tons." The newspaper editors blamed not Hawkins but the Corps, which had "assured Congress the Tennessee-Tombigbee would carry 27-million tons of cargo of all kinds during its first year of operation. The actual result: Only 1.7-million tons, or barely 6 percent of the Corps' projection." They continued: "When will Congress learn? The Corps and its Western counterpart, the Bureau of Reclamation, are notorious for overstating the expected economic benefits of projects they want to build. But these bureaucracies are not themselves wholly to blame, because the Congress, as keeper of the pork barrel, wants it that way."[5]

An eight-barge tow, the largest able to pass through the Tenn-Tom's locks, travels along the river section of the waterway. *(Courtesy of U.S. Army Corps of Engineers, Mobile District.)*

In San Francisco, editors of the *Examiner* proclaimed that the $2 billion waterway "that survived every budgetary ax in Congress, has become more a playground for the boating set than a major commercial route."[6] The *Washington Post*, too, kept a watchful eye on the project. In a January 1987 editorial, the *Post* called the waterway "the nation's largest wet elephant." With regard to the Tenn-Tom's projected impact, the editors wrote that "the waterway has provided passage for only 4.8 million tons of cargo in almost its first two years of operation, and a fifth of that was stone for its own banks."[7] Moreover, the majority of craft plying the canal were pleasure boats; in 1985, for example, 433 commercial vessels passed through the Tenn-Tom's Aberdeen Lock, compared to 1,280 pleasure craft.[8] The *Atlanta Journal and Constitution*, which labeled the Tenn-Tom "a 234-mile broken promise to the rural poor who live along the border of Mississippi and Alabama," reported that during the first seven years of operation "the waterway has seen more bass boats than barges."[9]

Asked why the barge traffic on the waterway was so low, Tom Frazier, an American Commercial Barge Line senior vice president, said that "it's cheaper to use the Mississippi, unless you're coming from east of Tenn-Tom and going right to Mobile. Coming off the Ohio River, it's better to

Lock D on the Tenn-Tom canal section *(Courtesy of U.S. Army Corps of Engineers, Mobile District.)*

go via the Mississippi, even if you're headed for Mobile."[10] The problem, just as critics had claimed, was the huge competitive advantage of the Mississippi River, which—even if a longer haul for some shippers—offered the cost savings of allowing for much larger tows (thirty-to-forty-barge tows, versus only eight-barge tows on Tenn-Tom, and *six*-barge tows south of Tenn-Tom where it joins the smaller Black Warrior River) and a lock-free passage that permitted faster travel speeds (eight to nine miles per hour versus three to four on the Tenn-Tom).[11]

Like the projected commercial traffic, regional economic development also fell far short of predictions. The commerce that did move on the waterway tended to be through traffic—its origins and destinations located outside the project area—giving the barge captains little reason to stop within the waterway's bounds, save to pass through the ten locks. One growth industry that generated mixed reactions among the residents of the Tenn-Tom corridor was the establishment of commercial garbage dumps to handle wastes from states outside the South, mainly from the Northeast. In 1989, three large, privately-owned, for-profit garbage dumps were planned along the waterway, the largest being a 5,000-acre facility in Greene County, Alabama, which was expected to be able to accommodate

The lone bass boat passing through the Aliceville Lock is a common sight on the Tenn-Tom, where pleasure craft far outnumber commercial barges. *(Courtesy of Marcel C. LaFollette.)*

10,000 tons of hazardous wastes. Just north of this site in Emelle (Sumter County), Alabama, was the largest commercial hazardous waste dump in America.[12] The siting of these dumps in economically disadvantaged areas with substantial minority-group populations raised the ironic prospect that a project, which had promised economic opportunity for African-Americans in rural Alabama, brought something far different.[13]

The history of the Tennessee-Tombigbee Waterway reflects many of the crucial changes in the development of large-scale public works in the United States. Perhaps more than any other single piece of legislation, the National Environmental Policy Act most directly shaped these changes. Its enactment into law in the very same year that Congress appropriated construction funds for Tenn-Tom assured that the waterway would be one of the first major projects built entirely under the aegis of the new legislation. As a result, a project already controversial on economic grounds was opened to public scrutiny in ways never before imagined by the Corps or its allies.

Organized public participation came from many quarters at different times. The most vocal were the environmentalists, railroads, fiscal conservatives, and minority advocates, all of whom challenged the engineering and bureaucratic arrogance of the federal government. And, although the limits to change were substantial, all groups—but especially the environmental and minority interests—succeeded in bringing about important alterations in the project and its construction. Their actions demonstrated the continuing importance of individuals and non-governmental organizations in pressing for change within a democracy.

In a decade punctuated with environmental protests and crises, the Tenn-Tom controversy also attracted unwanted national attention to the army engineers and their Congressional supporters throughout the 1970s and early 1980s, opening other Corps activities to reassessment. To an extent impossible to measure but apparent in press coverage and subsequent political debates, the Corps's credibility with the public was severely damaged by the fight over the Tenn-Tom. The issues raised by the waterway's challengers helped focus and frame the terms of further debates over all federally-funded water resources projects. For example, there were no omnibus water bills passed into law between 1976-1986; the Corps thus went without any new project starts during the last decade of the Tenn-Tom's construction.[14] Certainly, other factors also influenced the passage of a rivers and harbors bill during this period, but the intense controversy over the Tenn-Tom significantly strengthened the positions of that bill's opponents. Only after the waterway was finished did Congress pass another omnibus water bill. And, even more important, this legislation changed many of the rules of the game, including new cost-sharing provisions (in the form of user fees) and demands for greater accountability by the Corps for post-authorization changes in projects. The Tenn-Tom fight drew public attention to the social aspects of engineering, and it also served as the catalyst for formation of a new coalition of opponents, one that increased opportunities for significant and meaningful public participation in these debates.[15]

Although environmentalists, railroads, and fiscal conservatives lost their battle to stop the Tennessee-Tombigbee Waterway, they indirectly contributed to these changes in law and policy. A more tangible victory for the coalition was the requirement that expansion of the navigational course south of the Tenn-Tom to Mobile Bay could precede only with Congressional authorization. The Black Warrior-Tombigbee Waterway connects the Tenn-Tom near Demopolis to the Gulf of Mexico, and the Corps planned to expand that project to justify expenditures on the Tenn-Tom. Eight years after the Tenn-Tom's completion, Congress has yet to authorize that expansion; and with the low traffic volume on the Tenn-Tom, there appears little likelihood of an economic justification for so doing in the near future.

It is not unusual for proposed Corps projects to sit on the shelves for many years, authorized by Congress but dormant for want of appropriated funds for construction. Often, these projects are ignored because of the belief that their outrageous financial, political, or environmental costs will keep them from ever being built. The building of the Tenn-Tom, however, speaks to another possibility, one where regional political clout can, under the right circumstances, override otherwise formidable odds. This outcome indicates that, as standard policy, the contributions of projects like the Tenn-Tom should be more strictly evaluated after as well as before construction. Debate over these justifications too often occurs only during planning and construction, then stops after the ribbon is cut and the project complete. Public accountability, however, demands otherwise, and it is critical—for the reputation of the profession of engineering and for making wise choices on future public works projects.

The Tenn-Tom exemplifies the difficulties associated with building a major engineering project whose conception, purpose, and design reflected and embodied the values of an earlier time, even though its engineering incorporated the technical advances of its own era. The social consensus of what is appropriate and desirable for all sorts of things changes over time. For this particular project, regional and national environmental organizations helped to identify and emphasize some of these changes. However, they also failed to push others. One significant failure was in the inability to forge a strong coalition with the African-American communities in Mississippi and Alabama, or with the national civil rights organizations. This omission was a principal shortcoming of the entire U.S. environmental movement in the 1970s and early 1980s. From the perspective of many minority groups, environmentalists and environmental organizations seemed far more concerned with environmental quality as it pertained to the amenities and lifestyles of the white middle class, than with the quality of life concerns of America's poor or its racial and ethnic minorities. Although James Williams and other members of CLEAN attempted to reach out to the local African-American community, they were unable to prove themselves convincing or sincere. Just as the history of the Tenn-Tom illustrated the importance of expanding the environmental coalition to include fiscal conservatives, good government groups, and organized economic interests, the implications for the future are that environmental coalitions must expand even further to embrace questions of economic equity and social justice. This linkage stands as a major challenge to environmentalists in the United States, as well as internationally, and it will require a new thinking about the balance between preserving nature, maintaining human dignity, and enabling economic opportunity.

Unlike the other critics of the waterway—the environmentalists, railroads, and fiscal conservatives—the minority activists saw something to be gained through *completion* of the project. The failure of the Corps of Engineers and the supporters of the Tennessee-Tombigbee Waterway to anticipate the minority community's demand for a share of the project benefits clearly damaged the Corps's standing with the Congress in ways that went far beyond the Tenn-Tom. As the agency was accused of practicing employment discrimination within its Mobile District, it came under fire and was pushed by both the Carter Administration and the Congress to amend that situation. The heated and steady involvement of activists became necessary to force the agency to comply with the legal requirements associated with the expenditure of public funds, and to fulfill the justification that the agency itself had presented for the waterway. Advocacy was an essential ingredient. Without this constant push, the friction caused by racism slowed programs on affirmative action and civil rights.

There was nothing inevitable about the Tenn-Tom. There were many turning points, specific moments when either small elements of the project or the entire waterway itself could have been scotched or altered substantially. Not only was the Tenn-Tom the nation's largest water project when measured by earth moved and dollars spent, it was also the largest when measured in political and economic complexity and controversy.

Environmental litigation, sophisticated and hard-fought Congressional lobbying, active participation by grass-roots minority groups, and many other factors, all worked to create a dynamic tension that shaped the project's outcome and which helped to blend politics and technology on a monumental scale.

It was power politics from start to finish. During the course of this fight, careers were made and lost, extensive coalitions brought together, the media engaged, public education (and miseducation) fostered, a lot of money spent, and a natural river lost to an economic dream. As such, the Tenn-Tom story reflects the course of engineering and environmental politics in late twentieth-century America. The engineers who built the waterway did their job and did it well. Whether they should ever have been asked to do that job is a political question, an environmental question, an economic question. It may be the fate of modern engineering that these questions are inextricably entwined in the whirlpool of national policy debates over the appropriate relationship of technology and the environment.

Notes

Introduction

1. The Waxler Towing Company of Memphis, Tennessee, had gained the privilege of making the first commercial passage through the Tenn-Tom by winning a Corps of Engineers lottery. The *Eddie Waxler* pushed its 4-barge, 64,000-barrel petroleum shipment into the southern end of the Tenn-Tom on 10 January, opening day. See "Waterway Links Tennessee River to Gulf: First Commercial Cargo Shipped on $2 Billion Passage," *Washington Post*, 11 January 1985.

2. The early history of the Tenn-Tom is most fully recounted in James H. Kitchens III, "An Outlet to the Gulf: The Tennessee-Tombigbee Waterway, 1571–1971," unpublished manuscript, Office of History, U.S. Army Corps of Engineers, 1985.

3. When Congress appropriated the initial construction funds in 1970, the Tenn-Tom was designed to extend 253 miles in length. Subsequent realignments and river bend cutoffs shortened the watercourse to 234 miles.

4. For an analysis of the economic justifications advanced through the early 1960s, see Joseph L. Carroll, "A Critique of Waterway Benefit-Cost Analysis: The Tennessee-Tombigbee," Ph.D. dissertation, Indiana University, 1962.

5. Tennessee-Tombigbee Waterway Development Authority, *The Truth about Tenn-Tom* (Columbus, MS: Tennessee-Tombigbee Waterway Development Authority, August 1980), p. 2.

6. See Duane A. Thompson, "Factors Affecting Coal Traffic on the Tennessee-Tombigbee Waterway," Report of the Environment and Natural Resources Division, Congressional Research Service, Library of Congress, 17 July 1981, p. 4.

7. For a penetrating analysis of how support for the waterway rested largely upon the perpetration of false promises and unrealistic expectations by the Corps and the project's Congressional and regional backers, see Johnny Greene, "The Corps of Engineers in Fantasy Land," *Inquiry*, 2 (May 14, 1979), 13–17. A more sympathetic assessment is provided in George Werneth, "Trouble Along the Tenn-Tom," *Atlanta Magazine*, 22 (October 1982), 76–80, 107–108, 110.

8. Attempts to promote industrial development in the South are addressed in James C. Cobb, *The Selling of the South: The Southern Crusade for Industrial Development, 1936–1980* (Baton Rouge: Louisiana State University Press, 1982); and James C. Cobb, *Industrialization and South-*

ern Society, 1877–1984 (Lexington: University Press of Kentucky, 1984). For treatments of the general social and economic transformations of the South during the 20th century, see David R. Goldfield, *Cotton Fields and Skyscrapers: Southern City and Region, 1607–1980* (Baton Rouge: Louisiana State University Press, 1982); Pete Daniel, *Standing at the Crossroads: Southern Life since 1900* (New York: Hill and Wang, 1986); and Bruce J. Schulman, *From Cotton Belt to Sunbelt: Federal Policy, Economic Development, and the Transformation of the South, 1938–1980* (New York: Oxford University Press, 1991). For insight into Congressional pork-barrel politics, see John A. Ferejohn, *Pork Barrel Politics: Rivers and Harbors Legislation, 1947–1968* (Stanford, CA: Stanford University Press, 1974); and Brian Kelly, *Adventures in Porkland: How Washington Wastes Your Money and Why They Won't Stop* (New York: Villard Books, 1992).

9. Regional variations in the U.S. environmental movement are discussed in Samuel P. Hays, *Beauty, Health, and Permanence: Environmental Politics in the United States, 1955–1985* (Cambridge: Cambridge University Press, 1987); and Riley E. Dunlap and Angela G. Mertig (eds.), *American Environmentalism: The U.S. Environmental Movement, 1970–1990* (Philadelphia: Taylor & Francis, 1992).

Chapter I/Marshalling Support

1. See George Rogers Taylor, *The Transportation Revolution, 1815–1860* (New York: Harper and Row, 1951); and Ronald E. Shaw, *Canals for a Nation: The Canal Era in the United States, 1790–1860* (Lexington: University Press of Kentucky, 1990).

2. For the preconstruction history of the Tennessee-Tombigbee Waterway, see Joseph L. Carroll, "A Critique of Waterway Benefit-Cost Analysis: The Tennessee-Tombigbee," Ph.D. dissertation, Indiana University, 1962; William H. Stewart, Jr., *The Tennessee-Tombigbee Waterway: A Case Study in the Politics of Water Transportation* (University, AL: Bureau of Public Administration, University of Alabama, 1971); James William Jones II, "An Analytical History of the Tennessee-Tombigbee Waterway," M.A. thesis, University of Mississippi, 1982; James H. Kitchens III, "An Outlet to the Gulf: The Tennessee-Tombigbee Waterway, 1571–1971," unpublished manuscript, Office of History, U.S. Army Corps of Engineers, 1985; and Glover Wilkins, "My Association with the Tennessee-Tombigbee Waterway Project," unpublished manuscript, Mobile District, U.S. Army Corps of Engineers, 1985. For a general history of the people and geography of the waterway area, see James F. Doster and David C. Weaver, *Tenn-Tom Country: The Upper Tombigbee Valley in History and Geography* (University, AL: University of Alabama Press, 1987); and David S. Brose, *Yesterday's River: The Archaeology of 10,000 Years along the Tennessee-Tombigbee Waterway* (Cleveland, OH: Cleveland Museum of Natural History, 1991).

3. See *Survey from Tombigbee River by Way of Big Bear Creek to the Tennessee River to Connect Water Communications*, Executive Doc. No. 1, Part 2; Report of the Secretary of War, Engineers Report (44th Cong., 1st sess., 1876).

4. See U.S. Congress, House of Representatives, *Waterway between Tennessee and Tombigbee Rivers in the State of Mississippi*, House Doc. No. 218 (63rd Cong., 1st sess., 1913).

5. See U.S. Congress, House of Representatives, *Estimate of Cost of Examinations, Etc., of Streams Where Water Power Appears Feasible*, House Doc. 308 (69th Cong., 1st sess., 1926); and U.S. Congress, House of Representatives, *Warrior and Tombigbee Rivers and Tributaries, Alabama and Mississippi*, House Doc. No. 56 (73rd Cong., 1st sess., 1933).

6. See Kitchens, "An Outlet to the Gulf," p. 302. For a discussion of the political context surrounding public works during the New Deal, see Bonnie Fox Schwartz, *The Civil Works Administration, 1933–1934: The Business of Emergency Employment in the New Deal* (Princeton, NJ: Princeton University Press, 1984).

7. See Kitchens, "An Outlet to the Gulf," pp. 308–309. The idea of an "earth-moving revolution" is presented in Eugene Ferguson, "Technology and Its Impact on Society," in *Technology and Its Impact on Society: Teckniska Museet Symposia* (Stockholm: Teckniska Museet, 1979), p. 276. For the broader development of earth-moving equipment, see Peter N. Grimshaw, *Excavators* (Poole, Dorset, U.K.: Blandford Press, 1985).

8. See U.S. Congress, House of Representatives, *Waterway Connecting the Tombigbee and Tennessee Rivers, Alabama and Mississippi*, House Doc. No. 269 (76th Cong., 1st sess., 1939), p. 4.

9. See Bureau of Railway Economics, Association of American Railroads, *Government Expenditures for Construction, Operation, and Maintenance of Transport Facilities by Air, Highway, and Waterway and Private Expenditures for Construction, Maintenance of Way, and Taxes on Railroad Facilities* (Washington: Association of American Railroads, March 1965). Although not all railroads joined the AAR, its member railroads carried 95% of the nation's intercity rail freight.

10. U.S. House of Representatives, *Waterway Connecting the Tombigbee and Tennessee Rivers* (1939), p. 33.

11. See U.S. Congress, House of Representatives, Committee on Rivers and Harbors, *Improvement of Waterway Connecting the Tombigbee and Tennessee Rivers, Alabama and Mississippi* (77th Cong., 1st sess., 1941); U.S. House of Representatives, Committee on Rivers and Harbors, *Improvement of Waterway Connecting the Tombigbee and Tennessee Rivers, Alabama and Mississippi* (78th Cong., 1st sess., 1943); R. V. Fletcher, "Legislation Affecting Shippers and the Railroads," *Proceedings of the National Association of Shippers Advisory Board*, 5 (November 10–11, 1941), 36–38; and "Tennessee-Tombigbee Up Again," *Railway Age*, 115 (November 27, 1943), 880.

12. See U.S. Congress, House of Representatives, *Waterway Connecting the Tombigbee and Tennessee Rivers*, House Doc. No. 486 (79th Cong., 2nd sess., 1946).

13. Public Law 525.

14. See Kitchens, "An Outlet to the Gulf," p. 458.

15. Ibid., pp. 462–463.

16. See U.S. Congress, House of Representatives, Committee on Appropriations, Subcommittee on Deficiencies and Army Civil Functions, Hearings, *Investigation of Corps of Engineers Civil Works Program*, Part 2, *Proposed Tennessee-Tombigbee Waterway* (82nd Cong., 1st sess., 1951); Carroll, "A Critique of Waterway Benefit-Cost Analysis," pp. 22–26; and Kitchens, "An Outlet to the Gulf," p. 480.

17. See Kitchens, "An Outlet to the Gulf," pp. 501–504; and interview, John T. Greenwood with Representative Jamie Whitten (D-Mississippi), 11 December 1986 (transcript in Office of History archives, Headquarters, U.S. Army Corps of Engineers, Fort Belvoir, VA).

18. U.S. Army Corps of Engineers, Mobile District, *Tennessee-Tombigbee Waterway, Alabama and Mississippi: Design Memorandum No. 1—General Design* (Mobile, AL: U.S. Army Corps of Engineers, Mobile District, 30 June 1960 [approved by the Chief of Engineers, 12 April 1962]).

19. Doswell Gullatt and Associates, *An Evaluation of the Economic Justification for the Tennessee-Tombigbee Waterway Claimed by the Corps of Engineers in Its Restudy Report Dated June 30, 1960* (Washington: Doswell Gullatt and Associates, no date [probably 1961]), p. 2.

20. Ibid., p. 3.

21. Ibid.

22. See Col. Robert C. Marshall to South Atlantic Division Engineer, 12 April 1962 (misc. letters folder, file 1501–07, Tenn-Tom Litigation Management Unit, Mobile District, Corps of Engineers, Mobile, AL [hereafter referred to as "SAMDL"]). For the railroads' response to the Tenn-Tom's reclassification, see Adam Yarmolinsky to T. J. Reardon, Jr., 14 December 1962 (Tenn-Tom microfilm, set I, reel 65, frame 57, Office of History archives, Corps Headquarters).

23. U.S. Army Corps of Engineers, Mobile District, *Tennessee-Tombigbee Waterway, Alabama and Mississippi: Supplement to Design Memorandum No. 1—General Design. Reevaluation of Project Economics* (Mobile, AL: U.S. Army Corps of Engineers, Mobile District, 30 June 1966).

24. Ibid., p. 12.

25. Ibid., p. 13.

26. Jim J. Tozzi, memorandum for the record, 7 March 1967 (folder 2, box 34, James J. Tozzi papers, Office of History archives, Corps Headquarters [hereafter referred to as "Tozzi papers"]). Tozzi raised other serious questions about the reliability of the Corps's cost estimate in a follow-up document: Jim J. Tozzi, memorandum for the record, 14 March 1967 (Navigation Facilities folder, Project Files—Tenn-Tom, Office of the Assistant Secretary of the Army for Civil Works,

Corps Headquarters office files, Washington, D.C.). See also, interview, author with Steven Dola (Office of the Assistant Secretary of the Army for Civil Works), 9 January 1987.

27. [Donald G.] Waldon, memorandum for the record, 4 April 1967 (Alabama folder, box 6, file NN–3–51–86–24, Natural Resources Division—Water Resources, Office of Management and Budget [OMB], Record Group 51, National Archives, Suitland, MD). Waldon joined the Tennessee-Tombigbee Waterway Development Authority professional staff in 1975.

28. Stanley R. Resor to George H. Fallon, 30 March 1967 (Alabama folder, box 17, file NN–3–51–86–24, Natural Resources Division—Water Resources, OMB, RG 51, National Archives).

29. Stanley R. Resor, memorandum for the Chief of Engineers, 30 March 1967 (Alabama folder, box 17, file NN–3–51–86–24, Natural Resources Division—Water Resources, OMB, RG 51, National Archives). See also, Dola interview, 9 January 1987.

30. See Robert W. Farrell to J. E. Gobrecht, 16 July 1965; and P. A. Hollar to E. A. Brautigam, et al., 20 July 1965 (TTWW litigation file, Office of Counsel, Mobile District, U.S. Army Corps of Engineers [hereafter referred to as "MDOC"]).

31. See unsigned list, "Project No. 25, Tennessee-Tombigbee Rivers Committee," 28 November 1966 (TTWW litigation file, MDOC).

32. Association of American Railroads, "Tennessee-Tombigbee Waterway: Supplement to General Design Memorandum, Reevaluation of the Project Economics," preliminary review submitted to the Office of the Chief of Engineers on 15 December 1966 (copy in TTWW litigation file, MDOC). See also, James G. Tangerose to P. A. Hollar, 16 November 1966 (TTWW litigation file, MDOC).

33. Personnel files, Board of Engineers for Rivers and Harbors, U.S. Army Corps of Engineers, Fort Belvoir, VA.

34. See James G. Tangerose to members of AAR Tennessee-Tombigbee Waterway Project Committee, 10 February 1967, 20 April 1967, and 19 September 1968 (TTWW litigation file, MDOC).

35. See Burton N. Behling to Senator William Proxmire, 6 August 1968 (TTWW litigation file, MDOC); and Tangerose's testimony against the Tenn-Tom in U.S. Congress, House of Representatives, Committee on Appropriations, Subcommittee on Public Works, Hearings, *Public Works Appropriations for 1969* (90th Cong., 2nd sess., 1968), Part 4, pp. 174–181. For the Corps's response to these economic criticisms, see South Atlantic Division Planning Division, "SAD Comments on American Association of Railroads Review of Tennessee Tombigbee Report", 20 June 1967 (Tenn-Tom microfilm, set I, reel 66, frames 584–585); Nashville District Planning and Reports Branch, "Nashville District Comments on AAR Analysis of Excavation Costs, Canal and Divide Sections, Tennessee-Tombigbee Waterway," 16 June 1967 (Tenn-Tom microfilm, set I, reel 66, frames 589–590); and Mobile District, "MDO Comments on American Association of Railroads Review of Tennessee-Tombigbee Report," 16 June 1967 (Tenn-Tom microfilm, set I, reel 66, frames 593–602). Within the Office of the Assistant Secretary of the Army, the economic analysts were less optimistic about the Tenn-Tom. See Jim J. Tozzi, memorandum for the record, 7 March 1967 (folder 2, box 34, Tozzi papers).

36. See, for example, Burton N. Behling to Stanley R. Resor, 4 October 1968 (Alabama folder, box 17, file NN–3–51–86–24, Natural Resources Division—Water Resources, OMB, RG 51, National Archives).

37. See William C. Wagner, memorandum for the file, 5 December 1968 (TTWW litigation file, MDOC).

38. See James G. Tangerose to members of the Tennessee-Tombigbee Waterway Project Committee, 7 April 1969; and James G. Tangerose to Maj. Gen. Thomas J. Hayes III, 7 April 1969 (TTWW litigation file, MDOC).

39. For a detailed account of the TTWDA's early development, see Tennessee-Tombigbee Waterway Development Authority, *The Tennessee-Tombigbee Waterway Story* (Columbus, MS: Tennessee-Tombigbee Waterway Development Authority, January 1969); Wilkins, "My Association with the Tennessee-Tombigbee Waterway Project"; and Kitchens, "An Outlet to the Gulf,"

pp. 496–580. The TTWDA's voluminous files are contained in the Glover Wilkins Tennessee-Tombigbee Waterway Archives, Archives and Museums Department, Mississippi University for Women, Columbus, MS [hereafter cited as "GWTTW Archives"].

40. See William Histaspas Stewart, Jr., "The Tennessee-Tombigbee Waterway: A Case Study in the Politics of Water Transportation," Ph.D. dissertation, University of Alabama, 1968, pp. 459–461; interview, author with Lambert C. Mims (Mobile, AL, City Commissioner, 1965–1985), 25 March 1986; interview, author with Glover Wilkins, 15 and 16 November 1986; interview, Helen Pilkinton with Glover Wilkins, 1 and 8 July 1975 (transcripts of this interview are in file 2–1001–19, GWTTW Archives); and Wilkins, "My Association with the Tennessee-Tombigbee Waterway Project."

41. Its influence was not wholly positive, however, as critics in the 1970s and 1980s questioned the appropriateness of a promotional/lobbying organization receiving its entire funding from public revenues. See, for example, interview, author with Brent Blackwelder, 12 April 1988. For a general discussion of the politics surrounding Corps of Engineers projects, see John A. Ferejohn, *Pork Barrel Politics: Rivers and Harbors Legislation, 1947–1968* (Stanford, CA: Stanford University Press, 1974).

42. See, for example, Joseph R. Hartley, "Analysis of the Tennessee-Tombigbee Waterway Reevaluation of Project Economics, Tennessee-Tombigbee Waterway, Corps of Engineers," December 1966 (copy in 1966 Reanalysis folder, Tenn-Tom Litigation Management Unit files, SAMDL).

43. Statement of Glover Wilkins, in U.S. Congress, House of Representatives, Committee on Appropriations, Subcommittee on Public Works, Hearings, *Public Works for Water, Pollution Control, and Power Development and Atomic Energy Commission Appropriation Bill, 1971* (91st Cong., 2nd sess., 1970), Part 5, pp. 943–945.

44. Quoted in James M. Perry, "The Great Washington Pork Feast," *Audubon,* 81 (July 1979), 106.

45. Wilkins interview, 15 and 16 November 1986; and interview, author with Donald G. Waldon (TTWDA Administrator, 1984–), 27 and 28 May 1986. See also, Perry, "The Great Washington Pork Feast," pp. 102–107.

46. Glover Wilkins to John Duncan, 1 May 1970 (Correspondence and Statements/1971 folder, file 2–1000–344, GWTTW Archives). TTWDA staff also drafted a statement for Alabama Governor Albert P. Brewer with the same conditions. See Glover Wilkins to Albert P. Brewer, 1 May 1970 (file 4–B, carton 34, Jack Edwards papers, University of South Alabama Archives, Mobile, AL [hereafter referred to as "Edwards papers"]).

47. See Glover Wilkins to Jack Edwards, 9 April 1973 (file 4–B, carton 53, Edwards papers).

48. Pilkinton interview with Wilkins, 8 July 1975, pp. 43–44.

49. See Glover Wilkins to Don Waldon, 5 January 1970 (Correspondence/Point Clear Meeting folder, file 2–1000–301, GWTTW Archives). Waldon, who joined the TTWDA in 1975, succeeded Wilkins as administrator upon Wilkins's retirement in 1984.

50. Donald G. Waldon to Glover Wilkins, 29 January 1970 (Correspondence/Point Clear Meeting folder, file 2–1000–301, GWTTW Archives).

51. See, for example, Glover Wilkins to Jamie Whitten, 15 October 1970 (Correspondence/Weinberger Hearings—Nov. 1970 folder, file 2–1000–230, GWTTW Archives).

52. Jamie L. Whitten to Glover Wilkins, 16 November 1970 (Correspondence/Weinberger Hearings—Nov. 1970 folder, file 2–1000–230, GWTTW Archives). An account of the Weinberger meeting of 9 November 1970 is contained in the unsigned memorandum, "Budget Hearing Meetings with Deputy Director Caspar Weinberger," 13 November 1970 (same folder).

53. Glover Wilkins to Jamie Whitten, 20 November 1970 (file 2–1000–534, GWTTW Archives).

54. Wilkins interview, 15 and 16 November 1986; interview, author with F. L. (Les) Currie (Executive Director, Mobile District), 15 May 1986; interview, author with Vernon S. Holmes (Engineering Division, Mobile District), 30 May 1986; and interview, author with John Jeffrey Tidmore (Engineering Division, Mobile District), 20 May 1986.

55. See Ward Sinclair, "Party Climaxes Tennessee-Tombigbee Success Story," Louisville (KY) *Courier-Journal & Times,* 4 May 1975; Whitten interview, 11 December 1986; and interview, author with Representative Tom Bevill (D-Alabama), 20 May 1987.

56. For an analysis of Southern voting patterns after World War II, see Numan V. Bartley and Hugh D. Graham, *Southern Politics and the Second Reconstruction* (Baltimore: Johns Hopkins University Press, 1975). Also useful for a discussion of Southern politics is Earl Black and Merle Black, *The Vital South: How Presidents Are Elected* (Cambridge, MA: Harvard University Press, 1992).

57. See Jack Edwards, biographical sketch, 23 September 1986 (Jack Edwards's personal files, Mobile, AL); statement by W. Brevard Hand (Chairman, Mobile County Republican Executive Committee), 21 January 1969 (file 4–B, carton 34, Edwards papers); and interview, author with Jack Edwards, 20 November 1986.

58. For a brief discussion of the outcome of the 1968 election in the South, see Bartley and Graham, *Southern Politics and the Second Reconstruction,* pp. 127–135.

59. Virginia Hooper to Harry Dent, 7 January 1970 (file 2–1000–494, GWTTW Archives).

60. See Jeffrey K. Stine, "Environmental Politics and Water Resources Development: The Case of the Army Corps of Engineers during the 1970s," Ph.D. dissertation, University of California at Santa Barbara, 1984, pp. 58–62; and Brent Blackwelder, "Water Resource Development," in James Rathlesberger (ed.), *Nixon and the Environment: The Politics of Devastation* (New York: The Village Voice, 1972), pp. 59–69.

61. The regional differences in the appeal of environmental issues is examined in Samuel P. Hays, *Beauty, Health, and Permanence: Environmental Politics in the United States, 1955–1985* (Cambridge: Cambridge University Press, 1987); and Samuel P. Hays, "Environmental Political Culture and Environmental Political Development: An Analysis of Legislative Voting, 1971–1989," *Environmental History Review,* 16 (Summer 1992), 1–22. See also, Jeffrey K. Stine, "Environmental Politics in the American South: The Fight over the Tennessee-Tombigbee Waterway," *Environmental History Review,* 15 (Spring 1991), 1–24.

62. See "Construction Funds for the Tennessee-Tombigbee Waterway," *Congressional Record—Senate,* 116 (January 21, 1970), 533–539; "President Nixon Gives Tenn-Tom Construction Nod in Budget Request," *Tenn-Tom-Topics,* 1 (March 1970), 1; and William Greider, "Ditch Diggers' Delight: A New Mississippi River," *Washington Post,* 15 March 1970.

63. Robert E. Jordan III to George Pratt Shultz, 1 July 1970 (1973 Civil Works Budget folder, Assistant Secretary of the Army for Civil Works, Corps Headquarters office files, Washington, D.C.).

64. See press release, statement by the President, 7 October 1970 (copy in Correspondence/BOB Jan. 1970–Oct. 1971 folder, file 2–1000–227, GWTTW Archives).

65. Glover Wilkins to Mississippi State Board of Water Commissioners, 28 December 1970 (Comprehensive Correspondence on 1970 Planning folder, file 2–1000–200, GWTTW Archives).

Chapter II/Design

1. For general accounts of the waterway design, see Julian A. Greer, "The Tennessee-Tombigbee Waterway," *Military Engineer,* 65 (September–October 1973), 336–339; Frank C. Denning, "The Tennessee-Tombigbee Waterway," *Bulletin of the Permanent International Association of Navigation Congresses,* no. 23 (1976), 16–20; K. A. Godfrey, Jr., "Waterway Is Public Works Landmark," *Civil Engineering,* 56 (July 1986), 42–45; American Society of Civil Engineers, Mobile (AL) Chapter, "Tennessee-Tombigbee Waterway: Nomination for Outstanding Civil Engineering Award (OCEA)—1986," unpublished nomination report, 1986; and Emory L. Kemp, "The Engineering Aspects of the History of the Tennessee-Tombigbee Waterway," two parts (unpublished background report, History Associates Inc., Rockville, MD, March 1987 and April 1987).

2. For a lengthier discussion, see Jeffrey K. Stine, "United States Army Corps of Engineers," in Donald R. Whitnah (ed.), *Government Agencies,* Greenwood Encyclopedia of American Insti-

tutions (Westport: Greenwood Press, 1983), pp. 513–416. The number of division and district offices within the Corps has fluctuated throughout the twentieth century; at the time of this writing, the agency is considering another reorganization likely to eliminate some of those offices.

3. See Maj. Gen. George H. Walker to Ohio River Division Engineer, 12 May 1967 (Tenn-Tom microfilm, set I, reel 66, frames 578–580, Office of History archives, Corps Headquarters, Fort Belvoir, VA); and James H. Kitchens III, "An Outlet to the Gulf: The Tennessee-Tombigbee Waterway, 1571–1971," unpublished manuscript, Office of History, U.S. Army Corps of Engineers, 1985, pp. 458–459.

4. See Brig. Gen. H. G. Woodbury, Jr., memorandum to South Atlantic and Ohio River Division Engineers, 28 June 1967 (Tenn-Tom microfilm, set I, reel 66, frames 581–582); Maj. Gen. D. A. Raymond, memorandum to Ohio River Division Engineer, 20 March 1973 (Tenn-Tom files, Executive Office, Nashville District, Nashville, TN); interview, author with Brig. Gen. Kenneth E. McIntyre (South Atlantic Division Engineer, 1976–1979), 14 November 1986; and interview, author with Vernon S. Holmes, 30 May 1986.

5. For an overview of the Corps's design process for the Tenn-Tom, see interview, author with Benton Wayne Odom, Jr. (Engineering Division, Mobile District), 20 May 1986; interview, author with Frank C. Deming (Engineering Division, Mobile District), 21 November 1986; Holmes interview, 30 May 1986; and Kemp, "The Engineering Aspects of the History of the Tennessee-Tombigbee Waterway," part I, p. 2.

6. See U.S. Army Corps of Engineers, Mobile District, *Tennessee-Tombigbee Waterway, Alabama and Mississippi: Design Memorandum No. 1—General Design* (Mobile, AL: U.S. Army Corps of Engineers, Mobile District, 30 June 1960 [approved by the Chief of Engineers, 12 April 1962]).

7. The voluminous design documents for the Tenn-Tom are kept as part of the Civil Works Projects Files in the Engineering Division, Corps's Headquarters, Washington, D.C.

8. Initially, the Corps reported the length of the river section as 168 miles. By 1977, the agency had shortened the watercourse by 20 miles, primarily through channel realignments and the inclusion of additional cutoffs. See U.S. Army Corps of Engineers, Mobile District, *Second Supplemental Environmental Report: Continuing Environmental Studies, Tennessee-Tombigbee Waterway, Alabama and Mississippi*, volume I, *Overall Study* (Mobile, AL: U.S. Army Corps of Engineers, Mobile District, October 1977), p. 1.

9. For a technical presentation of this approach, see Margaret S. Petersen, *River Engineering* (Englewood Cliffs, N.J.: Prentice-Hall, 1986).

10. See Nathaniel Dehass McClure IV, "Tennessee-Tombigbee Waterway and the Environment," in Steven R. Abt and Johannes Gessler (eds.), *Hydraulic Engineering* (New York: American Society of Civil Engineers, 1988), p. 331.

11. Jack C. Mallory, memorandum for the files, 30 August 1976 (TTW Cutoffs folder, file 1501-07, Inland Environment Section, Planning Division, Mobile District, U.S. Army Corps of Engineers [hereafter referred to as "SAMPD-EI"]).

12. See U.S. Fish and Wildlife Service, Division of Ecological Services, *Bendway Management Study, Tennessee-Tombigbee Waterway, Alabama and Mississippi: A Fish and Wildlife Coordination Act Report* (Daphne, AL: U.S. Fish and Wildlife Service, Division of Ecological Services, March 1984).

13. For a detailed discussion of dredging, see Petersen, *River Engineering*, pp. 225–263. Prior to 1970, Corps personnel referred to dredged material as "spoil," a name they later purged from their vocabulary in an attempt to reshape the rhetoric surrounding the environmental debate. For changing attitudes toward wetlands and the Corps's responsibility for monitoring their use, see Jeffrey K. Stine, "Regulating Wetlands in the 1970s: U.S. Army Corps of Engineers and the Environmental Organizations," *Journal of Forest History*, 27 (April 1983), 60–75.

14. See Mobile District, *Design Memorandum No. 1—General Design*; U.S. Army Corps of Engineers, Mobile District, *Environmental Statement: Tennessee-Tombigbee Waterway, Alabama and Mississippi Navigation* (Mobile, AL: U.S. Army Corps of Engineers, Mobile District, March

1971); interview, author with Nathaniel D. (Skeeter) McClure IV (Planning Division, Mobile District), 16 May 1986; and interview, author with Frederick G. Thompson (Engineering Division, Mobile District), 21 May 1986.

15. The design concept is outlined in Frank C. Deming to South Atlantic Division Engineer, 1 June 1976 (Tenn-Tom microfilm, set I, reel 1, frames 1772–1775); interview, author with Kenneth D. Underwood (Engineering Division, Mobile District), 21 May 1986; and Gerald J. McLindon, "Creative Spoil: Design Concepts, Construction Techniques, and Disposal of Excavated Materials," *Environmental Geology and Water Sciences*, 7 (nos. 1/2, 1985), 91–108.

16. See Tommie Pierce, memorandum for the files, 29 January 1976 (TTW Project/Jan. 1975 —Feb. 1976 folder, file 1503–03, Real Estate Division, Mobile District [hereafter referred to as "SAMRE"]); and U.S. Army Corps of Engineers, Mobile District, *First Supplemental Environmental Report: Continuing Environmental Studies, Tennessee-Tombigbee Waterway* (Mobile, AL: U.S. Army Corps of Engineers, Mobile District, August 1975).

17. See Jack C. Mallory, memorandum for the file, 12 October 1977 (Tenn-Tom microfilm, set I, reel 1, frame 56).

18. Underwood interview, 21 May 1986; and McClure interview, 16 May 1986.

19. See Kemp, "The Engineering Aspects of the History of the Tennessee-Tombigbee Waterway," part I, pp. 6–8.

20. See U.S. Army Corps of Engineers, Mobile District, *Gainesville Lock and Dam, Tombigbee River, Alabama and Mississippi: Design Memorandum No. 4, General Design* (Mobile, AL: U.S. Army Corps of Engineers, Mobile District, January 1969).

21. John R. Thoman to B. J. Christiansen, 25 November 1968, reprinted in *ibid.*, pp. IV-2 and 3.

22. U.S. Army Corps of Engineers, Mobile District, *Gainesville Lock and Dam, Tennessee-Tombigbee Waterway, Alabama and Mississippi: Design Memorandum No. 12, Instrumentation* (Mobile, AL: U.S. Army Corps of Engineers, Mobile District, June 1972), p. 10.

23. U.S. Army Corps of Engineers, Mobile District, *Gainesville Lock and Dam, Tennessee-Tombigbee Waterway, Alabama and Mississippi: Design Memorandum No. 11, Spillway* (Mobile, AL: U.S. Army Corps of Engineers, Mobile District, December 1971), p. I–13; McClure interview, 16 May 1986; and interview, author with Jack C. Mallory (Planning Division, Mobile District), 16 May 1986.

24. Mobile District, *Environmental Statement*, p. 19. See also, C. Edward Carlson to Mobile District Engineer, 31 December 1970 (James D. Williams papers; when I used this collection it was in Williams's U.S. Fish and Wildlife Service office in Arlington, VA; he subsequently donated his papers to the University of Alabama Archives, University, AL [hereafter referred to as "Williams papers"]); and Mallory interview, 16 May 1986.

25. See Ralph R. W. Beene, Trip Report, 2 September 1971 (Tenn-Tom Waterway, General/1970–1971 folder, file 1518–01, CW–601, South Atlantic Division [SAD], Atlanta, GA).

26. See Mallory interview, 16 May 1986; and interview, author with Robert L. Crisp, Jr. (Engineering Division, South Atlantic Division), 8 October 1986.

27. See James W. Erwin, Trip Report, 2 December 1971 (Tenn-Tom Waterway, General/1970–1971 folder, file 1518–01, CW–601, SAD).

28. Col. Harry A. Griffith to Clark Hubbs, 28 June 1971 (ALA/MISS—COE: TTWW folder, box 4, file 429–81–37, Council on Environmental Quality records, Washington National Records Center, Suitland, MD).

29. James D. Wall to Chief of Engineering Division (SAD), 19 March 1971 (Tenn-Tom Waterway—General/1970–1971 folder, file 1518–01, CW–601, SAD).

30. See Mobile District, "Use of Earth-Filled Slurry Trenches for Control of Groundwater on the Tennessee-Tombigbee Waterway Project," unpublished abstract, no date [probably autumn 1977] (Board of Environmental Consultants/January 1978 folder, file 1501–03, SAMPD-EI); Thompson interview, 21 May 1986; Crisp interview, 8 October 1986; interview, author with Jack H. Bryan (Engineering Division, Mobile District), 17 November 1986; and Deming interview, 21 November 1986.

31. See Petros P. Xanthakos, *Slurry Walls* (New York: McGraw-Hill Book Company, 1979), pp. 1–17. For a detailed technical description of slurry trench walls, see Xanthakos, pp. 194–221; Ronald A. Antonino, "Earth-Filled Slurry Walls Provide Economical Seepage Control," *Civil Engineering*, 52 (April 1980), 72–74; and Edith Iglauer, "The Biggest Foundation," *New Yorker*, 48 (November 4, 1972), 140.

32. See Supplement to Design Memorandum No. 7, Gainesville Lock and Dam, Ala., Underseepage Studies, attached to letter, Col. Robert E. Snetzer to Chief of Engineers, 26 March 1970 (Gainesville L&D folder, Correspondence subfile, file 1518–01, Civil Works Project files, Water Projects Section, Office of the Chief of Engineers, Washington, D.C. [hereafter referred to as "CWPF, WPS, OCE"]).

33. George H. Mittendorf, memorandum to Mobile District Engineer, 28 June 1971 (Gainesville L&D Conference folder, Conferences/Review Comments subfile, file 1518–01, CWPF, WPS, OCE).

34. See, for example, Summary of Discussions and Decisions, Conference on Alternative Studies for Gainesville Dam, South Atlantic Division, 22 July 1971 (Gainesville L&D Conference folder, Conferences/Review Comments subfile, file 1518–01, CWPF, WPS, OCE).

35. See U.S. Army Corps of Engineers, Mobile District, *Studies of Project Costs and Benefits: Tennessee-Tombigbee Waterway* (Mobile, AL: U.S. Army Corps of Engineers, Mobile District, October 1975), pp. 14–15.

36. See Mobile District, "Use of Earth-Filled Slurry Trenches"; Thompson interview, 21 May 1986; Bryan interview, 17 November 1986; interview, author with Freddy R. Jones (Area Engineer for Tennessee-Tombigbee Waterway, 1975–1982), 21 November 1986; American Society of Civil Engineers, "Tennessee-Tombigbee Waterway"; and Engineering Division, Mobile District, "Information Document and Instructions on Design Concepts, Lock A and Spillway," 10 August 1977 (Tenn-Tom microfilm, set I, reel 1, frames 301–302).

37. Prior to 1976, the Corps listed the canal section as 45 miles in length. Channel realignments made during the design phase reduced the overall length of this section by one mile. See Mobile District, *Second Supplemental Environmental Report*, volume I, p. 1.

38. See Mobile District, *Environmental Statement*, pp. 20–21; and U.S. Army Corps of Engineers, Mobile District, *Environmental Study, Tennessee-Tombigbee Waterway, Ala., Miss., Tenn.* (Mobile, AL: U.S. Army Corps of Engineers, Mobile District, August 1970).

39. Mobile District Engineering Division, "Status Report, Tennessee-Tombigbee Waterway," 13 August 1971 (Williams papers).

40. See McClure interview, 16 May 1986; interview, author with George H. Atkins (Engineering Division, Mobile District), 18 November 1986; Thompson interview, 21 May 1986; and Kemp, "The Engineering Aspects of the History of the Tennessee-Tombigbee Waterway."

41. Mallory interview, 16 May 1986; interview, author with John W. Rushing (Planning Division, South Atlantic Division), 6 October 1986; Odom interview, 20 May 1986; Deming interview, 21 November 1986; and interview, author with Richard T. Kimberl (Engineering Division, Mobile District), 21 November 1986. For additional background on Cronenberg, see William D. Jones to Congressman Jack Edwards, 5 October 1971 (file 1–C, carton 34, Jack Edwards papers, University of South Alabama Archives, Mobile, AL).

42. Background of this design evolution and details of the final chain-of-lakes design are provided in U.S. Army Corps of Engineers, Mobile District, *Tennessee-Tombigbee Waterway, Alabama and Mississippi, Supplement to the Project Design Memorandum, GDM for Canal Section: Design Memorandum No. 5, General Design* (Mobile, AL: U.S. Army Corps of Engineers, Mobile District, September 1976).

43. Mobile District, Engineering Division, "Status Report: Tennessee-Tombigbee Waterway," 16 March 1973 (Tenn-Tom microfilm, set I, reel 14, frame 229). Also useful is Holmes interview, 30 May 1986.

44. Walter E. Mussell, et al., Memorandum for [South Atlantic] Division Engineer, 30 November 1973 (Tenn-Tom files, Executive Office, Nashville District, Nashville, TN).

45. See Col. Drake Wilson to South Atlantic Division Engineer, 25 January 1974 (TTWW

Wildlife Mitigation Study folder, file 1517–01, Environmental Resources Planning Section, Planning Division, Mobile District [hereafter referred to as "SAMPD-ER"]).

46. See Mobile District, "Alternative Plan Studies for Canal Section, Tennessee-Tombigbee Waterway," no date [probably January 1974] (TTWW Wildlife Mitigation Study folder, file 1517–01, SAMPD-ER). Additional detail is provided in Rushing interview, 6 October 1986.

47. See discussion in *Memorandum of Support of Federal Defendants' Motion for Judgment on the Pleadings* . . . , Environmental Defense Fund, et al. v. Clifford Alexander, et al., U.S. District Court for the Northern District of Mississippi, 2 July 1980; and Atkins interview, 18 November 1986.

48. Glover Wilkins, Administrator's Report, April–June 1970 (1970 Working Correspondence folder, file 2–1000–18, Glover Wilkins Tennessee-Tombigbee Waterway Archives, Archives and Museums Department, Mississippi University for Women, Columbus, MS [hereafter referred to as "GWTTW Archives"]).

49. Glover Wilkins to Thomas Abernethy, 25 January 1971 (file 2–1000–651, GWTTW Archives).

50. See Engineering Division, OCE, disposition form, 18 July 1975 (Canal Section Conference folder, Conferences/Review Comments subfile, file 1518–01, CWPF, WPS, OCE).

51. See Robert L. Crisp, Jr., Minutes of Lock D Conference, Tennessee-Tombigbee Waterway, 26 May 1977 (Tenn-Tom microfilm, set I, reel 1, frame 452).

52. Fred Thompson, memorandum for chief of Mobile District Engineering Design Branch, 16 September 1975 (Elimination of Lock D folder, file 401–07, SAMPD-EI).

53. David H. Webb to Mobile District Engineer, et al., 22 April 1977 (Tenn-Tom microfilm, set I, reel 1, frame 776). Emphasis in original.

54. Mobile District, "Tennessee-Tombigbee Waterway, Preliminary Evaluation of Design Concepts for Canal Section Involving Lock D," January 1977 (Board of Environmental Consultants/January 1977 Meeting folder, file 1501–03, SAMPD-EI).

55. Crisp, Minutes of Lock D Conference.

56. See *ibid.*, frame 453; and Robert L. Crisp, Jr., memorandum for record, 8 February 1977 (Tenn-Tom microfilm, set I, reel 1, frame 1070).

57. Bryan interview, 17 November 1986; and Kemp, "The Engineering Aspects of the History of the Tennessee-Tombigbee Waterway," part I, pp. 8–9.

58. U.S. General Accounting Office, *To Continue or Halt the Tenn-Tom Waterway? Information to Help the Congress Resolve the Controversy* (Washington: U.S. General Accounting Office, 15 May 1981), pp. 2–3. The divide cut was located near the 806-foot Woodall Peak, the highest point in Mississippi. For fuller descriptions of this section of the waterway, see U.S. Army Corps of Engineers, Nashville District, *General Design Memorandum for Divide Cut N–1: Tennessee-Tombigbee Waterway, Mississippi and Alabama,* 2 volumes (Nashville, TN: U.S. Army Corps of Engineers, Nashville District, March 1973); Cynthia A. Drew, "The Tennessee-Tombigbee Waterway: Divide Section," *Military Engineer,* 77 (May–June 1985), 170–174; and Jimmy Bates, memorandum for record, 29 July 1971, p. 3 (Other Agencies Coordination—Nashville District/TTW folder, file 1501–07, SAMPD-EI).

59. The definitive history of Project Plowshare has yet to be written. Useful discussions of the program appear in Ralph Sanders, *Project Plowshare: The Development of the Peaceful Uses of Nuclear Explosives* (Washington: Public Affairs Press, 1962); Gerald W. Johnson and Gary H. Higgins, *Engineering Applications of Nuclear Explosives: Project Plowshare,* Contract Report W–7405–eng–48 (Livermore, CA: Lawrence Radiation Laboratory, May 19, 1964); Gerald W. Johnson, "Excavating with Nuclear Explosives," *Discovery,* 25 (November 1964), 16–21; Carl R. Gerber, Richard Hamburger, and E. W. Seabrook Hull, *Plowshare* (Oak Ridge, TN: U.S. Atomic Energy Commission, March 1966); Melvin W. Jackson, "Ten Years of Plowshare," *Civil Engineering,* 37 (December 1967), 34–38; and Trevor Findlay, *Nuclear Dynamite: The Peaceful Nuclear Explosions Fiasco* ([Sydney?]: Brassey's Australia, 1990). For a comprehensive list of the articles, books, and technical reports addressing the civilian applications of nuclear explosives, see Robert

G. West and Robert C. Kelly (compilers), *A Selected, Annotated Bibliography of the Civil, Industrial, and Scientific Uses for Nuclear Explosives* (Oak Ridge, TN: U.S. Atomic Energy Commission, Division of Technical Information Extension, 1971). The standard textbook on the subject, which addressed many of the Plowshare activities, is Edward Teller, Wilson K. Talley, Gary H. Higgins, and Gerald W. Johnson, *The Constructive Uses of Nuclear Explosives* (New York: McGraw-Hill, 1968). See also, Lynn E. Weaver (ed.), *Education for Peaceful Uses of Nuclear Explosives* (Tucson: University of Arizona Press, 1970).

60. Edward Teller, "We're Going to Work Miracles," *Popular Mechanics*, 113 (March 1960), 97–101, 278, 280, 282. For the promotion of Project Plowshare and other nuclear programs, see Michael Smith, "Advertising the Atom," in Michael J. Lacey (ed.), *Government and Environmental Politics: Essays on Historical Developments since World War Two* (Washington: The Wilson Center Press, 1989), pp. 233–262.

61. See Bernard C. Hughes, *History of the U.S. Army Engineer Nuclear Cratering Group* (Livermore, CA: Lawrence Radiation Laboratory, January 1969).

62. Louis J. Circeo, Jr., "Engineering Properties and Applications of Nuclear Excavations," in (Highway Research Board) *Highway Research Record*, 50 (1964), 13.

63. *Ibid.*, p. 26.

64. See A. D. Suttle, Jr., to Edward Teller, 20 June 1960 (Mississippi folder, Office of History and Historical Records, Lawrence Livermore National Laboratory, Livermore, CA [hereafter referred to as "LLNL"]).

65. See Dave Rabb to A. D. Suttle, Jr., 14 July 1960 (Mississippi folder, Office of History and Historical Records, LLNL).

66. In preparing the feasibility study, Brown and Root teamed with the Mississippi firm M. T. Reed Construction Company. See A. D. Suttle, Jr., to George J. Darneille, 14 September 1960 (Mississippi folder, Office of History and Historical Records, LLNL); Glover Wilkins, "My Association with the Tennessee-Tombigbee Waterway Project," unpublished manuscript, Technical Library, Mobile District, U.S. Army Corps of Engineers, 1985, p. 42; William Histaspas Stewart, Jr., "The Tennessee-Tombigbee Waterway: A Case Study in the Politics of Water Transportation," Ph.D. dissertation, University of Alabama, 1968, p. 466; and Tennessee-Tombigbee Waterway Development Authority *Newsletter*, 1 (September 26, 1963), 2 (copy in Greenough/Mobile Harbor folder, file 18081, Mobile Municipal Archives, Mobile, AL).

67. See E. Graves, Jr., memorandum to C. Bacigalupi et al., 27 September 1960 (Mississippi folder, Office of History and Historical Records, LLNL).

68. "An Atomic Blast to Help Build a U.S. Canal," *U.S. News & World Report*, 54 (May 20, 1963), 14. See also, L. J. Votrman, "Nuclear Excavation," in Weaver, *Education for Peaceful Uses of Nuclear Explosives*, p. 67; and Crisp interview, 8 October 1986. The Army's overall planning effort is reported in Hughes, *History of the U.S. Army Engineer Nuclear Cratering Group*. The flurry of interest in using nuclear explosives to build the Tenn-Tom is exemplified in Col. Loren W. Olmstead to Wilmot N. Hess, 28 March 1961 (Mississippi folder, Office of History and Historical Records, LLNL); "Nuclear Blasts Are Proposed To Cut Tombigbee Waterway," Memphis (TN) *Commercial Appeal*, 25 September 1961; and "South In Great Era, Barnett Declares," Clarion (MS) *Ledger*, 27 September 1961.

69. Glover Wilkins to Glenn T. Seaborg, 17 February 1964 (Mississippi folder, Office of History and Historical Records, LLNL).

70. Paul B. Johnson to Glenn T. Seaborg, 5 March 1964 (Mississippi folder, Office of History and Historical Records, LLNL). See also, Paul B. Johnson to Robert S. McNamara, 5 March 1964 (same file); Lt. Col. Ernest Graves, Jr., memorandum to Mobile District Engineer, 6 March 1964 (same file); and Col. D. A. Raymond to Robert F. Dye, 1 April 1964 (same file).

71. See, for example, W. B. Pieper to Gerald Johnson, 15 April 1964 (Mississippi folder, Office of History and Historical Records, LLNL); Gerald W. Johnson to Robert G. Gibson, 28 January 1965 (same file); M. L. Merritt to Robert E. Miller, 29 March 1965 (same file); and Merritt to Miller, 31 March 1965 (same file).

72. See Lt. Col. Maurice K. Kurtz, Jr., to Michael M. May, 4 May 1967 (Mississippi folder, Office of History and Historical Records, LLNL).

73. U.S. Congress, House of Representatives, Committee on Appropriations, Subcommittee on Public Works, Hearings, *Public Works Appropriations for 1968*, Part I (90th Cong., 1st sess., 1967), p. 1355. For the official report, see U.S. Army Corps of Engineers, Mobile District, *Tennessee-Tombigbee Waterway, Divide Cut: Nuclear Excavation Feasibility Study* (Mobile, AL: U.S. Army Corps of Engineers, Mobile District, 15 February 1966). For the discussion of the nuclear excavation option presented in the 1971 Tenn-Tom environmental impact statement, see Mobile District, *Environmental Statement*, p. 29.

74. See U.S. Army Corps of Engineers, Nashville District, *Feature Design Memorandum N–2: Divide Cut, Tennessee-Tombigbee Waterway, Mississippi and Alabama* (Nashville, TN: U.S. Army Corps of Engineers, Nashville District, January 1975).

75. See Geotechnical and Civil Branch, OCE, disposition form to Directorate of Engineering and Construction, et al., 8 January 1974 (Divide Cut Conference folder, Conferences/Review Comments subfile, file 1518–01, CWPF, WPS, OCE).

76. See interview, author with William Brandes (Nashville District Engineer, 1971–1974), 18 September 1986.

77. Joseph M. Caldwell, memorandum for South Atlantic Division Engineer, 21 May 1973 (Bay Springs/1971–1976 folder, file 1501–07, SAMPD-EI).

78. See Lt. Col. Terrence J. Connell, memorandum to South Atlantic Division Engineer, 8 June 1973 (Bay Springs/1971–1976 folder, file 1501–07, SAMPD-EI).

79. Geotechnical and Civil Branch, disposition form to Directorate of Engineering and Construction, 8 January 1974.

80. James D. Wall to Mobile District Engineer, 24 June 1974 (TTW correspondence/SAD and OCE folder, file 1501–07, SAMPD-EI).

81. See Nathaniel D. McClure, Jr., notes of telephone conversation with John W. Rushing and Robert Crisp, Jr., 30 August 1976 (TTW correspondence/SAD and OCE folder, file 1501–07, SAMPD-EI).

82. See interview, author with James G. Goad (Program Development Office, Nashville District), 19 September 1986; and interview, author with Maj. Gen. Henry J. Hatch (Nashville District Engineer, 1974–1977), 16 December 1987.

83. James J. Tozzi, memorandum for the record, 14 March 1967 (folder 2, box 34, James J. Tozzi papers, Office of History archives, Corps headquarters, Fort Belvoir, VA).

84. Mobile District, *Environmental Statement*, p. 23. Under Secretary of the Army Thaddeus R. Beal elaborated on these considerations in a letter to CEQ Chairman Russell E. Train dated 23 July 1971 (ALA/MISS—COE: TTWW folder, box 4, file 429–81–37, CEQ records, WNRC, Suitland, MD). See also, interview, author with H. Joe Cathey (Engineering Division, Nashville District), 15 September 1986; and Mallory interview, 16 May 1986.

85. Mobile District, Comments on Testimony by James D. Williams on Tenn-Tom Waterway, 6 June 1973 (TTW correspondence/SAD and OCE folder, file 1501–07, SAMPD-EI).

86. See Bob Bryan, "Evacuated Dirt in Waterway Construction to be Placed in Valleys along the Route," Tupelo (MS) *Journal*, 9 July 1974; Cathey interview, 15 September 1986; and interview, author with Daniel F. Hall (Construction Division, Nashville District), 15 September 1986.

87. The two locks east of the Mississippi River with a higher lift than the Bay Springs Lock were the Wilson Lock on the Tennessee River (built in 1927; 94-foot lift) and the Walter F. George Lock on the Chattahoochee River (built in 1963; 85-foot lift). West of the Mississippi River, six locks had higher lifts than Bay Springs, and all of them were on the Columbia-Snake Waterway. See Institute for Water Resources, *Status of the Inland Waterways* (Fort Belvoir, VA: Institute for Water Resources, September 1987), Appendix A.

88. See Official Minutes, Bay Springs Lock and Dam, Tennessee-Tombigbee Waterway, Meeting at WES, 11–12 September 1975 (Bay Springs L&D Conference folder, Conferences/Review

Comments subfile, file 1518–01, CWPF, WPS, OCE); and interview, author with Dennis R. Williams (Engineering Division, Nashville District), 18 September 1986.

89. See Charles H. Tate, Jr., *Bay Springs Canal Surge Study, Tennessee-Tombigbee Waterway, Mississippi and Alabama: Hydraulic Model Investigation,* Technical Report H–78–9 (Vicksburg, MS: U.S. Army Engineer Waterways Experiment Station, Hydraulics Laboratory, June 1978); and Henry A. Malec, memorandum for the record, 22 March 1974 (Bay Springs/1971–1976 folder, file 1501–07, SAMPD-EI).

90. See Jackson H. Ables, Jr., *Filling and Emptying System for Bay Springs Lock, Tennessee-Tombigbee Waterway, Mississippi: Hydraulic Model Investigation,* Technical Report H–78–19 (Vicksburg, MS: U.S. Army Engineer Waterways Experiment Station, Hydraulics Laboratory, November 1978).

91. See U.S. Army Corps of Engineers, Mobile District, *Gainesville Lock and Dam, Tennessee-Tombigbee Waterway, Alabama and Mississippi: Design Memorandum No. 8, Lock and Channels* (Mobile, AL: U.S. Army Corps of Engineers, Mobile District, September 1969); and U.S. Army Corps of Engineers, Nashville District, *Tennessee-Tombigbee Waterway, Mississippi and Alabama: Design Memorandum No. N–12, Divide Section, Bay Springs Lock and Dam* (Nashville, TN: U.S. Army Corps of Engineers, Nashville District, February 1977), p. I–8.

92. See Nashville District Engineering Division, memorandum for the record, 22 June 1976 (Bay Springs L&D Conference folder, Conferences/Review Comments subfile, file 1518–01, CWPF, WPS, OCE).

93. Nashville Engineering Division, memorandum for the record, 22 June 1976; Ables, *Filling and Emptying System for Bay Springs Lock*; interview, author with Herman Gray (Engineering Division, Nashville District), 16 September 1986; and interview, author with Howard Boatman (Operations and Readiness Division, Nashville District), 16 September 1986. The Bankhead Lock hydraulic system is discussed in N. R. Oswalt, J. H. Ables, Jr., and T. E. Murphy, *Navigation Conditions and Filling and Emptying System, New Bankhead Lock, Black Warrior River, Alabama: Hydraulic Model Investigation,* Technical Report H–72–6 (Vicksburg, MS: U.S. Army Engineer Waterways Experiment Station, Hydraulics Laboratory, September 1972).

94. Addendum, Bay Springs Lock and Dam, Tennessee-Tombigbee Waterway, Meeting at WES, 11–12 September 1975 (Bay Springs L&D Conference folder, Conferences/Review Comments subfile, file 1518–01, CWPF, WPS, OCE).

Chapter III/Construction

1. See Emory L. Kemp, "The Engineering Aspects of the History of the Tennessee-Tombigbee Waterway," two parts (unpublished background report, History Associates Inc., Rockville, MD, March 1987 and April 1987); American Society of Civil Engineers, Mobile (AL) Chapter, "Tennessee-Tombigbee Waterway: Nomination for Outstanding Civil Engineering Award (OCEA)—1986," unpublished nomination report, 1986; and interview, author with Freddy R. Jones, 21 November 1986.

2. See interview, author with Vernon S. Holmes, 30 May 1986.

3. See U.S. Congress, House of Representatives, Committee on Appropriations, Subcommittee on Public Works, Hearings, *Public Works for Water and Power Development and Atomic Energy Commission Appropriation Bill, 1974,* Part I (93rd Cong., 1st sess., 1973), pp. 533–536; Maj. Gen. Daniel A. Raymond to Maj. Gen. J. W. Morris, 16 March 1973 (Construction Schedule folder, file 1518–01, Nashville District Civil Works Project files, Environment and Resources Branch, Planning Division, Mobile District, U.S. Army Corps of Engineers [hereafter referred to as "SAMPD-EC"]; interview, John T. Greenwood with Representative Jamie Whitten, 11 December 1986; interview, author with Representative Tom Bevill, 20 May 1987; interview, author with Victor V. Veysey (Assistant Secretary of the Army for Civil Works, 1975–1976), 30 January 1986; interview, author with F. L. (Les) Currie, 15 May 1986; interview, author with John Jeffrey Tidmore, 20 May 1986; and interview, author with Bory Steinberg (Civil Works Programs Division, Corps Headquarters), 9 January 1987.

4. Maj. Gen. Daniel A. Raymond to Maj. Gen. J. W. Morris, 16 March 1973 (Construction Schedule folder, file 1518–01, Nashville District Civil Works Project files, SAMPD-EC).

5. Brig. Gen. James L. Kelly to Joe L. Evins, 10 April 1973 (Tenn-Tom microfilm, set I, reel 57, frames 480–482, Office of History archives, Corps Headquarters, Fort Belvoir, VA). When the Corps commissioned the San Francisco engineering firm Jacobs Associates to produce an independent "construction analysis and cost estimate" of the divide cut in summer 1974, the agency learned that, despite Kelly's remarks to the contrary, accelerated construction would unquestionably raise the cost of the project. See Jacobs Associates, "Construction Analysis and Cost Estimate," report prepared for the Corps of Engineers, Nashville District, August 1974 (Jacobs Associates Report folder, box 14, Nashville District Design Branch records, Design Branch, Engineering Division, Mobile District [hereafter referred to as "SAMEN-DN"]).

6. See U.S. Congress, House of Representatives, Committee on Appropriations, Subcommittee on Public Works, Hearings, *Public Works for Water and Power Development and Atomic Energy Commission Appropriation Bill, 1975*, Part I (93rd Cong., 2nd sess., 1974), pp. 274, 446.

7. See Brig. Gen. Wayne S. Nichols to Col. William F. Brandes, 18 June 1974 (Tenn-Tom microfilm, set I, reel 37, frame 1674); and Col. Daniel D. Hall to SAD Engineering Division and Office of Counsel, 14 July 1976 (Tenn-Tom microfilm, set I, reel 1, frame 1921).

8. For an example of the Corps's fast-track military construction methods, see Frank N. Schubert, *Building Air Bases in the Negev: The U.S. Army Corps of Engineers in Israel, 1979–1982* (Washington: GPO, 1992), passim.

9. See Holmes interview, 30 May 1986; interview, author with Brig. Gen. Forrest T. Gay III (South Atlantic Division Engineer, 1982–1985), 21 May 1987; and Jeffrey K. Stine, "The Tennessee-Tombigbee Waterway and the Evolution of Cultural Resources Management," *The Public Historian*, 14 (Spring 1992), 7–30.

10. For praise of the project's management, see Brig. Gen. Drake Wilson to South Atlantic Division Engineer, 28 June 1977 (Tenn-Tom microfilm, set I, reel 1, frame 269).

11. For a general discussion of programming for the Tenn-Tom, see Holmes interview, 30 May 1986; Tidmore interview, 20 May 1986; interview, author with James G. Goad, 18 and 19 September 1986; interview, author with Robert L. Crisp, Jr., 8 October 1986; interview, author with William H. Osborne (Program Development Office, South Atlantic Division [SAD]), 9 October 1986; and Steinberg interview, 9 January 1987.

12. For a discussion of these meetings and the role of Robert Crisp, see Tidmore interview, 20 May 1986; interview, author with Kenneth D. Underwood, 21 May 1986; Holmes interview, 30 May 1986; interview, author with Harold C. Eaton (Affirmative Action Office, Mobile District), 30 May 1986; interview, author with Richard J. Connell, Jr. (Planning Division, SAD), 6 October 1986; interview, author with James Wall (Engineering Division, SAD, 1950–1980), 7 October 1986; Jones interview, 21 November 1986; and interview, author with Maj. Gen. Henry J. Hatch, 16 December 1987.

13. Crisp interview, 8 October 1986; and Tidmore interview, 20 May 1986.

14. Holmes interview, 30 May 1986. See also, Robert L. Crisp, Jr., trip report, 25 February 1974 (Quarterly Review Status folder, file 1501–07, Inland Environment Section, Environment and Resources Branch, Planning Division, Mobile District [hereafter referred to as "SAMPD-EI"]); minutes for quarterly status meeting on the Tennessee-Tombigbee Project, 31 May 1974; Robert L. Crisp, Jr., memorandum for South Atlantic Division Engineer, 18 December 1974 (Quarterly Review Status folder, file 1501–07, SAMPD-EI); Crisp interview, 8 October 1986; and interview, author with Richard T. Kimberl, 21 November 1986.

15. See Robert L. Crisp, Jr., trip report, 25 August 1975 (John Jeffrey Tidmore's personal papers, Civil Works Program Development and Management Branch, Engineering Division, Mobile District [hereafter referred to as "SAMEN-P"]).

16. Robert L. Crisp, Jr., trip report, 25 February 1974 (Quarterly Review Status folder, file 1501–07, SAMPD-EI). See also, interview, author with Daniel F. Hall, 15 September 1986.

17. U.S. Army Corps of Engineers, Mobile District, "Proposed Testing of Construction Meth-

ods, Federally Authorized Project, Portion of the Tennessee-Tombigbee Waterway within the Nashville District Jurisdiction," Public Notice No. 76–905, 29 April 1976 (Bay Springs/1971–1976 folder, file 1501–07, SAMPD-EI).

18. See Col. Robert K. Tener, memorandum for South Atlantic Division Engineer, 5 September 1978 (Tenn-Tom files, Executive Office, Nashville District, U.S. Army Corps of Engineers); and Brig. Gen. Kenneth E. McIntyre to John Bearden, 21 February 1979 (Conferences and Review Comments folder, file 1518–01, Civil Works Project Files, Water Projects Section, Directorate of Engineering and Construction, Corps Headquarters [hereafter referred to as "CWPF, WPS, OCE"]).

19. See Lee W. Miles, "Tenn Tom Constructors Equipment Department Final Report for Tennessee Tombigbee Waterway," unpublished in-house report (Boise, Idaho: Morrison-Knudsen Company, no date [probably 1984]). This document, a copy of which is in the Morrison-Knudsen Company archives in Boise, Idaho, was a "lessons learned" type report summarizing various aspects of the Tenn-Tom construction effort. See also, "Big Digging on 'Tenn-Tom' Waterway," *The eM-Kayan,* 38 (November 1979), 10–13; and "Major Shipping Route Unfolds: Earthmoving on Grand Scale along 'Tenn-Tom' Waterway," *The eM-Kayan,* 39 (August 1980), 3–7.

20. See David Proctor, "Boiseans Build Up Big Business Worldwide," *Idaho Statesman,* 1 May 1987. Besides Morrison-Knudsen, the Hoover Dam consortium included Bechtel Group, Kaiser Paving Company, J. F. Shea, Utah Construction Company, and Warren Brothers. See Joseph E. Stevens, *Hoover Dam: An American Adventure* (Norman: University of Oklahoma Press, 1988). The general activities of Morrison-Knudsen can be followed in its company magazine, *The eM-Kayan.*

21. See Miles, "Tenn Tom Constructors"; "Big Digging on 'Tenn-Tom' Waterway"; "Tenn Tom's Soil Headaches Get Heavy Doses of Hardhat Know-how," *Engineering News-Record,* 204 (January 17, 1980), 56–58; "Evacuation of 95-Million Yards Nears End on Tenn-Tom Project," *The eM-Kayan,* 41 (November 1982), 6–9; and Kemp, "The Engineering Aspects of the History of the Tennessee-Tombigbee Waterway," part II.

22. Miles, "Tenn Tom Constructors."

23. See Miles, "Tenn Tom Constructors"; and "Big Digging on 'Tenn-Tom' Waterway."

24. This work schedule was susceptible to rain delays, which caused only one or two missed days per month during the summer, but which could cost three or four lost weeks per month during the most adverse winter weather conditions. See Miles, "Tenn Tom Constructors."

25. Miles, "Tenn Tom Constructors."

26. See *ibid.*; "Major Shipping Route Unfolds"; "Big Digging on the Tenn-Tom Waterway," *The eM-Kayan,* 40 (January 1982), 4–5; Kemp, "The Engineering Aspects of the History of the Tennessee-Tombigbee Waterway," part II; American Society of Civil Engineers, "Tennessee-Tombigbee Waterway," p. 10; interview, author with Frederick G. Thompson, 21 May 1986; and Hall interview, 15 September 1986. Ultimately, 2.2 million square yards of filter fabric were laid under the riprap.

27. See "Riprap Machine to 'Armor' Tenn-Tom Slopes," *The eM-Kayan,* 40 (May 1981), 3; "Big Digging on the Tenn-Tom Waterway"; Miles, "Tenn Tom Constructors"; and Hall interview, 15 September 1986.

28. See John Van Brahana, "Beneath the Tenn-Tom," *Water Spectrum,* 7 (Winter 1975–76), 17–24; and Tombigbee River Valley Water Management District, *Annual Report,* 1985, p. 10.

29. See Crisp interview, 8 October 1986; and Hall interview, 15 September 1986.

30. See, for example, U.S. Army Corps of Engineers, Mobile District, *Environmental Statement: Tennessee-Tombigbee Waterway, Alabama and Mississippi Navigation* (Mobile, AL: U.S. Army Corps of Engineers, Mobile District, March 1971), p. 21; and interview, author with Frank B. Couch (Geotechnical Branch, Nashville District), 17 September 1986.

31. See Summary of Work-Level Meeting on Environmental Effects of Waterway on Groundwater and Mineral Resources, 2 October 1970, Mobile, Alabama (Ground Water and Mineral

Resources/TTW folder, file 1501–07, SAMPD-EI); Jimmy Bates, memorandum for record, 29 July 1971, p. 3 (Other Agencies Coordination—Nashville District/TTW folder, file 1501–07, SAMPD-EI); and Van Brahana, "Beneath the Tenn-Tom."

32. See "Tennessee-Tombigbee Waterway Project—Domestic Wells," unsigned and undated fact sheet [probably compiled by the Nashville District in autumn 1977], attached to letter, Maj. Ralph M. Danielson to Paul Roth, 5 December 1977 (General Reference/Misc. Letters folder, file 1501–07, Tenn-Tom Litigation Management Unit, Mobile District [hereafter referred to as "SAMDL"]).

33. See U.S. Army Corps of Engineers, Nashville District, *Report on Domestic Waterwells, Tennessee-Tombigbee Waterway Project* (Nashville, TN: U.S. Army Corps of Engineers, Nashville District, December 1977); "Public Meeting June 11 for Local Landowners," *Tishomingo County News*, 4 June 1981; Hamp Rogers, "Lowered Water Levels Prompt Public Meeting," *Northeast Mississippi Daily Journal*, 5 June 1981; and Hamp Rogers, "Low Water: Tishomingo Residents May File Suit," *Northeast Mississippi Daily Journal*, 12 June 1981.

34. Quoted in Rogers, "Lowered Water Levels Prompt Public Meeting."

35. Nashville District, *Report on Domestic Waterwells*. See also, Fred B. Shelton, discussion paper, no date [probably autumn 1977] (Domestic Well Report folder, Nashville District Geotechnical Branch files, box 16/12/19, Mobile District storage facility, Mobile, AL).

36. Interview, Henry Dahlinger, Herman Gray, and Robert Thomas with Marvin D. Simmons (Nashville District Geologist), 23 March 1989 (transcript available at Office of History archives, Corps Headquarters, Fort Belvoir, VA).

37. See Nathaniel D. McClure IV, memorandum for the files, 25 February 1977 (Tenn-Tom microfilm, set I, reel 1, frames 997–998).

38. Arundel-Atkinson-Ball, "Environmental Implementation Plan, Columbus Lock and Dam," no date [probably March 1975] (Columbus Lock & Dam Site folder, file 1501–07, SAMPD-EI).

39. See U.S. Army Corps of Engineers, Nashville District, "Water Quality Control Plan, Divide Section, Tennessee-Tombigbee Waterway," November 1978 (copy in Bay Springs Water Quality/1975–1977 folder, file 1501–07, SAMPD-EI).

40. Interview, author with H. Joe Cathey, 15 September 1986.

41. Lawrence R. Green, memorandum to chiefs of construction and engineering divisions, Mobile District, 1 November 1976 (Tenn-Tom microfilm, set I, reel 1, frame 1353).

42. C. E. White, Jr. to Col. Charlie L. Blalock, 6 December 1977 (TTWW litigation files, Office of Counsel, Mobile District).

Chapter IV/The Environmental Debate

1. Tennessee-Tombigbee Waterway Development Authority, *The Tennessee-Tombigbee Waterway Story* (Columbus, MS: Tennessee-Tombigbee Waterway Development Authority, January 1969), p. 13.

2. The environmental protest to the Corps's civil works program is discussed in Jeffrey K. Stine, "Environmental Politics and Water Resources Development: The Case of the Army Corps of Engineers during the 1970s," Ph.D. dissertation, University of California at Santa Barbara, 1984.

3. For a discussion of environmental concerns in the American South, see James C. Cobb, *The Selling of the South: The Southern Crusade for Industrial Development, 1936–1980* (Baton Rouge: Louisiana State University Press, 1982), pp. 229–253; Albert E. Cowdrey, *This Land, This South: An Environmental History* (Lexington: University Press of Kentucky, 1983); and Samuel P. Hays, *Beauty, Health, and Permanence: Environmental Politics in the United States, 1955–1985* (Cambridge: Cambridge University Press, 1987), passim. Of related interest is Thomas D. Clark, *The Greening of the South: The Recovery of Land and Forest* (Lexington: University Press of Kentucky, 1984).

4. Interview, author with Glenn H. Clemmer (CLEAN member), 15 April 1987. See also, interview, author with James D. Williams (CLEAN member), 7 January 1987.

5. CLEAN Charter, 10 March 1970 (CLEAN office files folder, Tennessee-Tombigbee Waterway Development Authority office files, Columbus, MS [hereafter referred to as "TTWDA"]).

6. The position paper, which CLEAN mailed to each member of the Mississippi Congressional delegation, was published in its entirety in "Tenn-Tom Project: Environmental Group Questions Waterway," West Point (MS) *Times Leader*, 7 May 1970. See also, Clemmer interview, 15 April 1987; and Boyd Gatlin to Michael J. Kerwin, 30 April 1970 (CLEAN office files folder, TTWDA). CLEAN's special committee on the Tenn-Tom was later made the sixth standing committee—the Committee on Conservation and Natural Resources. The original five committees dealt with industrial pollution, municipal pollution, agricultural pollution, population, and programs and public information.

7. The literature discussing the impact of NEPA is extensive. Some of the most important works are: Frederick R. Anderson, *NEPA in the Courts: A Legal Analysis of the National Environmental Policy Act* (Baltimore: Johns Hopkins University Press for Resources for the Future, 1973); Richard N. L. Andrews, *Environmental Policy and Administrative Change: Implementation of the National Environmental Policy Act* (Lexington, MA: D. C. Heath, 1976); Richard A. Liroff, *A National Policy for the Environment: NEPA and Its Aftermath* (Bloomington: Indiana University Press, 1976); Lynton K. Caldwell, *Science and the National Environmental Policy Act: Redirecting Policy through Procedural Reform* (University, AL: University of Alabama Press, 1982); and Serge Taylor, *Making Bureaucracies Think: The Environmental Impact Statement Strategy of Administrative Reform* (Stanford, CA: Stanford University Press, 1984).

8. See W. L. Towns to Mobile District Engineer, 27 December 1968 (Tenn-Tom file, Division of Ecological Services records, U.S. Fish and Wildlife Service, Washington, D.C.); and U.S. Fish and Wildlife Service, Division of Ecological Services, *Bendway Management Study, Tennessee-Tombigbee Waterway, Alabama and Mississippi: A Fish and Wildlife Coordination Act Report* (Daphne, AL: U.S. Fish and Wildlife Service, Division of Ecological Services, March 1984).

9. See U.S. Fish and Wildlife Service, *Bendway Management Study*; and U.S. Geological Survey project proposal, "Upper Cretaceous Geology of the Tennessee-Tombigbee Waterway" (Charles C. Smith, project chief), 19 June 1978 (TTWW June 1978 folder, file 1501–07, Inland Environment Section, Environment and Resources Branch, Planning Division, Mobile District [hereafter referred to as "SAMPD-EI"]). For sympathetic accounts of the wilderness values sacrificed as a result of the waterway's construction, see Sam Love, "A Native Returns to the Tombigbee," *Living Wilderness*, 37 (Summer 1973), 24–26, 28, 30–31; and Johnny Greene, "Selling Off the Old South," *Harper's*, 254 (April 1977), 39–46, 51–58.

10. See U.S. Army Corps of Engineers, Mobile District, *Environmental Statement: Tennessee-Tombigbee Waterway, Alabama and Mississippi Navigation* (Mobile, AL: U.S. Army Corps of Engineers, Mobile District, March 1971). Estimates for the divide cut excavations were later raised to 160 million cubic yards.

11. Glenn H. Clemmer, "The Tennessee-Tombigbee Waterway: Are We Sure of the Costs?," in Harry L. Sherman (ed.), *Environmental Quality II: An MSCW Leadership Forum* (Columbus, MS: Mississippi State College for Women, 1972), p. 106.

12. U.S. Fish and Wildlife Service, *Bendway Management Study*, p. 11. See also, C. A. Schultz, *Completion Report: Tombigbee River Basin Preimpoundment Studies, January 1, 1978—December 31, 1980*, North Mississippi Fisheries Investigations, Mississippi Project F–47 (Jackson, MS: Mississippi Department of Wildlife Conservation, 1981).

13. These three species were *Noturus munitus, Hybopsis aestivalis*, and *Percina lenticula*. See R. Dale Caldwell to Col. Harry A. Griffith, 23 June 1971 (Inquiries and Replies, TTW/1971 folder, file 1501–07, SAMPD-EI); Richard Dale Caldwell, "A Study of the Fishes of the Upper Tombigbee River and Yellow Creek Drainage Systems of Alabama and Mississippi," Ph.D. dissertation, University of Alabama, 1969; and Chief of the Office of Endangered Species, memorandum to Assistant Secretary for Fish and Wildlife and Parks, 23 March 1976 (James D. Williams papers, Arlington, VA [hereafter referred to as "Williams papers"]).

14. Caldwell to Griffith, 23 June 1971. See also, University of Alabama, Tuscaloosa, "Chemical and Biological Effects of the Mixing of the Tennessee and Tombigbee Water Systems," a project proposal submitted to the U.S. Army Corps of Engineers, Mobile District, March 1971 (Williams papers).

15. See, for example, John E. Cooper to Walter J. Hickel, 26 June 1970 (file 2–1000–932, Glover Wilkins Tennessee-Tombigbee Waterway Archives, Archives and Museums Department, Mississippi University for Women, Columbus, MS [hereafter referred to as "GWTTW Archives"]).

16. See University of Alabama, "Chemical and Biological Effects of the Mixing of the Tennessee and Tombigbee Water Systems."

17. Thaddeus R. Beal to Russell E. Train, 23 July 1971 (Miscellaneous Reference TTW folder, file 401–07, SAMPD-EI). See also, Ralph M. Perhac, *Heavy Metal Distribution in Bottom Sediment and Water in the Tennessee River-Loudon Lake Reservoir System*, Research Report No. 40 (Knoxville, TN: University of Tennessee, Water Research Center, December 1974).

18. See Randall Grace, "For the Betterment of the Nation: The Selling of the Tennessee-Tombigbee Waterway," in Barry Allen and Mina Hamilton Haefele (eds.), *In Defense of Rivers: A Citizens' Workbook on Impacts of Dam and Canal Projects* (Stillwater, NJ: Delaware Valley Conservation Association, 1976), p. 149.

19. See Roderick Nash, *Wilderness and the American Mind*, third edition (New Haven: Yale University Press, 1982); and Tim Palmer, *Endangered Rivers and the Conservation Movement* (Berkeley: University of California Press, 1986).

20. See Clemmer interview, 15 April 1987; and James E. Crowfoot and Julia M. Wondolleck, "Citizen Organizations and Environmental Conflict," in James E. Crowfoot and Julia M. Wondolleck (eds.), *Environmental Disputes: Community Involvement in Conflict Resolution* (Washington: Island Press, 1990), p. 4.

21. See, for example, U.S. Army Engineer District, Mobile, "Environmental Impact Study, Tennessee-Tombigbee Waterway, Mississippi, Alabama, Tennessee: Technical Studies Work Plan," unpublished draft, July 1970 (Jack Mallory's personal files, SAMPD-EI).

22. See, for example, "Isn't There Room for All?," Columbus (MS) *Commercial Dispatch*, 22 August 1971; and Glover Wilkins to Governor Louie B. Nunn, 23 April 1971 (file 2–1000–932, GWTTW Archives).

23. Harry A. Griffith, "Tennessee-Tombigbee Waterway: Audience Participation," in Sherman, *Environmental Quality II*, p. 114.

24. "Tenn-Tom Can Relieve Heavy Traffic on Mississippi," news release from the office of Senator John Sparkman, 17 September 1971 (file 2–1000–1063, GWTTW Archives).

25. Lt. Col. Paul D. Sontag to R. Dale Caldwell, 13 July 1971 (Inquiries and Replies, TTW/1971 folder, file 1501–07, SAMPD-EI).

26. Optimistic accounts of the Corps's response to the environmental challenges of the 1970s can be found in Daniel A. Mazmanian and Jeanne Nienaber, *Can Organizations Change? Environmental Protection, Citizen Participation, and the Corps of Engineers* (Washington: Brookings Institution, 1979); and Martin Reuss, *Shaping Environmental Awareness: The United States Army Corps of Engineers Environmental Advisory Board, 1970–1980* (Washington: GPO, 1983). The Corps's environmental critics, on the other hand, were legion. For examples, see Jeffrey K. Stine and Michael C. Robinson, *The U.S. Army Corps of Engineers and Environmental Issues in the Twentieth Century: A Bibliography* (Washington: GPO, 1984).

27. See, for example, John E. Cooper to Lt. Gen. Frederick J. Clarke, 9 June 1970; and Edward Lee Rogers to Clarke, 21 July 1970 (Williams papers).

28. John E. Cooper and Robert A. Kuehne to Dear Colleague, 6 July 1970 (Williams papers).

29. Cooper and Kuehne to Dear Colleague, 6 July 1970. Cooper became an assistant professor of biology at the Community College of Baltimore in fall 1970.

30. John E. Cooper to Col. Harry A. Griffith, 14 July 1970 (Tenn-Tom Waterway, General/1970–1971 folder, file 1518–01, CW–601, South Atlantic Division office, U.S. Army Corps of Engineers, Atlanta, GA [hereafter referred to as "SAD"]).

31. See Marjorie H. Carr, "The Fight to Save the Oklawaha," in Elizabeth R. Gillette (ed.), *Action for Wilderness* (San Francisco: Sierra Club, 1972), pp. 157–169.

32. "Scientists and Citizens Form Southeast Environmental Action Group," press release [Robert A. Kuehne and John E. Cooper], 1 August 1970 (Williams papers).

33. John E. Cooper to Edward Lee Rogers, 24 July 1970 (Williams papers).

34. See "Halt to Tenn-Tom Is First Objective," Birmingham (AL) *News*, 3 August 1970; "Tenn-Tom Waterway Is New Target of CLEAN," West Point (MS) *Times Leader*, 13 August 1970; and Martha R. Cooper and John E. Cooper, "Report on August 1, 1970 meeting at Mississippi State University, Starkville, Mississippi," no date (Williams papers).

35. Cooper and Cooper, "Report on August 1, 1970 meeting," pp. 2–3.

36. Ibid., p. 5.

37. Ibid., p. 6.

38. Ibid., p. 7.

39. John S. Ramsey to Dear Colleague, 7 September 1970. See also, John S. Ramsey to Mobile District Engineer, 1 September 1970 (Williams papers).

40. Quoted in "CLEAN Will Study Tenn-Tom Ecology," Columbus (MS) *Commercial Dispatch*, 18 August 1970. See also, "Tennessee-Tombigbee Waterway Hit," Nashville (TN) *Tennessean*, 13 August 1970; and Clemmer interview, 15 April 1987.

41. Boyd Gatlin to Thomas Abernethy, 3 November 1970 (file 2–1000–932, GWTTW Archives).

42. See, for example, Lt. Gen. F. J. Clarke to Edward Lee Rogers, 15 September 1970 (Inquiries and Replies/TTW 1970 folder, file 1501–07, SAMPD-EI); Stan Atkins, "Top Environmental Study Pledged," Mobile (AL) *Register*, 13 November 1970; and "Tenn-Tom and the Environment," Chattanooga (TN) *Times*, 27 November 1970.

43. John E. Cooper to E. Leo Koester, 11 March 1971 (TTWW litigation file, MDOC).

44. Undated form letter (probably early February 1971) from the Alabama Conservancy (Williams papers). See also, Charles S. Prigmore to James Allen, 18 October 1970 (Inquiries and Replies/TTW 1970 folder, file 1501–07, SAMPD-EI).

45. See Charles Prigmore's telegrams to the House and Senate Public Works Subcommittees, members of the Alabama Congressional Delegation, and President Richard Nixon, 19 May 1971; and John B. Scott, Jr., to E. A. Drago, 27 May 1971 (Alabama Conservancy folder, file 103–01, SAMPD-EI). These efforts were reported in Dan Gross, "Conservancy Raps Tenn-Tom," Birmingham (AL) *Post-Herald*, 20 May 1971.

46. Telegram, Jack Edwards to Charles S. Prigmore, 19 May 1971 (file 4–B, carton 37, Jack Edwards papers, University of South Alabama Archives, Mobile, AL [hereafter referred to as "Edwards papers"]). The Bankhead Wilderness area is near Birmingham.

47. Charles S. Prigmore to Jack Edwards, 3 June 1971 (file 4–B, carton 37, Edwards papers).

48. See Charles T. Traylor to Marion Edey, 29 July 1971 (Williams papers).

49. See U.S. Army Engineer District, Mobile, "Environmental Impact Study . . . Technical Studies Work Plan."

50. Col. R. P. Tabb to Mobile District Engineer, 19 June 1970 (Technical Studies Workplan TTW folder, file 1501–07, SAMPD-EI).

51. Lt. Gen. Frederick J. Clarke to Maj. Gen. Richard H. Free, 3 August 1970 (Tenn-Tom Waterway—General/1970–1971 folder, file 1518–01, CW–601, SAD). See also, interview, author with Frederick J. Clarke (Chief of Engineers, 1969–1973), 15 December 1987.

52. Col. Robert R. Werner, memorandum for record, 22 October 1970 (EAB Meetings/1970 folder, box 1, Environmental Advisory Board records, OCE Civil Works files, Office of History archives, Corps Headquarters, Fort Belvoir, VA). Martin Reuss covered this exchange and the entire meeting in his history of the EAB; see Reuss, *Shaping Environmental Awareness*, pp. 9–11.

53. See interview, Martin Reuss with Lynton K. Caldwell (EAB member), 13 May 1980 (transcript in Office of History archives, Corps Headquarters); and Reuss, *Shaping Environmental Awareness*, p. 81.

54. "Army Engineers Announce Plans for Detailed Environmental Study of Tennessee-Tombigbee Waterway," news release, U.S. Army Engineer District, Mobile, 12 November 1970 (Information Pamphlet/TTW folder, file 103–01, SAMPD-EI). See also, A. J. Chamberlin, memorandum for the record, 12 November 1970 (Press Conference 12 November 1970 folder, file 1501–07, SAMPD-EI); and U.S. Army Corps of Engineers, Mobile District, *Information Pamphlet: Environmental Study, Tennessee-Tombigbee Waterway, Mississippi, Alabama and Tennessee* (Mobile, AL: U.S. Army Corps of Engineers, Mobile District, November 1970), p. 3.

55. CEQ, which was established by NEPA in 1970, was charged with advising the President and monitoring executive branch compliance with NEPA through a review of environmental impact statements.

56. U.S. Army Corps of Engineers, Mobile District, *Environmental Study, Tennessee-Tombigbee Waterway, Mississippi, Alabama and Tennessee: Technical Studies Work Plan* (Mobile, AL: U.S. Army Corps of Engineers, Mobile District, 16 November 1970), p. 9.

57. Ibid., pp. 9–10.

58. Ibid., p. 10.

59. These government bodies included: the Bureau of Sport Fisheries and Wildlife, the Bureau of Outdoor Recreation, the National Park Service, and the Geological Survey, all within the U.S. Department of the Interior; the Federal Highway Administration, the Coast Guard, the Federal Aviation Administration, and the Federal Railroad Administration, all part of the U.S. Department of Transportation; the U.S. Environmental Protection Agency's Water Quality Office and Bureau of Water Hygiene; the Department of Commerce; the Department of Housing and Urban Development; the Tennessee Valley Authority; the Appalachian Regional Commission; the Department of Agriculture; and the states of Alabama, Mississippi, and Tennessee.

60. U.S. Army Corps of Engineers, Mobile District, *Environmental Statement: Tennessee-Tombigbee Waterway, Alabama and Mississippi Navigation* (Mobile, AL: U.S. Army Corps of Engineers, Mobile District, March 1971). See also, Thaddeus R. Beal to Russell E. Train, 20 April 1971 (Williams papers).

61. Mobile District, *Environmental Statement*, p. 2.

62. Russell E. Train to Stanley R. Resor, 4 June 1971 (Environmental Statement Agency Replies/TTW folder, file 1501–07, SAMPD-EI).

63. James D. Williams to Russell E. Train, 19 May 1971 (Williams papers).

64. Timothy Atkeson to James D. Williams, 17 June 1971 (Litigation/1971–1972 folder, file 410–01, TTWW litigation file, Office of Counsel, Mobile District [hereafter referred to as "MDOC"]).

65. James D. Williams to William D. Ruckelshaus, 17 June 1971 (file 2–1000–949, GWTTW Archives).

66. John R. Thoman to James D. Williams, 3 August 1971 (Williams papers). See also, interview, author with Gerald Miller (Environmental Assessment Branch, Region IV Office, EPA), 10 April 1986.

67. James D. Williams to Members of the Public Works and Appropriations Committees of the House and Senate, 25 May 1971 (Williams papers); reprinted in U.S. Congress, Senate, Subcommittee of the Committee on Appropriations, Hearings, *Public Works for Water and Power Development and Atomic Energy Commission Appropriations for Fiscal Year 1972* (92nd Cong., 1st sess., 1971), p. 731. See also, James D. Williams to Henry S. Reuss, 1 July 1971 (litigation—1971/1972 folder, file 410–01, MDOC).

68. James D. Williams and Glenn H. Clemmer to "Dear Environmentalist," 21 May 1971 (litigation—1971/1972 folder, file 410–01, MDOC).

69. See "Kentuckian Voices Tenn-Tom Canal Objection," Mobile (AL) *Press*, 7 May 1970.

70. Senate Subcommittee of the Committee on Appropriations, *Public Works for Water and Power Development . . . 1972*, pp. 672–673.

71. Robert A. Kuehne to Lt. Gen. F. J. Clarke, 17 June 1970 (litigation—1971/1972 folder, file 410–01, MDOC).

72. See, for example, William Partington to Persons interested in the Tennessee-Tombigbee Waterway, 4 June 1971 (Williams papers).
73. William M. Partington to Michael Frome, 4 June 1971 (Williams papers).
74. Form letter from Marion Edey, 6 July 1971 (Williams papers).
75. Glover Wilkins to Members of the Authority, 1 March 1971 (file 2–1000–981, GWTTW Archives).
76. See William E. Timmons, memorandum for Dwight Chapin, 18 June 1970 (General NR7–1/A–Z Projects/1969–70 folder, box 22, file group NR, Nixon Presidential Materials Project, Suitland, MD).
77. Louie B. Nunn to the President, 22 October 1970 (Correspondence/BOB Jan. 1970–Oct. 1971 folder, file 2–1000–227, GWTTW Archives).
78. Louie B. Nunn to Richard Nixon, 8 December 1970 (Correspondence/BOB Jan. 1970–Oct. 1971 folder, file 2–1000–227, GWTTW Archives). See also, Hugh W. Sloan, Jr., to Louie B. Nunn, 29 October 1970 (General NR7–1/A–Z Projects/1969–70 folder, box 22, file group NR, Nixon Presidential Materials Project). In February 1971, Tennessee Governor Winfield Dunn urged the President to attend. See Dwight L. Chapin to Governor Winfield Dunn, 25 February 1971 (Ex NR7–1/A–Z Projects/1971–72 folder, box 19, file group NR, Nixon Presidential Materials Project).
79. Louie B. Nunn to H. R. Haldeman, 10 February 1971 (file 2–1000–889, GWTTW Archives).
80. William E. Timmons, memorandum for Dwight Chapin, 22 February 1971 (Presidential Trip to Tennessee-Tombigbee Dedication folder, box 7, John Whitaker papers, Nixon Presidential Materials Project [hereafter referred to as "Whitaker papers"]). See also, interview, author with Alabama Governor George C. Wallace, 19 May 1986.
81. See Whitaker's handwritten comments on Dave Parker, memorandum for John Whitaker, et al., 1 March 1971 (Presidential Trip to Tennessee-Tombigbee Dedication folder, box 7, Whitaker papers). For a discussion of overall environmental policy making in the Nixon Administration, see John C. Whitaker, *Striking a Balance: Environment and Natural Resources Policy in the Nixon-Ford Years* (Washington: American Enterprise Institute for Public Policy Research, 1976).
82. See William E. Timmons to Jack Edwards, 31 March 1971 (Ex NR7–1/A–Z Projects/1971–72 folder, box 19, file group NR, Nixon Presidential Materials Project).
83. See Glover Wilkins to George C. Wallace, 22 April 1971 (file 2–1000–889, GWTTW Archives).
84. Russell E. Train, memorandum for John Whitaker, 2 April 1971 (Presidential Trip to Tennessee-Tombigbee Dedication folder, box 7, Whitaker papers).
85. See John C. Whitaker, memorandum for Dwight Chapin, 6 April 1971 (Presidential Trip to Tennessee-Tombigbee Dedication folder, box 7, Whitaker papers).
86. John C. Whitaker, memorandum for John D. Ehrlichman, 13 April 1971 (Presidential Trip to Tennessee-Tombigbee Dedication folder, box 7, Whitaker papers).
87. Russell E. Train, memorandum for the President, 10 May 1971 (Presidential Trip to Tennessee-Tombigbee Dedication folder, box 7, Whitaker papers).
88. Wesley K. Sasaki, memorandum to Mr. Rice and Mr. [Bill] Gifford, 6 April 1971 (Tennessee folder, box 9, file NN–3–51–86–24, Natural Resources Division—Water Resources, Office of Management and Budget [OMB] Record Group 51, National Archives, Suitland, MD).
89. John C. Whitaker, memorandum for Dwight Chapin, 13 May 1971 (Presidential Trip to Tennessee-Tombigbee Dedication folder, box 7, Whitaker papers).
90. Joe Browder to John Whitaker, 15 May 1971 (Presidential Trip to Tennessee-Tombigbee Dedication folder, box 7, Whitaker papers).
91. Telegram, members of CLEAN to the President, 19 May 1971 (Williams papers). See also, "CLEAN Telegraphs Nixon: 'Defer Work on Tenn-Tom'," Columbus (MS) *Commercial Dispatch*, 23 May 1971.
92. James D. Williams to Russell E. Train, 19 May 1971 (Williams papers).
93. See "Conservancy Raps Tenn-Tom," Birmingham (AL) *Post-Herald*, 20 May 1971; Charles

S. Prigmore, "Voice of the People: The Tenn-Tom Issue," Birmingham (AL) *News,* 26 May 1971; and "Tenn-Tom Project Denounced by League of Women Voters," Birmingham (AL) *Post-Herald,* 22 May 1971.

94. "Barge Boon, or a Threat to Ecology?," Louisville (KY) *Courier-Journal,* 25 May 1971.

95. Council on Environmental Quality, list of questions and answers for the Tennessee-Tombigbee Waterway ground breaking, no date [probably 15 May 1971] (Presidential Trip to Tennessee-Tombigbee Dedication folder, box 7, Whitaker papers).

96. See "Tennessee-Tombigbee Waterway: The President's Remarks at the Dedication Ceremony at Mobile, Alabama, May 25, 1971," *Weekly Compilation of Presidential Documents,* May 31, 1971, pp. 806–808.

97. John C. Whitaker, memorandum for John Ehrlichman, 18 May 1971 (Presidential Trip to Tennessee-Tombigbee Dedication folder, box 7, Whitaker papers).

98. See John C. Whitaker, memorandum for Clifford Hardin, 19 May 1971 (Presidential Trip to Tennessee-Tombigbee Dedication folder, box 7, Whitaker papers).

99. Elsie Carper, "Tennessee-Tombigbee Plan Criticized," Louisville (KY) *Courier-Journal,* 1 June 1971. For examples of similar coverage, see "So They're Saying . . . ," Huntsville (AL) *Times,* 28 May 1971; "Non-Political Politics," Tuscaloosa (AL) *News,* 29 May 1971; "Good and Bad Canals," Tampa (FL) *Tribune,* 27 June 1971; and "Tombigbee Waterway Needs Reappraisal," Louisville (KY) *Times,* 17 July 1971. Environmental criticism leveled at the President's support of the Tenn-Tom was reported in: Elsie Carper, "Big Waterway Plan Assailed by EPA Unit," *Washington Post,* 1 June 1971; "Waterway Work Goes on Despite Environmental Protest," *Engineering News-Record,* 186 (June 10, 1971), 14; and Leon Lindsay, "Nixon Caught in Middle of Environmental Dispute," *Christian Science Monitor,* 17 June 1971.

100. Rowland Evans and Robert Novak, "Nixon's Wallace Problem," *Washington Post,* 28 May 1971.

101. John D. Ehrlichman, confidential memorandum for John N. Mitchell, 24 September 1971 (Ex NR7–1/A–Z Projects/1971–72 folder, box 19, file group NR, Nixon Presidential Materials Project).

102. See Paul E. Roberts, "The Role of Economics in the Environmental Law Suit," *Proceedings of the 1971 DuPont Environmental Engineering Seminar,* 26 (November 1972), 37–42; and "Action Now," *Sierra Club Bulletin,* 56 (August 1971), 27.

103. See interview, author with James D. Williams, 11 February 1987; and interview, author with Paul E. Roberts, 5 August 1987.

Chapter V/In Court—Act I

1. The National Wildlife Federation subsequently joined the lawsuit as a plaintiff. See Lewis Nolen, "Suit Planned on Pesticide," Memphis (TN) *Commercial Appeal,* 28 July 1970; and Report of Committee on Agricultural Pollution Concerning CLEAN's Opposition to the Fire Ant Eradication Program, 22 October 1970 (James D. Williams papers, Arlington, VA [hereafter referred to as "Williams papers"]).

2. Interview, author with Glenn H. Clemmer, 15 April 1987.

3. See *Environmental Defense Fund v. Corps of Engineers,* 325 F. Supp. 728 (E. D. Ark. 1971).

4. Arnold ran unsuccessfully for the U.S. Congress (4th District, Arkansas) in 1966 and 1972. From 1975–1978, he served as legislative assistant for Senator Dale Bumpers (D—Arkansas), and in 1978 he was appointed a federal judge in Arkansas. Praise of Arnold's competence was unanimous. See, for example, interview, Jo Ann McCormick with Judge William C. Keady, 2 May 1986; interview, author with Hunter M. Gholson (attorney for TTWDA), 28 May 1986; and Clemmer interview, 15 April 1987. See also, Roderick A. Cameron to Richard S. Arnold, 13 June 1974 (Williams papers); Dennis Puleston to Richard S. Arnold, 19 June 1974 (Williams papers); interview, author with Richard S. Arnold, 1 June 1987; and *Who's Who in America,* 44th edition (Wilmette, IL: Marquis Who's Who, 1986), vol. I, p. 90.

5. Interview, author with James D. Williams, 11 February 1987.

6. Jim Williams and Glenn Clemmer to Persons attending Tenn-Tom Meeting, 6 July 1971 (Williams papers).

7. "Environmental Defense Fund and Others File Lawsuit to Stop Huge Federal Waterway Project in Alabama and Mississippi," Environmental Defense Fund news release, 14 July 1971 (Williams papers).

8. "Tombigbee Waterway Needs Reappraisal," Louisville (KY) *Times,* 17 July 1971.

9. *Complaint,* Environmental Defense Fund, Inc., et al. v. Corps of Engineers of the United States, et al., Civil Action No. 1395–71, U.S. District Court for the District of Columbia, 14 July 1971.

10. Arnold interview, 1 June 1987.

11. *Memorandum of Law in Support of Motion for Preliminary Injunction,* Environmental Defense Fund, Inc., et al. v. Corps of Engineers of the United States Army, et al., U.S. District Court for the District of Columbia, 21 July 1971, p. 5. See also, *Complaint,* 14 July 1971; and "Environmental Defense Fund and Others File Lawsuit."

12. *Motion for Preliminary Injunction,* Environmental Defense Fund, Inc., et al. v. Corps of Engineers of the United States Army, et al., U.S. District Court for the District of Columbia, 21 July 1971.

13. See Richard S. Arnold to James D. Williams, 29 July 1971 (Williams papers).

14. See interview, author with Paul E. Roberts, 5 August 1987; and interview, author with James D. Williams, 7 January 1987.

15. See John E. Cooper to Richard A. Arnold, 27 July 1971; and Cooper to Arnold, 1 August 1971 (Williams papers).

16. John E. Cooper and Martha R. Cooper, "A Critique of Some Aspects of U.S. Army Corps of Engineers Environmental Statement, Tennessee-Tombigbee Waterway, March 1971," submitted to Richard Arnold, 30 July 1971 (Williams papers).

17. U.S. Army Corps of Engineers, Mobile District, *Environmental Statement: Tennessee-Tombigbee Waterway, Alabama and Mississippi Navigation* (Mobile, AL: U.S. Army Corps of Engineers, Mobile District, March 1971), p. 2.

18. Cooper and Cooper, "A Critique of Some Aspects of the U.S. Army Corps of Engineers Environmental Statement," p. 1.

19. *Ibid.,* p. 11.

20. Richard S. Arnold to Dennis Puleston, 30 August 1971 (Williams papers).

21. See interview, author with Nathaniel D. (Skeeter) McClure IV, 16 May 1986.

22. Glover Wilkins to members of the Authority, 14 July 1971 (file 2–1000–981, Glover Wilkins Tennessee-Tombigbee Waterway Archives, Archives and Museums Department, Mississippi University for Women, Columbus, MS [hereafter referred to as "GWTTW Archives"]). See also, interview, author with Glover Wilkins, 15 and 16 November 1986.

23. See Glover Wilkins to members of the Authority, 20 July 1971 (file 2–1000–981, GWTTW Archives); and Wilkins interview, 15 and 16 November 1986.

24. Gholson interview, 28 May 1986. See also, John C. Stennis to Richard G. Kleindienst, 30 September 1971 (Ex NR–7–1/A–Z Projects/1971–72 folder, box 19, file group NR, Nixon Presidential Materials Project, National Archives and Records Administration, Washington, D.C.).

25. Jack Edwards to Clark MacGregor, 5 August 1971 (Presidential Trip to Tennessee-Tombigbee Dedication folder, box 7, John Whitaker papers, Nixon Presidential Materials Project [hereafter referred to as "Whitaker papers"]).

26. Richard M. Fairbanks, memorandum for John C. Whitaker, 30 August 1971 (Presidential Trip to Tennessee-Tombigbee Dedication folder, box 7, Whitaker papers).

27. Glover Wilkins to members of the Authority, 30 July 1971 (file 2–1000–981, GWTTW Archives).

28. Gholson interview, 28 May 1986. Gholson was a law partner with Burgin in 1971.

29. Glover Wilkins, memorandum, 5 August 1971 (file 2–1000–949, GWTTW Archives).

30. *Ibid.*

31. Glover Wilkins, memorandum for Tennessee-Tombigbee Waterway Development Authority Members, et al., 18 August 1971 (file 2–1000–5, GWTTW Archives).

32. Glover Wilkins to Frank Koisch, et al., 12 January 1972 (file 2–1000–982, GWTTW Archives).

33. Tennessee-Tombigbee Waterway Development Authority, Environmental Policy Statement (Columbus, MS: Tennessee-Tombigbee Waterway Development Authority, August 1971).

34. Glover Wilkins to members of the Authority, 30 July 1971 (file 2–1000–981, GWTTW Archives).

35. Glover Wilkins to members of the Authority, 8 September 1971 (file 2–1000–981, GWTTW Archives). See also, Wilkins interview, 15 and 16 November 1986. For his special appeal for additional financial assistance from the state of Alabama, see Glover Wilkins to George C. Wallace (letter marked "Personal and Confidential"), 9 November 1971 (file 2–1000–889, GWTTW Archives).

36. John W. Rushing to Dick Edwards, undated [probably early August 1971] (Williams papers). See also, interview, author with John W. Rushing, 6 October 1986; and interview, author with Willis E. Ruland (Planning Division, Mobile District), 19 November 1986.

37. See Rushing to Edwards, undated; and Alfred P. Holmes, Jr., memorandum to Gary Bohlke, 28 April 1972 (Litigation/1971–72 folder, file 410–01, Office of Counsel, Mobile District, U.S. Army Corps of Engineers [hereafter referred to as "SAMOC"]).

38. A detailed summary of the two-day hearing was provided in John H. Gullett to Glover Wilkins, 17 September 1971 (file 2–1000–19, GWTTW Archives).

39. See Findings of Fact, Conclusions of Law and Order, Environmental Defense Fund, Inc., et al. v. Corps of Engineers of the United States Army, et al., U.S. District Court for the District of Columbia, 21 September 1971. The case was later reported as *Environmental Defense Fund, Inc. v. Corps of Engineers*, 331 F. Supp. 925 (D.D.C. 1971). See also, "Federal Judge Rules against Southern Waterway Project," *Washington Post*, 22 September 1971; William M. Blair, "Judge Enjoins Tennessee Canal Plan," *New York Times*, 22 September 1971; "Tombigbee Tempest," *Washington Evening Star*, 30 September 1971; "A Two-Lane Mississippi," *New York Times*, 3 October 1971; and "EDF Wins Injunction against Tennessee-Tombigbee Waterway," *EDF Letter*, 2 (October 1971), 1.

40. Richard S. Arnold to Dennis Puleston, 11 January 1972 (Williams papers).

41. Richard S. Arnold to Dennis Puleston, 27 September 1971 (Williams papers).

42. Richard S. Arnold to Gaylord Nelson, 4 October 1971 (Williams papers).

43. See Glover Wilkins to members of the Authority, 21 September 1971 (file 2–1000–981, GWTTW Archives).

44. Gholson interview, 28 May 1986.

45. Arnold interview, 1 June 1987. Arnold, now a federal judge with the Court of Appeals of the 8th Circuit, said that he did not share this view. For the variances among District Courts, see Robert A. Carp and C. K. Rowland, *Policymaking and Politics in the Federal District Courts* (Knoxville: University of Tennessee Press, 1983).

46. Glover Wilkins, memorandum for Members of the Authority, 10 January 1972 (file 2–1000–982, GWTTW Archives). "We knew we would be sunk," Wilkins later recalled, "if the case was tried in Washington." Wilkins interview, 15 November 1986. See also, Glover Wilkins, "My Association with the Tennessee-Tombigbee Waterway Project," unpublished manuscript, Mobile District, U.S. Army Corps of Engineers, 1985, pp. 103–104.

47. See Alfred P. Holmes, Jr., memorandum to the file, no date [probably 4 or 5 November 1971] (TTW litigation files, SAMOC).

48. See John H. Gullett to Glover Wilkins, 16 November 1971 (file 2–1000–949, GWTTW Archives).

49. Holmes, memorandum to the file, no date [probably 4 or 5 November 1971].

50. Gullett to Wilkins, 16 November 1971.

51. Richard S. Arnold to Dennis Puleston, 11 January 1972 (Williams papers).

52. Quoted in Philip W. Smith, "Tenn-Tom Attorneys File Plea," Mobile (AL) *Register*, 15 December 1971.
53. *Defendants' Motion for Transfer of Venue*, Environmental Defense Fund, et al. v. Corps of Engineers of the United States Army, et al., U.S. District Court for the District of Columbia, 21 January 1972.
54. *Hearing on Motions for a Change of Venue*, Environmental Defense Fund, et al. v. Corps of Engineers of the United States Army, et al., U.S. District Court for the District of Columbia, 27 January 1972; and *Order*, Environmental Defense Fund, et al. v. Corps of Engineers of the United States Army, et al., U.S. District Court for the District of Columbia, 27 January 1972.
55. "Preliminary Injunction Affects Tennessee-Tombigbee Waterway," *Congressional Record—Senate*, 117 (September 21, 1971), 32614. See also, "Judge Stops Tombigbee Work," Chattanooga (TN) *News Free Press*, 21 September 1971.
56. This point is mentioned in Carroll Brumfield, "The Tenn-Tom," *Environmental Action*, 4 (July 22, 1972), 11. See also, Glover Wilkins to James O. Eastland, 23 February 1972 (file 2–1000–1049, GWTTW Archives); Wilkins, "My Association with the Tennessee-Tombigbee Waterway Project," p. 106; and Wilkins interview, 15 and 16 November 1986.
57. "Sparkman Supports Tenn-Tom Bid for Venue Change on Continuation of Waterway," press release, office of Senator John Sparkman, 24 January 1972 (copy in file 2–1000–1063, GWTTW Archives).
58. Arnold interview, 1 June 1987.
59. See William F. Brandes, "The Tennessee-Tombigbee Waterway Project v. The National Environmental Policy Act," term paper, EWRE 279, School of Engineering, Vanderbilt University, January 1975, p. 9.
60. *Ibid.*, p. 10.
61. Glover Wilkins to Authority members, 9 February 1972 (file 2–1000–982, GWTTW Archives).
62. *Memorandum Order*, Environmental Defense Fund, Inc., et al., v. Corps of Engineers of the United States Army, et al., U.S. District Court for the Northern District of Mississippi, Eastern Division, 13 April 1972.
63. Arnold interview, 1 June 1987.
64. T. Marx Huff to Richard S. Arnold, et al., 13 March 1972 (Williams papers).
65. See *Order*, Environmental Defense Fund, Inc., et al. v. Corps of Engineers of the United States Army, et al., U.S. District Court for the Northern District of Mississippi, 13 April 1972.
66. *Ibid.*; and *Memorandum Order*, Environmental Defense Fund, et al. v. Corps of Engineers of the United States Army, et al., U.S. District Court for the Northern District of Mississippi, 8 June 1972. See also, Carroll Brumfield, "Judge Deals Setback to Suit on Tennessee-Tombigbee Plan," Memphis (TN) *Commercial Appeal*, 14 April 1972; and Keady interview, 2 May 1986.
67. Richard S. Arnold to Jon T. Brown, 31 August 1971 (Williams papers).
68. Hunter M. Gholson, memorandum to Tennessee-Tombigbee Waterway Development Authority members, 10 June 1972 (file 2–1000–14, GWTTW Archives).
69. Keady interview, 2 May 1986. A detailed summary of the testimony is provided in Alfred P. Holmes, Jr. to Office of the Chief of Engineers, 30 June 1972 (Litigation/1973–1976 file, folder 410–01, SAMOC).
70. See Gholson interview, 28 May 1986; John H. Gullett to Thompson Pound, 4 February 1972 (file 2–1000–949, GWTTW Archives); and Holmes to Office of the Chief of Engineers, 30 June 1972.
71. Williams interview, 7 January 1987; and Clemmer interview, 15 April 1987.
72. See minutes of the CLEAN Board of Directors meeting, 5 July 1972 (Williams papers).
73. Arnold interview, 1 June 1987; and Gholson interview, 28 May 1986.
74. Brandes, "The Tennessee-Tombigbee Waterway Project v. The National Environmental Policy Act," p. 19. See also, interview, author with William Brandes (Nashville District Engineer, 1971–1974), 18 September 1986.

75. *Environmental Defense Fund, Inc. v. Corps of Engineers of the United States Army*, 348 F. Supp. 916 (N.D. Miss. 1972). The phrase *with prejudice* meant that the District Court decision was final and was subject only to appeal. Had Keady issued his decision *without prejudice*, either party could have come back to the District Court on the same issue.

76. *Motion for Injunction Pending Appeal before the Honorable William C. Keady*, Environmental Defense Fund, Inc., et al. v. Corps of Engineers of the United States Army, et al., U.S. District Court for the Northern District of Mississippi, 28 August 1972, pp. 46–52.

77. *Motion for Injunction Pending Appeal in the U.S. Court of Appeals for the Fifth Circuit*, Environmental Defense Fund, Inc., et al., v. Corps of Engineers of the United States Army, et al., U.S. Court of Appeals for the Fifth Circuit, 30 August 1972.

78. *Appeal from the United States District Court for the Northern District of Mississippi*, Environmental Defense Fund, Inc., et al. v. Corps of Engineers of the United States Army, et al., U.S. Court of Appeals for the Fifth Circuit, 14 September 1972.

79. See Richard S. Arnold to Dennis Puleston, 18 September 1972 (Williams papers).

80. See *Brief for Appellants*, Environmental Defense Fund, Inc., et al. v. Corps of Engineers of the United States Army, et al., U.S. Court of Appeals for the Fifth Circuit, 10 November 1972; *Environmental Defense Fund, Inc. v. Corps of Engineers of the United States Army*, 492 F. 2d 1123 (5th Cir. 1974); Richard S. Arnold to Edward Lee Rogers, 29 May 1973 (Williams papers); and "Court Told Action Would Intrude," Texarkana (AR) *Gazette*, 29 May 1973.

81. See *Environmental Defense Fund v. Corps of Engineers* (5th Cir. 1974); and *Brief for Appellees*, Environmental Defense Fund, Inc., et al. v. Corps of Engineers of the United States Army, et al., U.S. Court of Appeals for the Fifth Circuit, 22 December 1972.

82. *Environmental Defense Fund v. Corps of Engineers* (5th Cir. 1974), pp. 2884–2885.

83. *Ibid.*, p. 2908.

84. See Jon T. Brown to Richard S. Arnold, 26 April 1974 (Williams papers).

85. Richard S. Arnold to Dennis Puleston, 25 April 1974. See also, Richard S. Arnold to Jon T. Brown, 6 May 1974 (Williams papers).

86. Dennis Puleston to Richard S. Arnold, 7 May 1974 (Williams papers).

87. See Dennis Puleston to Richard S. Arnold, 19 June 1974 (Williams papers).

88. Interview, author with George H. Atkins, 18 November 1986.

89. Brandes, "The Tennessee-Tombigbee Waterway Project v. The National Environmental Policy Act," p. 25.

90. *Ibid.*, pp. 17–18.

91. Alfred P. Holmes, Jr., memorandum to Gary Bohlke, 28 April 1972 (Litigation/1971–72 folder, file 410–01, SAMOC).

Chapter VI/The Environmentalist-Railroad Coalition

1. See James G. Tangerose, "Budget Request—Civil Works Program, Corps of Engineers, Fiscal Year 1971," Association of American Railroads *Waterway News Letter*, no. 70–1 (February 19, 1970), 1–6.

2. See James G. Tangerose, memorandum to members of the AAR Tennessee-Tombigbee Waterway Committee, 19 March 1970 (TTWW litigation file, Office of Counsel, Mobile District, U.S. Army Corps of Engineers [hereafter referred to as "MDOC"]).

3. W. C. Wagner, memorandum for file, 17 March 1970 (TTWW litigation file, MDOC).

4. See Robert J. Schaefer to James G. Tangerose, 18 March 1970 (TTWW litigation file, MDOC).

5. E. Leo Koester to Wade M. Richards, 15 April 1970 (TTWW litigation file, MDOC). See also, W. G. Whitsett to Brooks H. Gordon, 20 March 1970 (TTWW litigation file, MDOC).

6. John F. Smith to E. A. Brautigam, et al., 21 April 1970 (TTWW litigation file, MDOC).

7. The testimonies were printed in U.S. Congress, House of Representatives, Committee on Appropriations, Subcommittee on Public Works, Hearings, *Public Works for Water, Pollution Control, and Power Development and Atomic Energy Commission Appropriation Bill, 1971* (91st

Cong., 2nd sess., 1970), Part 5, pp. 386–425; and U.S. Congress, Senate, Subcommittee of the Committee on Appropriations, Hearings, *Public Works for Water, Pollution Control, and Power Development and Atomic Energy Commission Appropriations for Fiscal Year 1971* (91st Cong., 2nd sess., 1970), pp. 444–456.

8. B. H. Gordon to W. G. Whitsett, et al., 8 May 1970 (TTWW litigation file, MDOC). Robert A. Kuehne, an associate professor of zoology at the University of Kentucky at Lexington, testified against the Tenn-Tom at the subcommittee hearings, although his participation was not sponsored by the railroad interests. See House Subcommittee on Public Works, *Public Works . . . Appropriations Bill, 1971*, Part 5, pp. 425–427; James G. Tangerose to N. E. White, 12 June 1970 (TTWW litigation file, MDOC); "L&N Leads Protest against Tenn-Tom," Mobile (AL) *Register*, 6 May 1970; and "Mistaken Dissent on Tenn-Tom Funds," Mobile (AL) *Register*, 8 May 1970.

9. W. Gavin Whitsett to Elvis J. Stahr, 19 May 1970 (TTWW litigation file, MDOC). See also, interview, author with Philip M. Lanier (Senior Vice President-Law, Louisville and Nashville Railroad), 23 November 1987. The Audubon Society thanked Whitsett for the information, but took no immediate action. See Chaplin B. Barnes to W. Gavin Whitsett, 15 June 1970 (TTWW litigation file, MDOC).

10. N. E. White to J. G. Tangerose, 8 June 1970 (TTWW litigation file, MDOC).

11. E. Leo Koester to Claude Ryan, 19 June 1970. See also, Ryan to Koester, 14 July 1970 (both letters are in the TTWW litigation file, MDOC).

12. See John E. Cooper to Leo Koester, 9 July 1970 (TTWW litigation file, MDOC); House Subcommittee on Public Works, *Public Works . . . Appropriations Bill, 1971*, Part 5, pp. 425–427; and Senate Subcommittee of the Committee on Appropriations, *Public Works . . . Appropriations for Fiscal Year 1971*, pp. 465–466.

13. Cooper to Koester, 9 July 1970.

14. John G. Mitchell (ed.), *Ecotactics: The Sierra Club Handbook for Environmental Activists* (New York: Simon and Schuster, 1970).

15. Cooper to Koester, 9 July 1970.

16. See E. Leo Koester to John E. Cooper, 17 July 1970 (Louisville and Nashville Railroad Collection, Record Group 123, University of Louisville Archives, Louisville, KY).

17. Brooks H. Gordon to E. A. Brautigam, et al., 10 August 1970 (TTWW litigation file, MDOC).

18. See "History of the Association of American Railroads," unpublished and undated manuscript in the Intermodal Policy Division, Association of American Railroads, Washington, D.C.

19. Brooks H. Gordon to J. W. Ingram, et al., 28 August 1970 (TTWW litigation file, MDOC).

20. Philip M. Lanier to Russell E. Train, 25 May 1971 (James D. Williams papers, Arlington, VA [hereafter referred to as "Williams papers"]). See also, Lanier interview, 23 November 1987; and Brooks H. Gordon to Col. Harry A. Griffith, 11 September 1970 (Inquiries and Replies/TTW 1970 folder, file 1501–07, Inland Environment Section, Environment and Resources Branch, Planning Division, Mobile District [hereafter referred to as "SAMPD-EI"]).

21. See, for example, E. Leo Koester to James Williams, 14 June 1971, 15 June 1971, and 16 June 1971 (file 2–1000–949, Glover Wilkins Tennessee-Tombigbee Waterway Archives, Archives and Museums Department, Mississippi University for Women, Columbus, MS [hereafter referred to as "GWTTW Archives"]); and E. Leo Koester to James A. Schultz, 21 June 1971 (Louisville and Nashville Railroad Collection).

22. Richard S. Arnold to Roderick Cameron, 27 July 1971 (Williams papers). See also, interview, author with James D. Williams, 7 January 1987; and E. Leo Koester to Jim Williams, 14 June 1971 (file 2–1000–949, GWTTW Archives).

23. Roderick A. Cameron to Philip M. Lanier, 20 August 1971 (Williams papers).

24. Philip M. Lanier to Roderick A. Cameron, 24 August 1971 (Williams papers). See also, Lanier interview, 23 November 1987.

25. Interview, author with James D. Williams, 11 February 1987.

26. See Richard S. Arnold to Paul Roberts, 18 August 1971 (Williams papers).

27. See Prime E. Osborn to John Breckinridge, 17 January 1973 (TTWW litigation file, MDOC); E. Leo Koester to Frank A. Stubblefield, 1 March 1973 (Williams Papers); and Brent Blackwelder, "Comments on National Water Commission's Final Report," (news report from Environmental Policy Center), 16 July 1973 (Williams papers). The Commission—which was established by Congress in 1968 to report in five years on the problems of national water resource availability, distribution, and conservation—released its final report in June 1973; it left the user charge recommendation intact. See National Water Commission, *Water Policies for the Future* (Washington: GPO, 1973).

28. See E. Leo Koester to James Williams, 20 March 1973 (Williams papers).

29. E. Leo Koester to James Williams, 24 August 1973 (Williams papers).

30. Statement of Association of Southeastern Biologists on the Tennessee-Tombigbee Waterway, 13 April 1973 (Williams papers).

31. Interview, author with Glenn H. Clemmer, 15 April 1987. See also, U.S. Congress, Congressional Budget Office, *Financing Waterway Development: The User Charge Debate*, CBO Staff Working Paper (Washington: GPO, July 1977); and American Enterprise Institute for Public Policy Research, *Waterway User Charges* (Washington: American Enterprise Institute for Public Policy Research, 30 September 1977).

32. See E. Leo Koester to Bob Kuehne, Jim Williams, and Glenn Clemmer, 16 April 1973 (attached to the letter was Koester's "Suggested Resolution for CLEAN"); E. Leo Koester to Glenn Clemmer, 8 May 1973; E. Leo Koester to James Williams, 24 August 1973; and E. Leo Koester to Jim Williams and Glenn Clemmer, 5 December 1973 (all letters in Williams papers).

33. See E. Leo Koester to Philip M. Lanier and James W. Hoeland, 1 June 1973 (TTWW litigation file, MDOC); and *Plaintiffs' Responses and Objections to Defendants' First Set of Interrogatories*, Environmental Defense Fund, et al. v. Clifford Alexander, et al., U.S. District Court for the Northern District of Mississippi, 11 July 1978. The travel expenses paid by the L&N Railroad were less than $275.

34. See James D. Williams's statement in U.S. Congress, House of Representatives, Committee on Appropriations, Subcommittee on Public Works, Hearings, *Public Works for Water and Power Development and Atomic Energy Commission Appropriation Bill, 1974* (93rd Cong., 1st sess., 1973), Part 6, pp. 1225–1230; and U.S. Congress, Senate, Subcommittee of the Committee on Appropriations, Hearings, *Public Works for Water and Power Development and Atomic Energy Commission Appropriations for Fiscal Year 1974* (93rd Cong., 1st sess., 1973), Part 6, pp. 7225–7231.

35. See Koester to Lanier and Hoeland, 1 June 1973.

36. E. Leo Koester to R. E. Bisha, 10 August 1973 (TTWW litigation file, MDOC).

37. *Ibid.*

38. E. Leo Koester to P. F. Osborn, et al., 8 November 1973 (TTWW litigation file, MDOC).

39. E. Leo Koester to Brent Blackwelder, 26 July 1974 (TTWW litigation file, MDOC).

40. See Glover Wilkins to Jack Edwards, 21 June 1973; Edwards to Wilkins, 25 June 1973 (both letters in file 4-B, carton 53, Jack Edwards papers, University of South Alabama Archives, Mobile, AL); and American Rivers Conservation Council, et al., *Disasters in Water Development: Some of the Most Economically Wasteful and Environmentally Destructive Projects of the Corps of Engineers, the Bureau of Reclamation, and the Tennessee Valley Authority* (n.p.: April 1973).

41. See Bob Bryan, "Evacuated Dirt in Waterway Construction to be Placed in Valleys along the Route," Tupelo (MS) *Journal*, 9 July 1974.

42. Brent Blackwelder to Roy Ash, 30 September 1974 (TTWW litigation file, MDOC).

43. Brent Blackwelder to Russell Peterson, 1 October 1974. See also, Brent Blackwelder to Russell Train, 1 October 1974 (TTWW litigation file, MDOC).

44. Brent Blackwelder to Philip M. Lanier, 2 October 1974 (TTWW litigation file, MDOC).

45. Philip M. Lanier to President Gerald Ford, 2 October 1974 (Tenn-Tom microfilm, set I, reel 65, frames 599–601, Office of History archives, Corps Headquarters, Fort Belvoir, VA).

46. Brent Blackwelder to Russell Peterson, 28 February 1975 (TTWW litigation file, MDOC).

47. See Sherrie Clemmer, "CLEAN Newsletter," March 1975 (copy in TTWW litigation file, MDOC).

48. See "Tenn-Tom: Review '75" (copy in TTWW litigation file, MDOC).

49. Mobile Bay Audubon Society, "Resolution," 1 July 1975 [signed by Myrt Jones, Society President] (correspondence/1966–1975 folder, file 1501–07, Environmental Resources Planning Section, Environment and Resources Branch, Planning Division, Mobile District [hereafter referred to as "SAMPD-ER"]).

50. See E. Leo Koester to Brent Blackwelder, 12 March 1975 (TTWW litigation file, MDOC); and "Rail Spokesman Labels 'Tenn-Tom' Waterway 'Waste of Tax Money,'" *Traffic World*, 161 (March 17, 1975), 31.

51. See "The Tennessee-Tombigbee Waterway," Alabama Conservancy *Action Report*, no. 13 (April 1975), 2. Grace had only recently completed his masters degree in geography, where his research focused on mussels in the Tennessee River. See Randall Grace, "The Fresh-Water Mussel Industry of the Lower Tennessee River: Ecology and Future," M.S. thesis, Western Kentucky University, 1974.

52. See U.S. Congress, House of Representatives, Committee on Appropriations, Subcommittee on Public Works, Hearings, *Public Works for Water and Power Development and Energy Research Appropriations Bill, 1976* (94th Cong., 1st sess., 1975), Part 9, pp. 1311–1315; and U.S. Congress, Senate, Subcommittee of the Committee on Appropriations, Hearings, *Public Works for Water and Power Development and Energy Research Appropriations for Fiscal Year 1976* (94th Cong., 1st sess., 1975), Part 7, pp. 6911–6918.

53. See Randall Grace, memorandum (subj.: Congressional Contacts, May 6–7, 1975, Concerning the Proposed Tenn-Tom Waterway), no date (TTWW litigation file, MDOC).

54. See E. Leo Koester to Cliff Massoth, 29 August 1975; E. Leo Koester to Glenn Clemmer, 29 August 1975; and Minutes of Railroad Tenn-Tom Waterway Meeting, Tuscaloosa, AL, 13 September 1975 (TTWW litigation file, MDOC). The Corps estimated Alabama's cost sharing contribution for the replacement of highway bridges to be $38 million.

55. Testimony of Brent Blackwelder before the Senate Public Works Committee, 8 October 1975, typescript (TTWW litigation file, MDOC).

56. Jon T. Brown to Dennis Puleston, 5 December 1974 (Williams papers).

57. Proposal—Tennessee Tombigbee Waterway Meeting, Louisville, KY, March 24, 1975 (Williams papers). The railway interests represented at the meeting included the L&N, the Southern Railway Company, the St. Louis-San Francisco Railway Company, the Illinois Central Gulf Railroad, the Seaboard Coast Line Industries, and the Brotherhood of Locomotive Engineers. Environmentalists were represented by officials from CLEAN, the Environmental Policy Center, the Environmental Defense Fund, and the Coalition on American Rivers. See also, E. Leo Koester to Glenn Clemmer, 16 April 1975 (TTWW litigation file, MDOC); and Lanier interview, 23 November 1987.

58. Interview, author with Joseph L. Carroll (Transportation Systems Program, Pennsylvania State University), 19 and 20 November 1987.

59. Arlie Schardt to The President, 14 November 1975 (ALA/MISS-COE/TTW folder, box 4, file 429–81–37, CEQ records, Washington National Records Center, Suitland, MD).

60. Jon T. Brown, memorandum to Philip Lanier, 28 June 1976 (Paul E. Roberts personnel papers, Washington, D.C.).

61. Glenn Clemmer to E. Leo Koester, 29 June 1975 (TTWW litigation file, MDOC). See also, Glenn Liming to E. Leo Koester, 30 June 1975 (TTWW litigation file, MDOC).

62. See E. Leo Koester to Randall Grace, 15 August 1975 (TTWW litigation file, MDOC).

63. The goals, purposes, and tactics of the TRCC were discussed at length in Randall Grace, memorandum to participants of the 25 September 1976 Starkville, MS, meeting, no date; and Tombigbee River Conservation Council "Annual Report, '76–'77," no date (TTWW litigation file, MDOC).

64. Randall Grace to Dear Friend of the Tombigbee, no date [probably September 1976] (TTW Litigation file, MDOC). See also, Randall Grace to Leo Koester, 7 October 1976 (TTWW litigation file, MDOC); and "A Vital Election Is Set for November 23, 1976: Alabama Voters Must Decide Whether to Invest Nearly $50 Million to Replace Bridge which Will Be Destroyed if the

Tenn-Tom Waterway Is Built," campaign literature distributed by the Alabama Conservancy, no date (copy in Alabama Conservancy folder, file 103–01, SAMPD-EI).

65. E. Leo Koester to Randall Grace, 9 November 1976 (TTWW litigation file, MDOC).

66. E. Leo Koester to Randall Grace, 7 January 1977 (TTWW litigation file, MDOC). See also, E. Leo Koester to Randall Grace, 13 October 1977 (TTWW litigation file, MDOC); and E. Leo Koester to Frank Owen, 20 June 1978 (TTWW litigation file, MDOC).

67. TRCC Action Alert, 20 December 1976 (TTWW litigation file, MDOC).

68. The TTWDA's pamphlet was entitled "Tennessee-Tombigbee Waterway—Short Cut to Progress: Facts for Press, Radio, and Television." The TRCC countered by titling its pamphlet "The Tenn-Tom Waterway—Short Cut to Disaster." The pamphlets were published in 1976 and 1977, respectively. Copies of both publications are in Charles B. Castner's personal files, CSX Transportation, Louisville, KY.

69. Copies of *Tenn-Tom Review* are contained in Castner's personal files.

70. See U.S. Army Corps of Engineers, Mobile District, "Fact Sheet, Tennessee-Tombigbee Waterway: Response to *Tenn-Tom: Review*, Tombigbee River Conservation Council," January 1977 (copy in litigation folder, file 1517–01, SAMPD-ER).

71. Glover Wilkins to Col. Charlie Blalock, 28 February 1977 (Tenn-Tom microfilm, set I, reel 1, frames 999–1000).

72. See Tombigbee River Conservation Council "Annual Report, '76–'77," no date (copy in TTWW litigation file, MDOC). The National Audubon Society later intervened as well.

73. Owen had previously volunteered his time to the TRCC. See CLEAN newsletter, February 1978 (copy in TTWW litigation file, MDOC).

74. Statement of E. Leo Koester before a public meeting sponsored by the Tennessee-Tombigbee Waterway Development Authority, Jackson, MS, 9 August 1978 (TTWW litigation file, MDOC).

75. E. Leo Koester to Richard Briggs, 17 August 1978 (TTWW litigation file, MDOC). See also, Lanier interview, 23 November 1987.

Chapter VII/Economic Accountability

1. See Diana Morgan, "Earth Day: Scientists Reflect on 20 Years of Activism," *The Scientist*, 4 (April 16, 1990), 1 and 12.

2. "Tennessee-Tombigbee," *Congressional Record—Senate*, 117 (June 4, 1971), 18176–18177. Representative Henry S. Reuss (D-Wisconsin), chairman of the House Government Operations Committee's Subcommittee on Conservation and Natural Resources, also questioned the Tenn-Tom at this time. See Elsie Carper, "Army Corps Promises Waterway Effects Study," *Washington Post*, 5 June 1971; and Dick Ritter, "Hard Look at Tenn-Tom Waterway Plan Urged," *Federal Times*, 7 July 1971.

3. "Public Works Appropriations," *Congressional Record—Senate*, 117 (July 31, 1971), 28480.

4. "Public Works for Water and Power Development and Atomic Energy Commission Appropriations," *Congressional Record—Senate*, 117 (July 31, 1971), 28480.

5. *Ibid.*, pp. 28480–28487.

6. *Ibid.*, p. 28488. See also, "Tombigbee's Foes Nipped," Chattanooga (TN) *News Free Press*, 1 August 1971.

7. John Stennis to Lt. Gen. Frederick J. Clarke, 18 May 1970 (James D. Williams papers, Arlington, VA [hereafter referred to as "Williams papers"]).

8. John Stennis to Caspar W. Weinberger, 9 December 1970 (Tennessee folder, box 9, file NN–3–51–86–24, Natural Resources Division—Water Resources, Office of Management and Budget [OMB], Record Group 51, National Archives, Suitland, MD).

9. "Conference Report on H.R. 10090, Public Works—AEC Appropriations," *Congressional Record—House*, 117 (September 22, 1971), 32719–32720.

10. *Ibid.*, p. 32722.

11. *Ibid.*, pp. 32726–32727. See also, "Tenn-Tom Injunction Incurs Blast," Birmingham (AL) *News*, 23 September 1971.

12. "Military Procurement Authorizations," *Congressional Record—Senate,* 117 (September 22, 1971), 32885.

13. See interview, author with Brent Blackwelder, 12 April 1988.

14. John C. Stennis to Glover Wilkins, 3 October 1972 (file 2–1000–385, Glover Wilkins Tennessee-Tombigbee Waterway Archives, Archives and Museums Department, Mississippi University for Women, Columbus, MS [hereafter referred to as "GWTTW Archives"]).

15. Ward Sinclair, "Party Climaxes Tennessee-Tombigbee Success Story," Louisville (KY) *Courier-Journal & Times,* 4 May 1975. See also, U.S. Congress, House of Representatives, Committee on Appropriations, Subcommittee on Public Works, Hearings, *Public Works for Water and Power Development and Energy Research Appropriation Bill, 1976* (94th Cong., 1st sess., 1975), Part 8.

16. "Stop This $3 Billion Boondoggle," Chicago *Sun-Times,* 27 January 1980.

17. See U.S. Army Corps of Engineers, Mobile District, *Tennessee-Tombigbee Waterway, Alabama and Mississippi: Design Memorandum No. 1—General Design* (Mobile, AL: U.S. Army Corps of Engineers, Mobile District, 30 June 1960); U.S. Army Corps of Engineers, Mobile District, *Tennessee-Tombigbee Waterway, Alabama and Mississippi: Supplement to Design Memorandum No. 1—General Design. Reevaluation of Project Economics* (Mobile, AL: U.S. Army Corps of Engineers, Mobile District, 30 June 1966); and "Army Engineers Will Soon Ask for Bids for Gainesville Lock—First Step in Constructing Long-Sought Tennessee-Tombigbee Waterway," news release, U.S. Army Engineer District, Mobile, 8 April 1971 (file 2–1000–981, GWTTW Archives).

18. See Mobile District Engineering Division, "Chronology of Estimate, 1966 to Date," 10 December 1975 (litigation folder, file 401–07, Inland Environment Section, Planning Division, Mobile District, U.S. Army Corps of Engineers [hereafter referred to as "SAMPD-EI"]); and William Fickel, "Briefing on Economic Reanalysis of Tennessee-Tombigbee Waterway," 15 June 1977 (Tenn-Tom microfilm, set I, reel 1, frame 563, Office of History archives, Corps Headquarters, Fort Belvoir, VA).

19. Maj. Gen. Daniel A. Raymond to Maj. Gen. J. W. Morris, 16 March 1973 (Construction Schedule folder, file 1518–01, Nashville District Civil Works Project records, Coastal Environment Section, Planning Division, Mobile District [hereafter referred to as "SAMPD-EC"]).

20. See Vernon S. Holmes, memorandum to District Engineer, 15 November 1974 (Tenn-Tom microfilm, set I, reel 1, frame 2230); interview, author with Vernon S. Holmes, 30 May 1986; and interview, author with John Jeffrey Tidmore, 20 May 1986.

21. See A. G. (George) Johnson, memorandum for record, 6 January 1975 (Tenn-Tom microfilm, set I, reel 1, frame 2242); and deposition of John Jeffrey Tidmore, 10–12 May 1978 (depositions/defendant folder, Tenn-Tom Litigation Management Unit files, Mobile District [hereafter referred to as "SAMDL"]).

22. Brig. Gen. Carroll N. LeTellier to the Mobile and Nashville District Engineers, 2 June 1975 (TTW correspondence/SAD and OCE folder, file 1501–07, SAMPD-EI).

23. See interview, author with Carroll N. LeTellier (South Atlantic Division Engineer, 1973–1976), 14 April 1988; deposition of Vernon Holmes, 22 June 1978, pp. 16–17 (depositions/defendant folder, Tenn-Tom Litigation Management Unit files, SAMDL); and Tidmore interview, 20 May 1986. For the Corps's account of these cost increases, see U.S. Army Corps of Engineers, Mobile District, *Studies of Project Costs and Benefits: Tennessee-Tombigbee Waterway* (Mobile, AL: U.S. Army Corps of Engineers, Mobile District, October 1975), pp. 1–2; and interview, author with Frank C. Deming, 21 November 1986.

24. Frank C. Deming to Mobile District Public Affairs Office, 19 November 1975 (litigation/L&N v. U.S. 1984 folder, file 410–01, Office of Counsel, Mobile District [hereafter referred to as "MDOC"]).

25. U.S. Army Corps of Engineers, Mobile District, news release, "Tennessee-Tombigbee Waterway in Alabama and Mississippi Re-evaluation," 21 January 1976 (copy in file 2–1000–943, GWTTW Archives).

26. See Environmental Policy Center, news release, 17 December 1975 (copy in Paul E. Roberts

personal papers, Washington, D.C.); and Committee for Leaving the Environment of America Natural, press release, "CLEAN Renounces Waterway," 26 January 1976 (TTWW litigation file, MDOC). For the lasting impact of the Corps's misrepresentation to Congress, see "Tenn-Tom Troubles Roll On," Birmingham (AL) News, 3 December 1978.

27. Maj. Gen. Ernest Graves to South Atlantic Division Engineer, 20 January 1976 (cost estimate folder, Tenn-Tom Litigation Management Unit files, SAMDL). See also, Department of the Army [Office of the Assistant Secretary of the Army for Civil Works], Information Paper, no date [probably April or May 1977] (Tenn-Tom microfilm, set I, reel 57, frame 93).

28. R. K. (Keith) Adams, memorandum for the record, no date [probably mid-February 1976] (TTLMU/Counts I & II folder, file 401–07, SAMDL).

29. See U.S. Congress, House of Representatives, Committee on Appropriations, Subcommittee on Public Works, Hearings, *Public Works for Water and Power Development and Energy Research Appropriation Bill, 1977* (94th Cong., 2nd sess., 1976), Part 1, pp. 576–754, passim.

30. Interview, author with Roger A. Burke (Planning Division, Mobile District), 17 November 1986; interview, author with F. L. (Les) Currie, 15 May 1986; interview, author with Lawrence R. Green (Planning Division, Mobile District), 21 November 1986; and Tennessee-Tombigbee Waterway Development Authority, "Facts and Information on the Tennessee-Tombigbee Waterway and Its Opponents' Allegations," 5 April 1979 (copy in TTLMU/Sen. Nelson folder, file 401–07, SAMDL).

31. See U.S. Army Corps of Engineers, Mobile District, *An Evaluation of the Transportation Economics of the Tennessee-Tombigbee Waterway, Final Report. Volume I: Executive Summary*, report prepared by A. T. Kearney Management Consultants, Inc. (Mobile, AL: U.S. Army Corps of Engineers, Mobile District, February 1976); and William Fickel, "Briefing on Economic Reanalysis of Tennessee-Tombigbee Waterway," 15 June 1977 (Tenn-Tom microfilm, set I, reel 1, frame 563).

32. U.S. Army Audit Agency, *Report of Audit: Tennessee-Tombigbee Waterway Project, U.S. Army Corps of Engineers*, Audit Report SO 76–408 (Atlanta, GA: U.S. Army Audit Agency, Southern District, 17 September 1976), p. 6.

33. Ibid., p. 27.

34. Ibid., pp. 31–32. A detailed discussion of this analysis is provided in U.S. Army Corps of Engineers, Mobile District, *An Evaluation of the Transportation Economics of the Tennessee-Tombigbee Waterway, Final Report. Volume II: Project Report*, report prepared by A. T. Kearney Management Consultants, Inc. (Mobile, AL: U.S. Army Corps of Engineers, Mobile District, April 1976). See also, Joseph L. Carroll and Chester L. Meade, *Review of Coal Forecasts on the Tennessee-Tombigbee Waterway*, Final Report to the Louisville and Nashville Railroad Co. (University Park, PA: Pennsylvania Transportation Institute, Pennsylvania State University, December 1977); and Robert F. Church and Charles J. Mundo, *Study Plan for Estimating the Impact on Railroads of the Tennessee-Tombigbee Waterway* (Cambridge, MA: U.S. Department of Transportation, Research and Special Programs Administrations, Transportation Systems Center, 15 May 1980).

35. U.S. Army Corps of Engineers, Mobile District, *An Evaluation of the Transportation Economics of the Tennessee-Tombigbee Waterway, Draft Final Report. Volume I: Executive Summary*.

36. [Maj. Gen. Ernest] Graves, memorandum to [Alex] Shwaiko, et al. (subj.: Tenn-Tom Cost/Benefit Reanalysis), 21 October 1975 (litigation/L&N v. U.S. 1984 folder, file 410–01, MDOC).

37. As quoted in Robert Crisp, disposition form to South Atlantic Division Engineer, 20 November 1975 (Exhibit #76, exhibit file, EDF v. Corps of Engineers, U.S. District Court for the Northern District of Mississippi, Oxford, MS).

38. See U.S. Congress, Senate, Subcommittee of the Committee on Appropriations, Hearings, *Public Works for Water and Power Development and Energy Research Appropriations for Fiscal Year 1977* (94th Cong., 2nd sess., 1976), Part 1, p. 172.

39. Victor V. Veysey, memorandum for Inspector General, 5 December 1975 (Tenn-Tom microfilm, set I, reel 1, frame 2334). See also, interview, author with Steven Dola, 9 January 1987.

40. U.S. Army Audit Agency, *Report of Audit: Tennessee-Tombigbee Waterway Project, U.S. Army Corps of Engineers,* Audit Report SO 76–408 (Atlanta, GA: U.S. Army Audit Agency, Southern District, 17 September 1976). A general account of the audit is presented in "Army Auditors Fault Corps on Tenn-Tom Estimating," *Engineering News-Record,* 200 (April 27, 1978), 11. See also, U.S. Army Corps of Engineers, Mobile District, *Briefing Data for Army Audit Agency: Tennessee-Tombigbee Waterway* (Mobile, AL: U.S. Army Corps of Engineers, Mobile District, January 1976).

41. U.S. Army Audit Agency, *Report of Audit: Tennessee-Tombigbee Waterway Project,* pp. 18–19.

42. Ibid., pp. 3–4.

43. Ibid., p. 11.

44. Ibid., p. 3. For similar criticisms of the Tenn-Tom, see Joseph L. Carroll and Srikanth Rao, "Economics of Public Investment in Inland Navigation: Unanswered Questions," *Transportation Journal,* 17 (Spring 1978), 27–54; and Joseph L. Carroll and Srikanth Rao, "The Role of Sensitivity in Inland Water Transport Planning: The Case of the Tennessee-Tombigbee," *Transportation Research Forum,* 19 (no. 1, 1978), 46–54.

45. See, for example, Brent Blackwelder, "Year End Report on Water Resource Boondoggles," Environmental Policy Center news release, 4 January 1974 (copy in TTWW litigation file, MDOC); and Randall Grace to Walter D. Huddleston, 18 July 1975 (TTWW litigation file, MDOC).

46. U.S. General Accounting Office, *Improvements Needed in Making Benefit-Cost Analyses for Federal Water Resources Projects* (Washington: U.S. General Accounting Office, 20 September 1974).

47. By 1989, neither the GAO Records Manager nor the GAO Historian were able to find the draft report in the agency's files, leading them to conclude that the document had been discarded. However, attorneys for the L&N Railroad and the Environmental Defense Fund subpoenaed GAO documents in their litigation of the Tenn-Tom in 1978, and they released samples of those documents—including the draft GAO report and related materials—to reporters. See, for example, Stewart Lytle, "Stennis Linked to Hiding of True Cost of Tenn-Tom," Birmingham (AL) *Post-Herald,* 1 May 1978.

48. Stennis later discussed his conversation with Staats at a press conference in Columbus, MS, on 13 December 1978. See Lt. Col. Donald R. Pope, memorandum for the record, 15 December 1978 (TTLMU/general reference folder, file 1501–07, SAMDL).

49. See Henry Eschwege to Secretary of the Army, 22 February 1977 (Tenn-Tom microfilm, set I, reel 57, frames 83–85). Additional discussion of these events are contained in Lawrence R. Green to Mobile District Engineer, 10 June 1977 (Tenn-Tom microfilm, set I, reel 56, frame 17); and Mobile District Office of Counsel, "Attorney Work Product," 1 September 1978 (TTLMU/litigation reference papers folder, file 401–07, SAMDL).

50. Quoted in Ward Sinclair, "$1.8 Billion Tennessee-Tombigbee Canal: Spurious Analysis Aids Water Project," *Washington Post,* 5 February 1979. Also reported in Ward Sinclair, "GAO Withheld Critical Report on Corps' Role in Barge Canal," *Washington Post,* 30 April 1978; Wayne King, "Documents Indicate Corps Misled Congress on Major Southern Canal," *New York Times,* 26 November 1978; Peter Kovacs and Brett Guge, "In Face of Critics, Lawsuit, Towering Cost, Tenn-Tom Has Potent Backers," Birmingham (AL) *News,* 30 November 1978; and Wayne King, "Army Engineers under Attack for Their 'Dig We Will' Zeal," *New York Times,* 4 February 1979. Representative Robert Edgar referred to this episode in a speech on the House floor: "Supplemental Appropriations and Rescission Bill," *Congressional Record—House,* 126 (June 18, 1980), 15327. See also Elmer B. Staats's letter to the editor, "That Missing G.A.O. Report," in *New York Times,* 26 December 1978.

51. U.S. General Accounting Office, *An Overview of Benefit-Cost Analysis for Water Resources Projects,* p. 1.

52. Ibid., p. 14. See also, Paul E. Roberts, "Benefit-Cost Analysis: Its Use (Misuse) in Evaluat-

ing Water Resource Projects," *American Business Law Journal*, 14 (Spring 1976), 73–84; and "Cost-Benefit Trips up the Corps," *Business Week*, no. 2573 (February 19, 1979), 96–97.

53. See "Who Will Finance Tenn-Tom Bridges Still in Question?," Birmingham (AL) *News*, 30 October 1974.

54. See Stephen Blaine Russell, "Tombigbee River Valley Water Management District and the Tennessee-Tombigbee Waterway," M.A. thesis, University of Mississippi, 1983, pp. 95–96.

55. See *ibid.*, pp. 93–115.

56. M. Barry Meyer, memorandum to James O. Eastland, 25 February 1975 (TTLMU/Counts III–IV folder, file 401–07, SAMDL).

57. "Federal-Aid Highway Act of 1975," *Congressional Record-House*, 121 (December 18, 1975), 41737. See also, Morris Cunningham, "Sleeper in Road Bill Offers States a Lift," Memphis (TN) *Commercial Appeal*, 13 January 1976; "Along Tenn-Tom Waterway: Measure Calls for $100 Million to Replace Bridges, Highways," Starkville (MS) *Daily News*, 14 January 1976; Ward Sinclair, "Legislation Contained $100 Million to Aid in Tombigbee Waterway Project," Louisville (KY) *Courier-Journal & Times*, 25 January 1976; and interview, John T. Greenwood with Jamie Whitten, 11 December 1986.

58. Environmental Policy Center fact sheet, "Federal-Aid Highway Act of 1975 & Local Cost Share on Corps of Engineers Navigation Projects," no date [probably late December 1975] (copy in TTWW litigation file, MDOC).

59. See Environmental Policy Center fact sheet, "House Inserts Provision Relieving Mississippi of Its Local Cost Share for the Tennessee-Tombigbee Waterway as Part of Federal-Aid Highway Act," 14 January 1976 (TTWW litigation file, MDOC).

60. Brent Blackwelder to James Lynn, 16 January 1976 (TTWW litigation file, MDOC).

61. Brock Evans (Sierra Club), Bill Painter (American Rivers Conservation Council), Brent Blackwelder (EPC), and Rafe Pomerance (Friends of the Earth) to Robert Jones, 25 January 1976 (statements folder, Tennessee-Tombigbee Waterway carton, Sierra Club Washington, D.C. Office Records, Bancroft Library, Berkeley, CA).

62. See E. Leo Koester to J. C. Laney, 13 February 1976 (TTWW litigation file, MDOC).

63. See R. K. Adams, memorandum for the record, no date [probably mid-February 1976] (TTLMU/Counts I and II folder, file 401–07, SAMDL); and Whitten interview, 11 December 1986.

64. Section 156 of the Federal Aid Highway Act of 1976, *Public Law* 94–280. See also, "Conference Report on H.R. 8235, Federal-Aid Highway Act," *Congressional Record-House*, 122 (April 13, 1976), 10811–10819; and "Federal-Aid Highway Act of 1976—Conference Report," *Congressional Record-Senate*, 122 (April 13, 1976), 10742–10748.

65. See Tombigbee River Conservation Council, *Tenn-Tom Review*, 4 (1978), 3.

66. See Glover Wilkins to Alabama and Mississippi Authority Members [marked "EXTREMELY CONFIDENTIAL"], 24 September 1976 (file 2–1000–926, GWTTW Archives); "Tenn-Tom Bridges Not Target of Ford's Funding Cutback," Columbus (MS) *Commercial Dispatch*, 27 September 1976; "Profiles in Pork: Tennessee-Tombigbee Waterway," Coalition for Water Project Review news release, 10 March 1980 (Water Resources Program files, National Wildlife Federation, Washington, D.C.); and Whitten interview, 11 December 1986.

67. George Reiger, "Let's Terminate the Tenn-Tom!," *Georgia Sportsman*, 2 (February 1977), 23.

68. Birmingham Audubon Society, "Resolution Opposing Uneconomic Water Projects, Including Tennessee-Tombigbee Waterway," 17 March 1977 (Tenn-Tom microfilm, set I, reel 57, frames 172–173).

69. Ray Blanton, George C. Wallace, Julian Carroll, Cliff Finch, and Glover Wilkins to Jimmy Carter, 22 November 1976 (file 2–1000–626, GWTTW Archives).

70. Jimmy Carter to the Congress of the United States, 21 February 1977 (folder SP–2–3–3, box SP–3, WHCF/Subject files, Jimmy Carter Library, Atlanta, GA). For the interactions of the Corps with the Carter Administration, see Jeffrey K. Stine, "Environmental Politics and Water

Resources Development: The Case of the Army Corps of Engineers during the 1970s," Ph.D. dissertation, University of California at Santa Barbara, 1984, pp. 170–207.

71. See Edward Walsh, "President Adds 14 Water Projects to 'Hit List'," *Washington Post,* 24 March 1977; and Douglas R. Sease, "$1.6 Billion Southern Canal on Carter Kill List Has Many Foes but Also Powerful Supporters," *Wall Street Journal,* 1 April 1977.

72. See Office of the White House Press Secretary, press release, 23 March 1977 (Water Resources Projects folder, box 94, Staff Offices [Press, Granum] file, Jimmy Carter Library); and Mobile District, "Announcement of Public Meeting on Tennessee-Tombigbee Waterway, Alabama and Mississippi," 23 March 1977 (Greenough—Mobile Harbor folder, file 18081, Mobile Municipal Archives, Mobile, AL). For a summary of Carter's water projects review, see Charles O. Jones, *The Trusteeship Presidency: Jimmy Carter and the United States Congress* (Baton Rouge: Louisiana State University Press, 1988), pp. 143–149; and Marc Reisner, *Cadillac Desert: The American West and Its Disappearing Water* (New York: Viking, 1986), pp. 317–333.

73. Glover Wilkins to "Dear Friend and Supporter," 17 March 1977 (Greenough—Mobile Harbor folder, file 18081, Mobile Municipal Archives). See also, Perry, "The Great Washington Pork Feast," p. 106; interview, author with Pat Ross (TTWDA), 27 May 1986; interview, author with Donald G. Waldon, 27 and 28 May 1986; and Glover Wilkins, "My Association with the Tennessee-Tombigbee Waterway Project," unpublished manuscript, Mobile District, U.S. Army Corps of Engineers, 1985, p. 150.

74. See Tombigbee River Conservation Council "Annual Report, '76–'77," no date (copy in TTWW litigation file, MDOC).

75. CLEAN newsletter, March 1977 (copy in Tenn-Tom files, Tom Bevill House of Representatives office papers, Washington, D.C.).

76. See Ernest W. Todd, Jr., and W. J. Hearin to Board of Directors, Tennessee-Tombigbee Advisory Committee, et al., 24 March 1977 (Greenough—Tenn-Tom folder, file 27078, Mobile Municipal Archives).

77. For an extended discussion of the public meeting and related events, see Currie interview, 15 May 1986; interview, author with Robert L. Crisp, Jr., 8 October 1986; Waldon interview, 27 and 28 May 1986; Ross interview, 27 May 1986; interview, author with Maj. Gen. Henry J. Hatch, 16 December 1987; and Johnny Greene, "The Corps of Engineers in Fantasy Land," *Inquiry,* 2 (May 14, 1979), 13–17. Examples of the numerous letters in support of the waterway from elected officials from all levels within Mississippi and Alabama can be found in NR7–1/Tenn-Tom folder, box NR–19, WHCF-Subject files, Jimmy Carter Library. See also, Senate of Alabama, Resolution, "Urging President Carter to Continue the Tennessee-Tombigbee Waterway Development Project," 22 March 1977; and James Free, "Bevill Denounces Carter's Cut in Water Projects," Birmingham (AL) *News,* 22 February 1977. For informal comments on the meeting by participants, see C. Ruel Ewing, Jr., to Clifford Alexander, 31 March 1977 (Tenn-Tom microfilm, set I, reel 31, frame 51); Sherrill M. Clemmer to Col. Henry Hatch, 31 March 1977 (Tenn-Tom microfilm, set I, reel 37, frame 125); and J. C. Laney to Col. Charlie Blalock, 2 August 1977 (litigation/1977 folder, file 410–01, MDOC; later published under the title "Tenn-Tom: More Faith than Fact," in the Anniston [AL] *Star,* 15 August 1977).

78. U.S. Army Corps of Engineers, Mobile District, *Public Meeting on the Tennessee-Tombigbee Waterway, Alabama and Mississippi, Held at Columbus, Mississippi, on 29 March 1977,* 5 volumes (Mobile, AL: U.S. Army Corps of Engineers, Mobile District 1977).

79. U.S. Congress, Senate, Subcommittee of the Committee on Appropriations, Hearings, *Public Works for Water and Power Development and Energy Research Appropriations for Fiscal Year 1978* (95th Cong., 1st sess., 1978), Part 7, p. 2318.

80. President Jimmy Carter, Statement on Water Projects, 18 April 1977 (copy in ALA/MISS—COE: TTWW folder, box 4, file 429–81–37, Council on Environmental Quality [CEQ] records, Washington National Records Center, Suitland, MD). See also, Walter Pincus, "Carter Narrows 'Hit List,' Seeks Halt to 17 Projects," *Washington Post,* 19 April 1977.

81. Randall Grace to President Jimmy Carter, 22 April 1977 (litigation-1977 folder, file 410–01,

MDOC). See also, "Public Works Appropriations," *Congressional Record—Senate*, 123 (June 30, 1977), 21806 and 21813.

82. Tennessee-Tombigbee Waterway Development Authority, "President Carter Approves Tenn-Tom" (Columbus, MS: Tennessee-Tombigbee Waterway Development Authority, 18 April 1977).

83. For a scathing attack on the project, see Coalition for Water Project Review, *Tennessee-Tombigbee Waterway: Background Report* (Washington: Coalition for Water Project Review, March 1979). By 1979, active membership in the Coalition for Water Project Review included: American League of Anglers, American Rivers Conservation Council, Coalition on American Rivers, Defenders of Wildlife, Environmental Action, Environmental Defense Fund, Environmental Policy Center, Environmentalists for Full Employment, Friends of the Earth, Fund for Animals, Izaak Walton League of America, National Audubon Society, National Wildlife Federation, Natural Resources Defense Council, Sierra Club, and Trout Unlimited.

84. See interview, author with Jill Lancelot, 6 August 1987; and Blackwelder interview, 12 April 1988. The growing dissatisfaction with the waterway among fiscal conservatives is addressed in Frank Morring, Jr., and Bill Stevens, "Many See Tenn-Tom Waterway as the Dream that Turned into a Nightmare," Birmingham (AL) *Post-Herald*, 24 January 1977.

85. James O. Eastland to Col. Charlie L. Blalock, 6 September 1977 (Tenn-Tom microfilm, set I, reel 1, frames 183–184). For the Senator's lobbying of President Carter, see James O. Eastland to the President, 27 October 1978 (file 2–1000–1049, GWTTW Archives).

86. For the background of this election, see James H. Kitchens's interview with Tom Bevill, 21 February 1985 (copy in Office of History archives, Corps Headquarters).

87. "The Tennessee-Tombigbee Waterway: A Lesson in Federal Mismanagement of Water Resources Development," *Congressional Record—Senate*, 124 (October 4, 1978), 33335. See also, James Nathan Miller, "Trickery on the Tenn-Tom," *Reader's Digest*, 113 (September 1978), 138–143. For Proxmire's earlier skepticism of the Tenn-Tom, see William Proxmire to E. Leo Koester, 25 April 1975 (TTW litigation file, MDOC).

88. Jimmy Carter to Clifford Alexander, 29 November 1978 (Water Policy/Decision Memo folder, box 313, DPS-Eizenstat file, Jimmy Carter Library).

89. Clifford L. Alexander, Jr., memorandum for the President, 5 December 1978 (Water Policy/Decision Memo folder, box 313, DPS-Eizenstat file, Jimmy Carter Library).

90. *Ibid*.

91. Stu Eizenstat, memorandum for the President, 6 December 1978 (Water Policy/Decision Memo folder, box 313, DPS-Eizenstat file, Jimmy Carter Library).

Chapter VIII/Social Justice: "A Cruel Hoax"

1. "West Alabama's Short Cut to Progress: Special Report, the Tenn-Tom Impact," *Alabama News Magazine*, 42 (June 1976), 18.

2. See Joseph J. Molnar, Leisle A. Ewing, Brenda Clark, and Macon Tidwell, *Developing the Tennessee-Tombigbee Corridor: Perceptions and Preferences of West Alabama Residents*, Rural Sociology Series No. 5 (Auburn, AL: Agricultural Experiment Station, Auburn University, 1981), p. 3.

3. William F. Brandes, "The Tennessee-Tombigbee Waterway Project v. The National Environmental Policy Act," term paper, EWRE 279, School of Engineering, Vanderbilt University, January 1975, p. 8. See also, Douglas C. Bachtel and Joseph J. Molnar, *Industrialization along the Tennessee-Tombigbee Waterway: Perceptions and Preferences of West Alabama Leaders*, Rural Sociology Series No. 4 (Auburn, AL: Agricultural Experiment Station, Auburn University, 1980); and "Time for Tenn-Tom?," *Chemical Week*, 106 (May 27, 1970), 23.

4. Tennessee-Tombigbee Waterway Development Authority, *The Tennessee-Tombigbee Waterway Story* (Columbus, MS: Tennessee-Tombigbee Waterway Development Authority, January 1969), p. 15.

5. Jack Edwards, news release, 21 January 1970 (Correspondence/January 1970 folder, file 2–1000–229, Glover Wilkins Tennessee-Tombigbee Waterway Archives, Archives and Museums

Department, Mississippi University for Women, Columbus, MS [hereafter referred to as "GWTTW Archives"]).

6. "Public Works and Atomic Energy Commission Appropriations, 1971," *Congressional Record—House*, 116 (June 24, 1970), 21248.

7. *Ibid.*, p. 21249.

8. Robert J. Brown, memorandum for Lt. Gen. F. J. Clarke, 14 July 1971 (Tenn-Tom microfilm, set I, reel 57, frames 644–645, Office of History archives, Corps Headquarters, Fort Belvoir, VA).

9. Thaddeus R. Beal to Robert J. Brown, 28 July 1971 (Tenn-Tom microfilm, set I, reel 57, frames 642–643).

10. See Bernard Garnett, "Black Judge Raps Foes of Waterway," *Race Relations Reporter*, 3 (January 17, 1972), 5–6. For a similar study of the perceived conflict between pollution and poverty in the Southeast during the early 1970s, see Oliver G. Wood, Jr., et al., "The BASF Controversy: Employment vs. Environment," *Essays in Economics*, no. 25 (November 1971), 1–75. The literature on environmental quality and social justice is growing rapidly. For the seminal studies of this relationship, see U.S. General Accounting Office, *Siting of Hazardous Waste Landfills and Their Correlation with Racial and Economic Status of Surrounding Communities* (Washington: U.S. General Accounting Office, 1983); Commission for Racial Justice, *Toxic Wastes and Race in the United States: A National Report on the Racial and Socio-Economic Characteristics of Communities with Hazardous Waste Sites* (New York: United Church of Christ, 1987); and Robert D. Bullard, *Dumping in Dixie: Race, Class, and Environmental Quality* (Boulder, CO: Westview Press, 1990).

11. Quoted in Garnett, "Black Judge Raps Foes of Waterway," p. 5. Branch was the first African-American probate judge to be elected in Alabama history.

12. Quoted in *ibid.*, p. 6. See also, Peter J. Bernstein, "Groundswell of Tenn-Tom Support Said Increasing," Mobile (AL) *Press Register*, 16 January 1972.

13. Quoted in Peter J. Bernstein, "Southern Hackles Raised over Tenn-Tom Injunction," Birmingham (AL) *News*, 16 January 1972.

14. Quoted in Garnett, "Black Judge Raps Foes of Waterway," p. 6.

15. Interview, author with James D. Williams, 11 February 1987. For an analysis of the poor job creation associated with Corps of Engineers water projects as compared to other federal programs, see Bruce M. Hannon and Roger H. Bezdek, "Job Impact of Alternatives to Corps of Engineers Projects," *Engineering Issues*, 99 (October 1973), 521–531.

16. See Federation of Southern Cooperatives Rural Training and Research Center, *Peoples Guide to Tennessee Tombigbee Waterway* (Epes, AL: Cooperator Press, 1975).

17. See South Atlantic Division, "Minority Business Enterprises Program," background paper, no date [probably summer 1977] (Tenn-Tom microfilm, set I, reel 31, frames 31–32); and U.S. Congress, Senate, Subcommittee of the Committee on Appropriations, Hearings, *Public Works for Water and Power Development and Energy Research Appropriations for Fiscal Year 1978* (95th Cong., 1st sess., 1978), Part 7, p. 2389. Also useful is interview, author with Tom P. Epps (Tenn-Tom Project Area Council), 27 May 1986.

18. See Col. Henry J. Hatch to "Prospective Bidder," 18 August 1976 (file 2–1000–942, GWTTW Archives). Resistance by the Associated General Contractors to the equal employment opportunities objectives on the Tenn-Tom is discussed in "AGC Challenges Government on EEO and Apprenticeship," *Engineering News-Record*, 205 (October 28, 1976), 41.

19. See Esther M. Harrison to Glover Wilkins, 26 August 1975 (file 2–1000–942, GWTTW Archives).

20. See minutes of the meetings of the Mississippi Economic Development Corporation/Tennessee-Tombigbee Division advisory board, 12 September 1975, 24 October 1975, and 25 March 1976; and Esther M. Harrison to Glover Wilkins, 20 November 1976 (file 2–1000–942, GWTTW Archives).

21. See Esther M. Harrison, Tennessee-Tombigbee Construction Assistance Center quarterly report, 1 July 1976 through 30 September 1976 (file 2–1000–942, GWTTW Archives).

22. See John Zippert to Col. [Harry A.] Griffith, 22 February 1973 (Tenn-Tom microfilm, set I, reel 65, frame 743); and Thomas A. Johnson, "Blacks Insist on Share of Canal Boom in South," *New York Times*, 12 October 1976.

23. Federation of Southern Cooperatives Rural Training and Research Center, *Peoples Guide to Tennessee Tombigbee Waterway*, pp. ii–iii.

24. *Ibid.*, pp. iii–v.

25. See Johnson, "Blacks Insist on Share of Canal Boom in South"; and "Equal Opportunity Employment on the Tennessee-Tombigbee Waterway Project," *Congressional Record—Extensions of Remarks* (June 15, 1976), 3351. For the general efforts of civil rights movement veterans during the 1970s to improve the economic conditions of Southern rural African-Americans, see Robert E. Anderson, Jr., *Poor, Rural, and Southern* (New York: Ford Foundation, 1978).

26. Mailgram, Wendell Paris to Joseph Evins, 6 May 1975 (James D. Williams papers, Arlington, VA [hereafter referred to as "Williams papers"]).

27. Testimony of Brent Blackwelder before the Senate Public Works Committee, 8 October 1975, typescript (TTWW litigation file, Office of Counsel, Mobile District, U.S. Army Corps of Engineers, Mobile, AL [hereafter referred to as "MDOC"]).

28. Victor V. Veysey to Senator Edward W. Brooke, 20 October 1975 (Tenn-Tom microfilm, set I, reel 57, frame 403).

29. Lamond Godwin to Glover Wilkins, 31 March 1976 (file 2–1000–942, GWTTW Archives).

30. Quoted in James Thompson, "Minorities May Oppose Waterway," Meridian (MS) *Star*, 31 January 1976. For similar statements, see "Corps' Employment Program Not Impressive to Blacks," Tupelo (MS) *Daily Journal*, 10 October 1977.

31. Glover Wilkins to Tom Bevill, 4 May 1977 (Tenn-Tom files, Tom Bevill House of Representatives office papers, Washington, D.C. [hereafter referred to as "Bevill papers"]).

32. See James Thompson, "Minority Group Seeks Assurances," Meridian (MS) *Star*, 24 October 1975; and John Zippert and Robert Valder, "The Tennessee Tombigbee Water Project—A 'White' Paper," report prepared under the auspices of the Minority Peoples Council on the Tennessee-Tombigbee Waterway and the NAACP Legal Defense and Educational Fund, August 1977, pp. 42–43.

33. Maj. Gen. Carroll N. LeTellier to Maj. Gen. Ernest Graves, 8 June 1976 (Tenn-Tom microfilm, set I, reel 31, frame 42).

34. Interview, author with Brig. Gen. Kenneth E. McIntyre, 14 November 1986.

35. Maj. Gen. Ernest Graves to Brig. Gen. Kenneth E. McIntyre, 7 July 1976 (Tenn-Tom microfilm, set I, reel 31, frame 41).

36. Howard H. Baker, Jr. to Victor V. Veysey, 29 March 1976 (Tenn-Tom microfilm, set I, reel 1, frames 1757–1758). Veysey's response, which detailed the Corps's program, is contained in Victor V. Veysey to Howard H. Baker, Jr., 2 June 1976 (Tenn-Tom microfilm, set I, reel 57, frames 142–144).

37. "Equal Opportunity Employment on the Tennessee-Tombigbee Waterway Project," *Congressional Record—Extensions of Remarks* (June 15, 1976), 3350–3351.

38. Victor V. Veysey to Augustus F. Hawkins, 15 July 1976 (Tenn-Tom microfilm, set I, reel 57, frame 136). See also, Victor V. Veysey to Augustus F. Hawkins, 18 October 1976 (Tenn-Tom files, Assistant Secretary of the Army for Civil Works—Pentagon Office files, Washington, D.C.).

39. Victor V. Veysey, memorandum for the Chief of Engineers, 27 September 1976 (Tenn-Tom microfilm, set I, reel 57, frame 124).

40. Zippert and Valder, "The Tennessee Tombigbee Water Project—A 'White' Paper," pp. 43–45.

41. *Ibid.*, pp. 45–46.

42. U.S. Army Corps of Engineers, Mobile District, news release, 18 August 1976 (copy in file 2–1000–943, GWTTW Archives).

43. "Special Bid Conditions, Affirmative Action Requirements, Equal Employment Opportunity," no date [probably August 1976] (copy in file 2–1000–942, GWTTW Archives).

44. Zippert and Valder, "The Tennessee Tombigbee Water Project—A 'White Paper,'" p. 48.

45. Charles R. Ford, memorandum for the Secretary of the Army, 16 May 1977 (Tenn-Tom microfilm, set I, reel 57, frame 181). Skilled workers were defined as operating engineers, plumbers/pipefitters, carpenters, iron workers, and cement masons.

46. W. F. McCraw, memorandum for the record, 2 June 1977 (Tenn-Tom files, Executive Office, Nashville District, U.S. Army Corps of Engineers, Nashville, TN). See also, Steven Riesenmy, "More Local, Minority Jobs Demanded," Tupelo (MS) *Daily Journal,* 2 June 1977; "Corps Hiring Policy Attacked," Meridian (MS) *Star,* 16 June 1977; and "Lack of Local Labor in Tenn-Tom Project Drawing Complaints," *South Mississippi Sun,* 27 June 1977.

47. Victor V. Veysey, memorandum for the Chief of Engineers, 27 September 1976 (Tenn-Tom microfilm, set I, reel 57, frames 125–126).

48. For examples of Alexander's exposure to the arguments of the African-American critics of the Tenn-Tom, see Wendell H. Paris to Clifford Alexander, 30 March 1977 (Tenn-Tom microfilm, set I, reel 57, frame 863); Vernon E. Jordan, Jr., to Clifford Alexander, 1 April 1977 (Tenn-Tom microfilm, set I, reel 57, frame 858); telegram, Director of NAACP Legal Defense and Education Fund to Secretary of the Army, 5 April 1977 (Tenn-Tom microfilm, set I, reel 57, frame 868); and Wendell Paris to Clifford L. Alexander, Jr., 10 May 1977 (Tenn-Tom microfilm, set I, reel 57, frames 845–846).

49. McIntyre interview, 14 November 1986.

50. Zippert and Valder, "The Tennessee Tombigbee Water Project—A 'White' Paper," p. 7.

51. *Ibid.,* pp. 10–16.

52. See Clifford L. Alexander, Jr., memorandum for the Acting Assistant Secretary of the Army for Civil Works, 28 April 1977 (Tenn-Tom microfilm, set I, reel 57, frame 186); Charles R. Ford, memorandum for the Chief of Engineers, 16 May 1977 (Tenn-Tom microfilm, set I, reel 57, frame 180); and interview, author with Roland E. Blanding (Affirmative Action Office, South Atlantic Division), 7 October 1986.

53. Jack H. Watson, Jr., to Clifford Alexander, 23 October 1978 (Tenn-Tombigbee/NR7–1 folder, box NR–19, WHCF-Subject files, Jimmy Carter Library, Atlanta, GA [hereafter referred to as "Carter Library"]).

54. Jack Watson and Larry Gilson, memorandum for members of the Interagency Coordinating Council, 3 January 1979 (Executive/FG–14–3 folder, box FG–109, WHCF-Subject files, Carter Library).

55. See White House Rural Development Initiatives, *Area Development from Large-Scale Construction: Planning and Implementation Guidelines* (n.p.: January 1980), p. I–3; and Leslie G. Range to Donald G. Waldon, 16 July 1980 (Tenn-Tom files, Bevill papers).

56. Jack H. Watson, Jr., to Clifford L. Alexander, Jr., 5 February 1980 (Executive/FG–14 folder, box FG–109, WHCF-Subject files, Carter Library). See also, Jack H. Watson, Jr., to Jack Edwards, 25 February 1980 (Tenn-Tom 1980 folder, file 228/20–3, Jack Edwards papers, University of South Alabama Archives, Mobile, AL [hereafter referred to as "Edwards papers"]); and Tom Bevill and John J. Rhodes to Dear Colleague, 23 June 1980 (Tenn-Tom 1980 folder, file 228/20–3, Edwards papers).

57. See Federation of Southern Cooperatives, Proposal to Provide Travel Assistance to Involve Low Income People in Policy Making Boards and Committees on the Tennessee Tombigbee Waterway, March 1981 (file 2–1000–1177, GWTTW Archives); and Thomas Till and Michael Welch, *A White House Rural Initiative in Transition: The Impact of Large-Scale Construction Projects on Rural Economic Development,* Final Report Prepared for the White House and the U.S. Community Services Administration (Washington: The National Rural Center, June 1981).

58. See Blanding interview, 7 October 1986.

59. Donald G. Waldon to Wendell Paris, 11 July 1980 (Minorities/1980 folder, Tennessee-Tombigbee Waterway Development Authority office files, Columbus, MS). For a discussion of events in the early 1980s, see interview, author with Nick J. Etheridge III (Small and Disadvantaged Business Utilization Office, Mobile District), 14 May 1986; Epps interview, 27 May 1986;

Eaton interview, 30 May 1986; and interview, author with William Hill (Office of Counsel, Nashville District), 19 September 1986.

60. Wendell Paris to Tom Bevill, 23 July 1980 (file 2–1000–1177, GWTTW Archives). See also, Donald G. Waldon to Wendell Paris, 31 July 1980 (Minorities/1980 folder, Tennessee-Tombigbee Waterway Development Authority office files); and Tom Bevill to Wendell Paris, 1 August 1980 (file 2–1000–1177, GWTTW Archives).

61. See interview, James H. Kitchens III with Louis Stokes, 26 February 1985 (transcript in Office of History archives, Corps Headquarters, Fort Belvoir, VA).

62. Maj. Gen. John F. Wall, "Affirmative Action on the Tenn-Tom," speech before Tenn-Tom contractors, Nashville, TN, 10 November 1982 (TT Contractors' Meeting/FY83 folder, file 601–07, Affirmative Action Office, Mobile District).

63. Judy Grande, "Should Stokes Have Backed 'Boondoggle' Canal?," Cleveland *Plain Dealer*, 22 August 1982.

64. Quoted in *ibid.*

65. Brent Blackwelder to Louis Stokes, 17 August 1982 [identical letters sent to all members of the Congressional Black Caucus] (Tenn-Tom files, Environmental Policy Center, Washington, D.C.).

66. Louis Stokes to Katie Hall, 8 December 1982 (Tenn-Tom files, Bevill papers). See also, Esther M. Harrison to Brent Blackwelder, 16 September 1982 (Tenn-Tom files, Environmental Policy Center).

67. Wall, "Affirmative Action on the Tenn-Tom."

68. Stephen R. Wilson, John Zippert, and Wendell Paris, *History and Accomplishments of Affirmative Action during the Construction of the Tennessee-Tombigbee Waterway* (Mobile, AL: U.S. Army Corps of Engineers, Mobile District, May 1986), p. ii. Detailed statistical tables for the total hours worked by minorities and women, broken down by occupation, are provided in Appendix 7 of this report, while Appendix 8 provides data on minority contracts and subcontracts on the Tenn-Tom.

69. For the African-American community's deflated expectations for the Tenn-Tom, see Molnar et al., *Developing the Tennessee-Tombigbee Corridor*, p. 88.

Chapter IX/In Court—Act II

1. For a detailed description of the litigation planning debates, see James T. B. Tripp, memorandum to Jon T. Brown, 15 October 1976 (Tenn-Tom correspondence/1975–1977 folder, box 20, Environmental Defense Fund papers, Department of Special Collections, State University of New York at Stony Brook).

2. See *Complaint for Declaratory and Injunctive Relief,* Louisville and Nashville Railroad v. Martin R. Hoffman, et al., U.S. District Court for the District of Columbia, 30 November 1976; and *Complaint for Declaratory and Injunctive Relief,* Environmental Defense Fund, Inc., et al. v. Martin R. Hoffman, et al., U.S. District Court for the District of Columbia, 30 November 1976.

3. *Ibid.* Except for descriptions of the plaintiffs and their direct interest in the suit, the wording of the two complaints was identical. For the Corps's descriptions of plans *A, B,* and *C,* see U.S. Army Corps of Engineers, Mobile District, *An Evaluation of the Transportation Economics of the Tennessee-Tombigbee Waterway, Final Report. Volume I: Executive Summary,* report prepared by A. T. Kearney Management Consultants, Inc. (Mobile, AL: U.S. Army Corps of Engineers, Mobile District, February 1976), pp. 6–9.

4. *Complaint,* L&N Railroad v. Hoffman. See also, *Brief of Appellants,* Environmental Defense Fund, et al. v. Clifford Alexander, et al., U.S. Court of Appeals for the Fifth Circuit, 10 September 1979.

5. *Complaint,* L&N Railroad v. Hoffman.

6. *Ibid.* See also, Philip W. Smith, "2 Suits Filed against Tenn-Tom Construction," Mobile (AL) *Register,* 2 December 1976; and the testimony of Brig. Gen. Kenneth E. McIntyre in U.S. Congress, Senate, *Public Works for Water and Power Development and Energy Research Appropriations for Fiscal Year 1978* (95th Cong., 1st sess., 1977), Part 2, pp. 1450–1451.

7. See Jeffrey K. Stine, "Environmental Politics and Water Resources Development: The Case of the Army Corps of Engineers during the 1970s," Ph.D. dissertation, University of California at Santa Barbara, 1984; Lettie M. Wenner, *The Environmental Decade in Court* (Bloomington: Indiana University Press, 1982); and Jeffrey G. Miller, *Citizen Suits: Private Enforcement of Federal Pollution Control Laws* (New York: John Wiley & Sons, 1987).

8. *Atchison, Topeka and Santa Fe Railway Co., et al. v. Callaway, et al.*, 382 F. Supp. 610 (D.C. 1974).

9. For the parallels between the Locks and Dam 26 case and the Tenn-Tom litigation, see David H. Webb, "Preliminary Analysis of the Two Lawsuits," no date [probably early 1977] (copy in TTW litigation file, Office of Counsel, Mobile District, U.S. Army Corps of Engineers [hereafter referred to as "MDOC"]).

10. See J. A. Austin to R. S. Hamilton, H. Emerson, and H. J. Bruce, 26 July 1976 (TTWW litigation file, MDOC). The Missouri Pacific Railroad Company did not join the Tenn-Tom litigation.

11. Interview, author with Lance D. Wood (Office of Counsel, Corps Headquarters), 19 May 1987.

12. See interview, author with F. L. (Les) Currie, 15 May 1986.

13. "L&N Greed Motivates Tenn-Tom Challenge," Mobile (AL) *Register*, 3 December 1976. Not all regional newspapers agreed. For the opposite view, see the series of articles in the Birmingham (AL) *Post-Herald* by Frank Morring, Jr., and Bill Steverson: "Many See Tenn-Tom Waterway as the Dream that Turned into a Nightmare," 24 January 1977; "Whether Tenn-Tom Pays off Is $2-Billion Question," 25 January 1977; "Tombigbee Is Too Valuable to Alton, Tenn-Tom Foes Say," 26 January 1977; and "Waterway's Foes Get Aid from L&N," 27 January 1977.

14. Sid McDonald to Prime F. Osborn, 22 January 1977 (TTWW litigation file, MDOC). McDonald sent copies of this letter to the governors and Congressional delegations of Alabama, Mississippi, Tennessee, and Kentucky. See also, "Lawmakers Seek Dismissal of Anti-Waterway Suit," Columbus (MS) *Commercial Dispatch*, 17 February 1977.

15. Prime F. Osborn to Sid McDonald, 8 February 1977 (TTWW litigation file, MDOC). Osborn sent copies of this letter to the same list used by McDonald. See also, Prime F. Osborn to Jack Edwards, 14 December 1976 (Tenn-Tom 1976 folder, file 228/20–3, Jack Edwards papers, University of South Alabama Archives, Mobile, AL); City of Mobile, "Resolution: Tennessee-Tombigbee Waterway," 21 December 1976 (Greenough—Mobile Harbor folder, file 18081, Mobile Municipal Archives, Mobile, AL); and Prime F. Osborn to Gary A. Greenough, 12 January 1977 (Greenough—Mobile Harbor folder, file 18081, Mobile Municipal Archives).

16. See interview, author with William L. Robertson (Office of Counsel, Corps Headquarters), 9 June 1986; Wood interview, 19 May 1987; and interview, author with Lt. Gen. John W. Morris (Chief of Engineers, 1976–1980), 15 December 1987.

17. See Affidavit of David H. Webb, 8 September 1977 (litigation/1977 folder, file 410–01, MDOC); and Robertson interview, 9 June 1986.

18. John H. Gullett to Hunter Gholson, 29 December 1976 (Tenn-Tom files, National Waterways Conference, Washington, D.C.).

19. Joseph R. Hartley to Glover Wilkins, et al., 24 February 1976 (file 2–1000–239, Glover Wilkins Tennessee-Tombigbee Waterway Archives, Archives and Museums Department, Mississippi University for Women, Columbus, MS [hereafter referred to as "GWTTW Archives"]). The Tombigbee River Valley Water Management District, the State of Alabama, and the Tombigbee Valley Development Authority joined the TTWDA as defendant-intervenors. Later, the Alabama Conservancy, the Birmingham Audubon Society, and the National Audubon Society entered the litigation as plaintiff-intervenors. See Stephen E. Roady to Guy Arello, 7 May 1980 (TAC/Tenn-Tom file, Alabama Conservancy, Birmingham, AL).

20. See interview, author with Hunter Gholson, 28 May 1986; and transcript of Hunter Gholson's "Lawsuit Update" to the TTWDA Board of Directors, no date [probably November 1978] (Authority memorandums/Jan–Dec 1978 folder, file 2–1000–357, GWTTW Archives).

21. Gholson interview, 28 May 1986; Currie interview, 15 May 1986; interview, author with Nathaniel D. (Skeeter) McClure, 16 May 1986; Robertson interview, 9 June 1986; and Wood interview, 19 May 1987.

22. Transcript of Gholson's "Lawsuit Update."

23. Ibid.; *Affidavit of Thomas S. Martin*, Environmental Defense Fund, et al. v. Clifford Alexander, et al., U.S. District Court for the Northern District of Mississippi, 11 September 1978; and Robertson interview, 9 June 1986. Christenbury left the Department of Justice in August 1979 and was replaced by Glenn V. Whitaker. See *Praecipe*, Environmental Defense Fund, et al. v. Clifford Alexander, et al., U.S. District Court for the Northern District of Mississippi, 14 August 1979.

24. See *Order Consolidating Actions*, Environmental Defense Fund, et al. v. Martin Hoffman, et al., U.S. District Court for the District of Columbia, 21 December 1976.

25. See *Order*, Environmental Defense Fund, et al. v. Martin Hoffman, et al., U.S. District Court for the District of Columbia, 31 March 1977.

26. *Plaintiffs' Request for Production of Documents* and *Plaintiffs' Interrogatories to Defendants*, Environmental Defense Fund, et al. v. Clifford Alexander, et al., U.S. District Court for the Northern District of Mississippi, 1 July 1977. The plaintiffs subsequently filed five more requests for production of documents with the U.S. District Court for the Northern District of Mississippi: on 9 September 1977, 28 July 1978, 13 July 1979, 8 July 1980, and 11 August 1980.

27. *Federal Defendants' Answers and Objections to Plaintiffs' Interrogatories (First Set)*, Environmental Defense Fund, et al. v. Clifford Alexander, et al., U.S. District Court for the Northern District of Mississippi, 11 October 1977. On the overall discovery effort, see interview, author with Stephen E. Roady (attorney for the plaintiffs), 20 May 1987.

28. See *Points and Authorities in Support of Federal Defendants' Motion to Enforce Discovery Agreement, or, in the Alternative, for Sanctions*, Environmental Defense Fund, et al. v. Clifford Alexander, et al., U.S. District Court for the Northern District of Mississippi, 30 June 1978; *Plaintiffs' Memorandum in Response to Federal Defendants' Motion to Enforce a So-Called "Discovery Agreement,"* Environmental Defense Fund, et al. v. Clifford Alexander, et al., U.S. District Court for the Northern District of Mississippi, 3 July 1978; and interview, author with Willis E. Ruland, 19 November 1986. Those microfilm reels are now held by the Office of History archives, Corps Headquarters, Fort Belvoir, VA.

29. *Amended Complaint*, Environmental Defense Fund, et al. v. Clifford Alexander, et al., U.S. District Court for the Northern District of Mississippi, 26 January 1978.

30. *Ibid.*

31. *Ibid.* The charge of illegally widening the channel from 170 to 300 feet was not made in the initial complaint filed in November 1976, because that information was only uncovered during the course of discovery.

32. *Ibid.*

33. *Ibid.*

34. *Ibid.*

35. *Ibid.*

36. Robertson interview, 9 June 1986.

37. See *Environmental Defense Fund, Inc. v. Alexander*, 467 F. Supp. 885, 887 (N.D. Miss. 1979).

38. Maj. Gen. F. J. Clarke to Special Assistant to the Secretary of the Army for Civil Functions, 17 February 1967 (Tenn-Tom microfilm, set I, reel 66, frame 631).

39. Stanley R. Resor, memorandum for the Chief of Engineers, 30 March 1967 (Defendants' Exhibit No. 1, volume 8, EDF v. Alexander, U.S. District Court for the Northern District of Mississippi, Oxford, MS).

40. See *Amended Complaint*, Environmental Defense Fund, et al. v. Clifford Alexander, et al., U.S. District Court for the Northern District of Mississippi, 26 January 1978.

41. Affidavit of Richard A. Hertzler, 14 July 1978 (TTLMU/Counts I–XI folder, file 401–07, Tenn-Tom Litigation Management Unit, Mobile District [hereafter referred to as "SAMDL"]).

See also, "Stewart Udall Says He's Shocked at Corps Chief's Waterway Data," *New York Times*, 5 February 1979.

42. See *Brief of Appellants*, Environmental Defense Fund, et al. v. Clifford Alexander, et al., U.S. Court of Appeals for the Fifth Circuit, 10 September 1979. Nor was Resor's memorandum mentioned in the environmental impact statement.

43. CLEAN newsletter, December 1976 (copy in TTWW litigation file, MDOC).

44. "The Engineering of Deceit," *New York Times*, 4 December 1978. For other examples, see "Congress Should Ditch Tennessee-Tombigbee Waterway," Louisville (KY) *Times*, 5 May 1978; James Nathan Miller, "Trickery on the Tenn-Tom," *Reader's Digest*, 113 (September 1978), 138–143; Wayne King, "Documents Indicate Corps Misled Congress on Major Southern Canal," *New York Times*, 26 November 1978; Peter Kovacs and Brett Guge, "The Tenn-Tom Waterway: Ten Years of Controversy," Birmingham (AL) *News*, 27 November 1978; "The Cost of Public Works," Sacramento (CA) *Bee*, 5 December 1978; Wayne King, "Army Engineers under Attack for Their 'Dig We Will' Zeal," *New York Times*, 4 February 1979; Johnny Greene, "The Corps of Engineers in Fantasy Land," *Inquiry*, 2 (May 14, 1979), 13–17; Marvin Bailey, "Dreams on the Tenn-Tom," Memphis (TN) *Commercial Appeal*, 27 June 1979; and James J. Kilpatrick, "Why this $3-billion Boondoggle?," *Washington Star*, 29 July 1980.

45. "The Tennessee-Tombigbee Waterway Project," *Congressional Record—House*, 125 (February 28, 1979), 3489.

46. See Wood interview, 19 May 1987; Robertson interview, 9 June 1986; and Alfred Holmes, disposition form, "Response DF to Readers Digest Article—Tenn-Tom," 1 September 1978 (TTLMU/litigation reference papers folder, file 401–07, SAMDL).

47. Glover Wilkins to Sam Green, 29 August 1978 (TTLMU/general reference folder, file 1501–07, SAMDL). See also, Miller, "Trickery on the Tenn-Tom."

48. Tennessee-Tombigbee Waterway Development Authority, "Facts and Information on the Tennessee-Tombigbee Waterway and Its Opponents' Allegations," 5 April 1979 (copy in TTLMU/Sen. Nelson folder, file 401–07, SAMDL).

49. *Environmental Defense Fund, Inc. v. Alexander*, 467 F. Supp. 885 (N.D. Miss. 1979).

50. 467 F. Supp. at 906.

51. 467 F. Supp. at 806, 910–911.

52. 467 F. Supp. at 886.

53. 467 F. Supp. at 908.

54. *Petition for Writ of Mandamus*, Environmental Defense Fund, et al. v. Clifford Alexander, et al., U.S. Court of Appeals for the Fifth Circuit, 22 March 1979. See also, Roady interview, 20 May 1987.

55. *Brief of Appellants*, Environmental Defense Fund, et al. v. Clifford Alexander, et al., U.S. Court of Appeals for the Fifth Circuit, 10 September 1979.

56. *Environmental Defense Fund, Inc. v. Alexander*, 467 F. Supp. 885 (N.D. Miss. 1979), aff'd, 614 F. 2d 474 (5 Cir. 1980).

57. *Environmental Defense Fund, Inc. v. Alexander*, 467 F. Supp. 885 (N.D. Miss. 1979), aff'd, 614 F. 2d 474 (5th Cir. 1980), cert. denied, 49 USLW 3289 (Oct. 21, 1980). See also, Jon T. Brown and Stephen E. Roady to Guy Arello, 29 May 1980 (TAC/Tenn-Tom file, Alabama Conservancy).

58. See *Plaintiffs' Pre-Trial Conference Memorandum*, Environmental Defense Fund, et al. v. Clifford Alexander, et al., U.S. District Court for the Northern District of Mississippi, 28 May 1980; and *Memorandum of Authorities in Support of the Plaintiffs' Renewed Motion for Summary Judgment*, Environmental Defense Fund, et al. v. Clifford Alexander, et al., U.S. District Court for the Northern District of Mississippi, 3 July 1980.

59. *Memorandum Opinion*, Environmental Defense Fund, et al. v. Clifford Alexander, et al., U.S. District Court for the Northern District of Mississippi, 1 October 1980.

60. *Memorandum in Support of Federal Defendants' Motion for Judgment on the Pleadings, or in the Alternative, for Summary Judgment*, Environmental Defense Fund, et al. v. Clifford Alexander, et al., U.S. District Court for the Northern District of Mississippi, 2 July 1980.

61. *Environmental Defense Fund, Inc. v. Alexander,* 501 F. Supp. 742 (N.D. Miss. 1980).
62. *Memorandum of Authorities in Support of Plaintiffs' Motion for Injunction Pending Appeal,* Environmental Defense Fund, et al. v. Clifford Alexander, et al., U.S. District Court for the Northern District of Mississippi, 26 November 1980.
63. *Order,* Environmental Defense Fund, et al. v. Clifford Alexander, et al., U.S. Court of Appeals for the Fifth Circuit, 23 January 1981.
64. *Environmental Defense Fund, Inc. v. Marsh,* 651 F. 2d 983 (5th Cir. 1981).
65. *Ibid.*
66. See Wood interview, 19 May 1987.
67. Gholson interview, 28 May 1986; Robertson interview, 9 June 1986; Wood interview, 19 May 1987; and *Petition of Defendants-Intervenors-Appellees for Panel Rehearing,* Environmental Defense Fund, et al. v. John Marsh, et al., U.S. Court of Appeals for the Fifth Circuit, 24 July 1981.
68. *On Petition for Rehearing,* Environmental Defense Fund, et al. v. Clifford Alexander, et al., U.S. Court of Appeals for the Fifth Circuit, 17 August 1981.
69. *Motion for Reformation of Order on Petition for Rehearing and Mandate,* Environmental Defense Fund, et al. v. John Marsh, et al., U.S. Court of Appeals for the Fifth Circuit, 19 August 1981. (Emphasis in original.)
70. *First Order on Remand,* Environmental Defense Fund, et al. v. Clifford Alexander, et al., U.S. District Court for the Northern District of Mississippi, 2 September 1981.
71. See N. D. McClure, disposition form to Mobile District Engineer, et al., 3 August 1981 (TTLMU/general reference folder, file 1501–07, SAMDL).
72. See *Brief of Appellee, Tennessee-Tombigbee Waterway Development Authority, Defendant-Intervenor,* 26 October 1982.
73. Interview, author with Daniel R. Burns (Construction Division, Mobile District), 22 May 1986; Robertson interview, 9 June 1986; Currie interview, 15 May 1986; McClure interview, 16 May 1986; Roady interview, 20 May 1987; and Wood interview, 19 May 1987. For the development of the Corps's initial bargaining positions on each aspect of the waterway, including its "fall back" positions, see Skeeter McClure, memorandum on "TTW Potential Impact of Proposed Injunction on Contracts Underway," 30 July 1981 (TTLMU/general reference folder, file 1501–07, SAMDL).
74. "Comments of the Louisville and Nashville Railroad Company on the Draft Supplement to the Environmental Impact Statement for the Tennessee-Tombigbee Waterway," 3 February 1982 (copy in L&N v. U.S. litigation folder, file 410–01, MDOC). For the Environmental Defense Fund's critique of the Tenn-Tom DSEIS, see James T. B. Tripp to Col. Robert H. Ryan, 3 February 1982 (Supplement & Final EIS folder, file 1518–01, Inland Environment Section, Planning Division, Mobile District [hereafter referred to as "SAMPD-EI"]).
75. U.S. Army Corps of Engineers, Mobile and Nashville Districts, *Final Supplement to the Environmental Impact Statement: Tennessee-Tombigbee Waterway, Alabama and Mississippi Navigation,* volume I—Environmental Impact Statement, volume II—Appendices (Mobile, AL, and Nashville, TN: U.S. Army Corps of Engineers, Mobile and Nashville Districts, April 1982).
76. *Plaintiffs' Motion for Continuation of Work Suspension,* Environmental Defense Fund, et al. v. John Marsh, et al., U.S. District Court for the Northern District of Mississippi, 28 June 1982.
77. *Plaintiffs' Objections to the Final Supplement to the Environmental Impact Statement for the Tennessee-Tombigbee Waterway,* Environmental Defense Fund, et al., v. John Marsh, et al., U.S. District Court for the Northern District of Mississippi, 28 June 1982.
78. Maj. Gen. E. R. Heiberg III, "Record of Decision: Tennessee-Tombigbee Waterway, Alabama and Mississippi Navigation," 30 June 1982 (copy in Record of Decision folder, file 1501–07, SAMDL).
79. *Plaintiffs' Response to Federal Defendants' Motion for Protective Order and Motion in Limine,* Environmental Defense Fund, et al. v. John Marsh, et al., U.S. District Court for the Northern District of Mississippi, 28 February 1983.
80. *Memorandum Order,* Environmental Defense Fund, et al. v. Clifford Alexander, et al.,

U.S. District Court for the Northern District of Mississippi, 20 July 1982. The Fifth Circuit Court of Appeals affirmed Keady's decision in December; see *Judgment,* Environmental Defense Fund, et al. v. John O. Marsh, et al., U.S. Court of Appeals for the Fifth Circuit, 15 December 1982. *Memorandum Order,* Environmental Defense Fund v. John D. Marsh, et al., U.S. District Court for the Northern District of Mississippi, 9 May 1983, formally ended the litigation. Judge Keady ordered the record of this case kept at the U.S. District Court in Oxford, Mississippi, for a period of 20 years. In so instructing, Keady was responding to advice given him by Norman Gillespie, the Clerk of the U.S. District Court for the Northern District of Mississippi. Gillespie, like Keady, was a native Mississippian, and he argued successfully that this case had historical significance and its record should therefore be preserved. Interview, Jo Ann McCormick with Norman Gillespie, 7 May 1986.

Chapter X/The Last Congressional Challenge

1. See "DRAFT—Tenn-Tom Deauthorization Bill," no date [probably early March 1979] (Tenn-Tom Deauthorization Bill folder, box 90, file M80–626, Gaylord Nelson papers, State Historical Society of Wisconsin, Madison, WS [hereafter referred to as "Nelson papers"]). Also reported in Ward Sinclair, "Sen. Nelson Says He'll Try Again to Kill Waterway," *Washington Post,* 11 February 1979.

2. See memorandum, Rebecca [Wodder] to Jeff [Nedelman], 7 February 1979 (Tenn-Tom Deauthorization Bill folder, box 90, file M80–626, Nelson papers).

3. Interview, author with Rebecca R. Wodder, 15 April 1988. Also useful are memorandums of 8, 20, and 21 March 1979 from Randall Grace to Rebecca Wodder; Steve Roady to Rebecca Wodder, 5 July 1979 and 18 September 1979; notes from Tenn-Tom meeting, 14 May 1980 (all items in Tenn-Tom Misc. folder, box 90, file M80–626, Nelson papers); and interview, author with Brent Blackwelder, 12 April 1988.

4. Office of Senator Gaylord Nelson, press release, "Nelson Calls for End to Tennessee-Tombigbee 'Boondoggle,'" 10 February 1979 (Tenn-Tom Hearings folder, box 90, file M80–626, Nelson papers).

5. "Tennessee-Tombigbee Waterway Deauthorization Act of 1979," *Congressional Record—Senate,* 125 (March 27, 1979), 3471–3476. See also, Office of Senator Gaylord Nelson, press release, "Nelson Introduces Bill to Kill Tenn-Tom Waterway," 27 March 1979 (Tenn-Tom Hearings folder, box 90, file M80–626, Nelson papers); and Gaylord Nelson to President Jimmy Carter, 27 March 1979 (Tennessee-Tombigbee Correspondence folder, box 90, file M80–626, Nelson papers).

6. "The Attempt to Deauthorize the Tennessee-Tombigbee Waterway," *Congressional Record—Senate,* 125 (April 4, 1979), 3935–3936.

7. Quoted in "Vow for Anti-Waterway Bill Gets Fiery Blasts from South," Memphis (TN) *Commercial Appeal,* 12 February 1979.

8. Interview, author with Gaylord Nelson, 15 April 1988.

9. Senate Committee on Environment and Public Works, news release, 12 April 1979 (Tenn-Tom folder, box 90, file M80–626, Nelson papers).

10. Quoted in Ward Sinclair, "Stennis Advice Kills Hearing on Waterway," *Washington Post,* 12 May 1979. As Sinclair reported: "Gravel denied that Stennis had threatened or suggested that the Appropriations Committee, of which he is a senior member, would stop the funding of any of Gravel's pet projects in Alaska."

11. Gaylord Nelson to Jennings Randolph, 11 May 1979 (Tenn-Tom file, Gaylord Nelson personal papers, Washington, D.C.).

12. Gaylord Nelson to Mike Gravel, 23 May 1979 (Tenn-Tom Hearings folder, box 90, file M80–626, Nelson papers). For an elaboration of these points, see Gaylord Nelson to Gary Hart, 23 May 1979 (Tenn-Tom file, Gaylord Nelson personal papers); and Nelson interview, 15 April 1988.

13. See unsigned staff working notes, "Press Release announcing hearings of SBC on Tenn-

Tom for release on July 24, 1979," no date [probably mid-July 1979] (Tenn-Tom Hearings/Aug. 16–17, 1979 folder, box 90, file M80–626, Nelson papers); and Nelson interview, 15 April 1988. Nelson speculated that Gravel backed down because Stennis had strong-armed him.

14. See unsigned staff working notes, "Hearing Outline for the Tenn-Tom Waterway," no date [probably mid-July 1979] (Tenn-Tom Hearings/Aug. 16–17, 1979 folder, box 90, file M80–626, Nelson papers).

15. Unsigned staff working notes, "Tenn-Tom Hearings by the Senate Committee on Small Business," no date [probably mid-July 1979] (Tenn-Tom Hearings/Aug. 16–17, 1979 folder, box 90, file M80–626, Nelson papers).

16. Jon T. Brown to Jeff Nedelman, 26 June 1979 (Tennessee-Tombigbee Correspondence folder, box 90, file M80–626, Nelson papers).

17. Gaylord Nelson to Mike Gravel, 30 July 1979 (Tenn-Tom Misc. folder, box 90, file M80–626, Nelson papers).

18. Jennings Randolph to Gaylord Nelson, 18 September 1979 (Tenn-Tom file, Nelson personal papers).

19. Nelson interview, 15 April 1988; and Wodder interview, 15 April 1988.

20. Gaylord Nelson to Alan Simpson, 13 March 1980 (Tenn-Tom folder, box 90, file M80–626, Nelson papers). Bills that die in committee, such as S. 769, are often reintroduced as amendments to other legislation. Nelson also lobbied President Carter again, and was again rebuffed. See Gaylord Nelson to President Jimmy Carter, 13 March 1980 (Tenn-Tom Floor Statement folder, box 90, file M80–626, Nelson papers); and Hubert L. Harris, Jr., to Gaylord Nelson, 18 April 1980 (Tenn-Tom Misc. folder, box 90, file M80–626, Nelson papers).

21. William S. Cohen to Gaylord Nelson, 8 May 1980 (Tenn-Tom Misc. folder, box 90, file M80–626, Nelson papers). For the growing opposition to the Tenn-Tom within the House of Representatives, see "Water Resources Development Act of 1979," *Congressional Record—House*, 126 (January 29, 1980), 976–980.

22. See Office of the Chief of Engineers, news release, "Army Corps of Engineers Suspends Award of Civil Works Construction and Maintenance Contracts," 25 March 1980 (copy in Corps Funding Shortfall/1980 folder, Heavy-Industrial Division, Associated General Contractors of America, Washington, D.C. [hereafter referred to as "AGC"]); Office of the Chief of Engineers, news release, "Army Corps of Engineers Announces Criteria for Continued Contract Actions," 4 April 1980; interview, author with Joseph K. Bratton (Chief of Engineers, 1980–1984), 22 May 1987; and Maj. Gen. E. R. Heiberg III to Hubert Beatty, 22 July 1980 (Corps Funding Shortfall/1980 folder, Heavy-Industrial Division, AGC).

23. Interview, author with Tom Bevill, 20 May 1987. See also, "Edgar's Fight against Pork," *Philadelphia Inquirer*, 23 May 1980; and "Congressman Tom Bevill's Involvement in Building the Tennessee-Tombigbee Waterway," unsigned and undated [probably written in 1985] publicity paper distributed by the Congressional office of Tom Bevill.

24. Gaylord Nelson and Daniel Patrick Moynihan to Dear Colleague, 16 May 1980 (Tenn-Tom folder, box 90, file M80–626, Nelson papers).

25. "Supplemental Appropriations and Rescission Bill," *Congressional Record—House*, 126 (June 18, 1980), 15320.

26. For the full debate, see *ibid.*, pp. 15320–15334.

27. *Ibid.*, p. 15321.

28. George E. Snyder to Dear Congressman, 13 June 1980 (Tenn-Tom 1980 folder, file 128/11–2, Jack Edwards papers, University of South Alabama Archives, Mobile, AL [hereafter referred to as "Edwards papers"]).

29. "Supplemental Appropriations and Rescission Bill," p. 15324.

30. See *ibid.*, p. 15334.

31. Glover Wilkins to Jack Edwards, 19 June 1980 (Tenn-Tom 1980 folder, file 228/20–3, Edwards papers).

32. Tennessee-Tombigbee Waterway Development Authority, *The Truth about Tenn-Tom* (Columbus, MS: Tennessee-Tombigbee Waterway Development Authority, August 1980).

33. "Supplemental Appropriations," *Congressional Record— Senate*, 126 (June 28, 1980), 17611.
34. *Ibid.*, p. 17612.
35. *Ibid.*.
36. *Ibid.*, p. 17614. See also, Office of Senator Gaylord Nelson, press release, "Nelson Sponsors Amendment to Cut Funds for Tenn-Tom Waterway," 27 June 1980 (Tenn-Tom folder, box 90, file M80–626, Nelson papers).
37. Martin Tolchin, "Senate Votes, 47–36, to Go on Financing Waterway in South," *New York Times*, 29 June 1980. See also, Charles R. Babcock, "Senate Approves a Spending Bill for $16.2 Billion," *Washington Post*, 29 June 1980.
38. "Supplemental Appropriations," (June 28, 1980), p. 17636.
39. "More Support for Waterway Not Justified," Louisville (KY) *Courier-Journal*, 17 July 1971.
40. See interview, author with Roger A. Burke, 17 November 1986; interview, author with John Jeffrey Tidmore, 20 May 1986; and interview, author with Paul E. Roberts, 5 August 1987.
41. See U.S. Army Corps of Engineers, Mobile District, *Tennessee-Tombigbee Waterway Briefing Information* (Mobile, AL: U.S. Army Corps of Engineers, Mobile District, May 1981), p. 9.
42. Robert K. Dawson to Henry Eschwege, 6 May 1981 (misc. folder, Tenn-Tom Litigation Management Unit files, Mobile District [hereafter referred to as "SAMDL"]). For an extended discussion of the sunk-cost justification for the Tenn-Tom, see U.S. House of Representatives, Committee on Appropriations, Subcommittee on Energy and Water Development, Hearings, *Energy and Water Development Appropriations for 1984* (98th Cong., 1st sess., 1983), Part 7, pp. 1–118.
43. John C. Stennis, et al. to Dear Senator, 19 March 1981 (printed in *Congressional Record— Senate*, 127 [March 19, 1981], 4715). See also, Senator John C. Stennis, news release, 19 March 1981 (copy in file 2–1000–1189, Glover Wilkins Tennessee-Tombigbee Waterway Archives, Archives and Museums Department, Mississippi University for Women, Columbus, MS [hereafter referred to as "GWTTW Archives"]); and Brian Kelly, *Adventures in Porkland: How Washington Wastes Your Money and Why They Won't Stop* (New York: Villard Books, 1992), p. 83.
44. See Frank C. Deming to South Atlantic Division Engineer, 11 May 1976 (cost reduction measures/TTW folder, file 1501–07, Environment and Resources Branch, Planning Division, Mobile District [hereafter referred to as "SAMPD-EI"]).
45. U.S. Army Corps of Engineers, Mobile District, *Studies of Project Costs and Benefits: Tennessee-Tombigbee Waterway* (Mobile, AL: U.S. Army Corps of Engineers, Mobile District, October 1975), p. 14.
46. *Ibid.*, p. 16.
47. See interview, author with Frank B. Couch, 17 September 1986; and interview, author with Maj. Gen. Henry J. Hatch, 16 December 1987.
48. U.S. Army Corps of Engineers, Mobile District, *Briefing Data for Army Audit Agency: Tennessee-Tombigbee Waterway* (Mobile, AL: U.S. Army Corps of Engineers, Mobile District, January 1976), p. V–45.
49. The perspective of Corps designers is described in interview, author with Kenneth D. Underwood, 21 May 1986; and interview, author with George H. Atkins, 18 November 1986. For criticism of this approach, see Representative Joel Pritchard, news release, "GAO Study Questions Tenn-Tom Project," 21 May 1981 (Tenn-Tom files, National Taxpayers Union [NTU], Washington, D.C.); and William H. Dempsey to Members of the House of Representatives, 13 July 1981 (Association of American Railroads office files, Washington, D.C.).
50. Joel Pritchard and Robert Edgar to "Dear Colleague," 23 June 1980 (Tenn-Tom 1980 folder, file 228/20–3, Edwards papers). For the quickly issued retort, see Tom Bevill and John J. Rhodes to "Dear Colleague," 23 June 1980 (Tenn-Tom 1980 folder, file 228/20–3, Edwards papers).
51. "Energy and Water Development Appropriations," *Congressional Record—House*, 126 (June 25, 1980), 16775.
52. *Ibid.*, p. 16791. See also, Michael Kilian, "From Your Congress—Mississippi River No. 2!," Chicago *Tribune*, 22 July 1980.

53. "Energy and Water Development Appropriations," p. 16791.

54. Jamie L. Whitten to Hunter M. Gholson, 1 August 1980 (file 2–1000–1190, GWTTW Archives).

55. "The $3,000,000,000 Ditch," *Washington Post,* 28 June 1980.

56. See "Another Place to Cut the Budget," Champaign-Urbana (IL) *News Gazette,* 22 June 1980; "River of No Financial Return," *New York Times,* 24 June 1980; "Stop This $3 Billion Boondoggle," Chicago *Sun-Times,* 27 June 1980; J. B. Collins, "Who Needs Tombigbee?," Chattanooga (TN) *News Free Press,* 29 June 1980; John Bennett, "Tenn-Tom Critics Become Embarrassing," Memphis (TN) *Commercial Appeal,* 3 July 1980; "Robbers without Guns," Atlanta (GA) *Constitution,* 3 July 1980; "In the Tenn-Tom Ditch," Raleigh (NC) *News & Observer,* 14 July 1980; "$3 Billion Boondoggle Gets Three of Four Idaho Votes," Idaho Falls *Post-Register,* 17 July 1980; "Here's a Plan for Spewing $3 Billion Down the Drain," Fort Myers (FL) *News-Press,* 19 July 1980; Kilian, "From Your Congress—Mississippi River No. 2!"; "Good Money for a Bad Project," *Los Angeles Times,* 27 July 1980; "Money Down the Tenn-Tom," Wichita (KS) *Eagle,* 29 July 1980; and James J. Kilpatrick, "Why This $3 Billion Boondoggle?," *Washington Star,* 29 July 1980.

57. Harry N. Cook to Glover Wilkins, 28 July 1980 (Tenn-Tom files, National Waterways Conference, Washington, D.C. [hereafter referred to as "NWC"]). See also, Tom Bevill and J. Bennett Johnston to Harry N. Cook, 3 September 1980 (Tenn-Tom files, Tom Bevill House of Representatives office papers, Washington, D.C. [hereafter referred to as "Bevill papers"]).

58. See James C. Jordan to Harry N. Cook, 8 August 1980 (Tenn-Tom file, NWC); James C. Jordan to David A. Johnston, 8 August 1980 (Corps Funding Shortfall/1980 folder, Heavy-Industrial Division, AGC); interview, author with James C. Jordan, 11 June 1986; and interview, author with Glover Wilkins, 15 and 16 November 1986.

59. "Personal and confidential" letter, Glover Wilkins to John C. Stennis, 8 August 1980 (file 2–1000–1189, GWTTW Archives).

60. Bevill interviews, 21 February 1985 and 20 May 1987; Jordan interview, 11 June 1986; and interview, author with Forrest T. Gay III (South Atlantic Division Engineer, 1982–1985), 21 May 1987.

61. "Energy and Water Development Appropriations," *Congressional Record—Senate,* 126 (September 10, 1980), 25010.

62. Stephen E. Roady to Debra Gordon-Hellman, 1 August 1980 (TAC/Tenn-Tom file, Alabama Conservancy, Birmingham, AL).

63. Prime F. Osborn to Dear Shareholder, 1 August 1980 (Tenn-Tom files, Environmental Policy Center, Washington, D.C. [hereafter referred to as "EPC"]). In addition to the 3,500 shareholders residing in Florida, Osborn sent similar requests to the shareholders living in New Hampshire (200) and Michigan (700). See Philip M. Lanier to Jon T. Brown, 8 August 1980 (Tenn-Tom files, EPC).

64. "Energy and Water Development Appropriations," p. 25040. See also, Lance Gay, "Senate Votes to Continue Canal Project," *Washington Star,* 10 September 1980.

65. "Strategy for Senator Moynihan's Hearings on Tenn-Tom," unsigned and undated memorandum marked *confidential* [compiled by TTWDA in early July 1980] (file 2–1000–1189, GWTTW Archives). With Randolph unable to attend the hearings, Stennis stood in his stead, hoping to ensure Moynihan did not deal too harshly with the Corps. See interview, author with Steven Dola, 9 January 1987.

66. "Energy and Water Development Appropriations," pp. 25024–25026.

67. See "Conference Report on H.R. 7590, Energy and Water Development Appropriations," *Congressional Record—House,* 126 (September 24, 1980), 27111.

68. See *ibid.,* p. 27115.

69. Environmental advocate and long-time Tenn-Tom opponent Gaylord Nelson lost his Senate seat in November 1980 after three terms, one of the Democratic casualties of the election.

70. That early strategy was discussed in Len Rippa and E. R. Osann to President Reagan, 3 August 1982 (Tenn-Tom files, NTU).

71. "The Hat Is on the other Head," *Los Angeles Times,* 4 March 1981.

72. "Sacred Cows that the President Missed," St. Petersburg (FL) *Times,* 22 February 1981. See also, George Neavoll, "Water Project Rip-Offs: Butchering the Sacred Cows," Wichita (KS) *Eagle-Beacon,* 15 February 1981; "Untouchable Pork?," Chicago *Tribune,* 2 March 1981; and "The Growing Clamor to Halt the Tenn-Tom," *Business Week,* no. 2711 (October 26, 1981), 58–59.

73. See Lance Gay, "Senate Panel Recommends Ending Work on Waterway," *Washington Star,* 14 March 1981; Joanne Omang, "Senate Panel Votes to Halt Controversial $2 Billion Tenn-Tom Canal," *Washington Post,* 14 March 1981; and interview, author with Senator Pete V. Domenici (R-New Mexico), 3 August 1987. For the justification of completing the waterway only to Columbus, Mississippi, see Steve Roady and Bill Bonvillian to Woodruff Price and Ed Whitfield, 27 March 1981 (Charles B. Castner's personal files, CSX Transportation, Louisville, KY). For Domenici's opposition to the waterway, see Pete Domenici to Myrt Jones, 31 January 1983 (correspondence/1983–1985 file, Mobile Bay Audubon Society papers [when I used this collection, it was in Society president Myrt Jones's house in Mobile, AL; the Society subsequently donated its papers to the University of South Alabama Archives, Mobile]).

74. See National Taxpayers Union, et al., "Tennessee-Tombigbee . . . the Facts," 21 July 1981 (copy of this two-page flyer in Tenn-Tom files, NTU); Jay D. Hair to Associate Members, 8 July 1981 (Water Resources Program files, National Wildlife Federation, Washington, D.C. [hereafter referred to as "NWF"]); and interview, author with Jill Lancelot, 6 August 1987.

75. "Tenn-Tom Boondoggle," *Congressional Record—House,* 127 (June 23, 1981), 13492.

76. "Flowing Rivers," *Wall Street Journal,* 29 July 1981. See also, Tennessee-Tombigbee Waterway Development Authority, "Analysis of July 23, 1981, House Vote on Tenn-Tom," no date [probably 30 July 1981] (Tenn-Tom 1983–1984 folder, file 228/20–3, Edwards papers); Margot Hornblower, "House Votes Funds to Continue Work on Big Waterway," *Washington Post,* 24 July 1981; Seth S. King, "Waterway Barely Survives in House," *New York Times,* 24 July 1981; John B. Oakes, "Pork—U.S. Prime," *New York Times,* 29 July 1981; and R. Jeffrey Smith, "The Waterway That Cannot Be Stopped," *Science,* 213 (August 14, 1981), 741–742, 744. The entire floor debate and the recorded vote are printed in "Energy and Water Development Appropriation," *Congressional Record—House,* 127 (July 23, 1981), 16983–17014.

77. See hand-signed list of attenders at the "Tenn-Tom meeting," 4 August 1981 (Tenn-Tom files, NTU). The groups represented included: Association of American Railroads, CSX, KCS Railway, Burlington, National Wildlife Federation, Environmental Policy Center, Sierra Club, American Rivers Conservation Council, Common Cause, National Taxpayers Union, and Brown & Roady. Not represented, yet part of the coalition, were the League of Women Voters, the Conservative Caucus, the National Audubon Society, Friends of the Earth, and the Coalition for Water Projects Review. See "Tennessee-Tombigbee . . . the Facts," 2 November 1981; and Len Rippa to Howard H. Baker, Jr., 27 October 1981 (Tenn-Tom files, NTU).

78. See agenda, 15 September 1981 (Tenn-Tom files, NTU).

79. See National Campaign to Stop the Tenn-Tom, "Why Fight the Tenn-Tom Ditch?," no date [probably April 1980] (copy in Tenn-Tom file, Alabama Conservancy); Tennessee-Tombigbee agenda, 13 October 1981 (Tenn-Tom files, NTU); Tennessee-Tombigbee agenda, 30 October 1981 (Tenn-Tom files, NTU); Len Rippa, memo (subj.: NTU Press Calls), no date [probably 27 October 1981] (Tenn-Tom files, NTU); and "NTU to Fight Tenn-Tom in Senate," *Dollars & Sense,* 12 (September 1981), 5.

80. See National Campaign to Stop the Tenn-Tom, "Why Fight the Tenn-Tom Ditch?"; American Rivers Conservation Council, et al., "Tennessee-Tombigbee . . . The Facts," 22 October 1981 (copy in Tenn-Tom files, EPC); and Blackwelder interview, 12 April 1988. For the role of the Association of American Railroads in lobbying against the waterway, see William H. Dempsey to Members of the House of Representatives, 13 July 1981 (Association of American Railroads office files, Washington, D.C.); and Frank N. Wilner to Forrest Hood James, Jr., 2 January 1981 (Association of American Railroads office files). For the Sierra Club's role, see Joe Fontaine to William L. Armstrong, 29 October 1981 (Tenn-Tom files, NTU). For editorial opposition to continued

Tenn-Tom appropriations, see "Now the Pork," *Boston Globe*, 18 July 1981; "Saying No to Pork-Barrelism," *Philadelphia Inquirer*, 22 July 1981; "$6 Billion Pork Barrel," Sacramento (CA) *Bee*, 22 July 1981; and "Flowing Rivers," *Wall Street Journal*, 29 July 1981.

81. Senators Charles H. Percy, Daniel P. Moynihan, Gordon J. Humphrey, Dave Durenberger, Richard G. Lugar, Alan K. Simpson, John H. Chafee, and William Proxmire to Dear Colleague, 28 October 1981 (Tenn-Tom files, NTU). See also, U.S. General Accounting Office, *To Continue or Halt the Tenn-Tom Waterway? Information to Help the Congress Resolve the Controversy* (Washington: U.S. General Accounting Office, 15 May 1981); Duane A. Thompson, *Factors Affecting Coal Traffic on the Tennessee-Tombigbee Waterway* (Washington: Library of Congress, Congressional Research Service, Environment and Natural Resources Division, 17 July 1981); "Tombigbee's Goal for Coal Movements: A 'Pipe Dream'," *Coal Week* (February 12, 1979), 3; and Christopher J. Dodd to Len Rippa, 29 September 1981 (Tenn-Tom files, NTU).

82. Coverage of the floor debate and the results of the rollcall vote are printed in "Energy and Water Development Appropriations Act," *Congressional Record—Senate*, 127 (November 4, 1981), 26410–26433.

83. Len Rippa to Brent Blackwelder, 30 November 1981 (Tenn-Tom files, EPC).

84. Quoted in Tom Scarritt, "Heflin Quarterbacks Secret Huddle to Call Next Year's Tenn-Tom Play," Birmingham (AL) *News*, 15 December 1981. See also, Senator Howell Heflin, news release, week of 16–20 November 1981 (copy in file 2–1000–1191, GWTTW Archives).

85. Senator Howell Heflin, newsletter, March 1982 (copy in file 2–1000–1191, GWTTW Archives).

86. [Jon T.] Brown and [Stephen E.] Roady to Tenn-Tom Legislative Group, 27 January 1982 (Tenn-Tom files, EPC).

87. Sierra Club Washington, D.C. Office, "Tennessee-Tombigbee Strategy 1982," no date [probably spring 1982] (Strategy folder, Tennessee-Tombigbee Waterway carton, Sierra Club Washington, D.C. Office records, Bancroft Library, Berkeley, CA [hereafter referred to as "Sierra Club records"]).

88. Sierra Club Washington, D.C. Office, Voting Profile on Tenn-Tom, no date [probably June 1982] (Voting Lists folder 4, Tennessee-Tombigbee Waterway carton, Sierra Club records).

89. See, for example, Tom Bevill to R. R. Johnston, Jr., 26 April 1982 (Tenn-Tom files, Bevill papers).

90. Len Rippa to Wendell Rawls (*New York Times*), 18 June 1982 (Tenn-Tom files, NTU).

91. Denny Shaffer to Editorial Page Editors, June 1982 (Strategy folder, Tennessee-Tombigbee Waterway carton, Sierra Club records). For examples of subsequent newspaper coverage, see "Pritchard again Challenging Biggest Pork-Barrel Project," Seattle (WA) *Times*, 4 July 1982; Robert McHugh, "Dismantle the Tenn-Tom and Eliminate Unemployment," Gulfport (MS) *Daily Herald*, 12 September 1982; Ruth Marcus, "Election Year Seen Aiding Environmentalists," *Washington Post*, 16 August 1982; and "Die, Monster, Die," *Arizona Daily Star*, 18 September 1982. For a longer, stinging economic criticism of the Tenn-Tom, see Joseph L. Carroll, "Tennessee-Tombigbee Waterway Revisited," *Transportation Journal*, 22 (Winter 1982), 5–20.

92. Joe Trainor, et al., to Dear Representative, 20 August 1982 (Tenn-Tom files, EPC). For additional examples of the anti-Tenn-Tom lobbying, see Rose McCullough to Tenn-Tom Lobbyists/Activists, 14 September 1982 (Tenn-Tom files, NTU); Jay D. Hair to Thomas P. O'Neill, Jr., 16 September 1982 (Water Resources Program file, NWF); and David A. Larsen and Ann C. Tate to presidents and conservation chairmen of Audubon Society chapters, 17 November 1982 (correspondence/1983–1985 file, Mobile Bay Audubon Society papers).

93. Len Rippa to Dear Editor, 30 August 1982 (Tenn-Tom files, NTU).

94. Statement of Congressman Bob Edgar on Tenn-Tom Amendment on FY 1983 Energy and Water Appropriations, 30 November 1982 (Tenn-Tom files, NTU).

95. Statement of Rep. Tom Bevill, Tenn-Tom News Conference, the Capitol, 29 November 1982 (Statements folder, Tennessee-Tombigbee Waterway carton, Sierra Club records). For its part, the Tennessee-Tombigbee Waterway Development Authority signed a $10,000 contract

with the Washington, D.C., public affairs firm Wagner and Baroody in October to undertake "a coordinated campaign to generate favorable public opinion in support of completion of the Tenn-Tom Waterway." See Joseph D. Baroody to Glover Wilkins, 1 October 1982 (July-October 1982 folder, Board Minutes files, Tennessee-Tombigbee Waterway Development Authority office records, Columbus, MS).

96. "Congressman Tom Bevill's Involvement in Building the Tennessee-Tombigbee Waterway," p. 8. No evidence is given to suggest who the "someone in authority" was, although the implication is it was Bevill. For the implications of this action, see ". . . And Continuing Irresolution," *Washington Post*, 13 December 1982.

97. See "Energy and Water Development Appropriation Bill," *Congressional Record—House*, 129 (June 7, 1983), 3669–3672; U.S. Congress, Environmental and Energy Study Conference *Floor Brief*, Senate Floor Brief #1 (June 20, 1983), p. 7; and interview, John T. Greenwood with Representative Jamie Whitten, 11 December 1986 (transcript in Office of History archives, Corps Headquarters, Fort Belvoir, VA).

98. See Bevill interview, 20 May 1987; and "Congressman Tom Bevill's Involvement in Building the Tennessee-Tombigbee Waterway," p. 8.

99. Bratton interview, 22 May 1987.

100. John C. Stennis to Tom Bevill, 11 July 1983 (Tenn-Tom files, Bevill papers).

101. Blackwelder interview, 12 April 1988.

Conclusion

1. Samuel R. Green, memorandum for [Mobile] District Engineer, 31 August 1983 (TTW Dedication/1983 folder, file 401–07, Public Affairs Office, Mobile District, U.S. Army Corps of Engineers, Mobile, AL). See also, John LoDico, "Historic Tenn-Tom Waterway Opens Amid Cautious Forecasts for Growth," *Traffic World*, 202 (June 10, 1985), 23–26.

2. Fred Grimm, "Waterway Boosters Roll Out Pork Barrel," Miami (FL) *Herald*, 3 June 1985.

3. Robert M. Press, "New Waterway: Boon for Economy or Boondoggle for Taxpayers?," *Christian Science Monitor*, 3 June 1985.

4. David Rogers, "Rivaling Cleopatra, A Pork-Barrel King Sails the Tenn-Tom," *Wall Street Journal*, 31 May 1985. See also, "Coal on Tenn-Tom below Expectations," *Tennessean*, 1 March 1985; David Treadwell, "Economic Reality Clouding Dreams for Canal's Impact," *Los Angeles Times*, 27 May 1985; "Conservationists Say Tenn-Tom Battle Not over Yet," Florence (AL) *Times Daily*, 2 June 1985; and "Toward the Bottom of the Barrel," *Washington Post*, 9 June 1985.

5. "A $2-Billion White Elephant," St. Petersburg (FL) *Times*, 23 February 1986.

6. David Tortorano, "Yachts Outnumber Industrial Ships on New Tenn-Tom Waterway," San Francisco *Examiner*, 1 June 1986. For other criticisms of the poor usage of the waterway, see Ron Martz, "Small Towns See Big Bucks Flowing Down Tenn-Tom," Atlanta (GA) *Journal and Constitution*, 2 June 1985; William E. Schmidt, "Traffic on New Tombigbee Waterway Falls Far Short of Army Engineers' Forecast," *New York Times*, 16 February 1986; Raad Cawthon, "Tenn-Tom Is Not Delivering Wealth that Some Expected," Atlanta (GA) *Journal and Constitution*, 2 March 1986; Eugene Carlson, "For Tenn-Tom, Traffic Is Only a Trickle," *Wall Street Journal*, 18 March 1986; "Tenn-Tom More Recreational Than Commercial," Birmingham (AL) *Post-Herald*, 12 May 1986; Marshall Ingwerson, "Banking on Tenn-Tom Waterway to Spur Region's Economy: Although Project Opened Early, Barge Traffic Remains a Trickle," *Christian Science Monitor*, 30 September 1986; David Treadwell, "'Tenn-Tom' Isn't a Draw for Commercial Traffic," *Los Angeles Times*, 21 December 1986; Cass Peterson, "The Fizzling of 200-Year-Old Dream," *Washington Post*, 26 December 1986; Tim Roberts, "Tenn-Tom Benefits Lagging Far Behind Rosy Predictions," Louisville (KY) *Courier-Journal*, 9 February 1987; and David Field, "Traffic Only 7% of Projections on the 'Tenn-Tom'," *Washington Times*, 2 October 1987.

7. "Wet Elephant," *Washington Post*, 5 January 1987.

8. See David Treadwell, "'Tenn-Tom' Isn't a Draw for Commercial Traffic, But Pleasure Boaters Love It," *Los Angeles Times*, 21 December 1986.

9. Jim Yardley, "Poverty, Not Prosperity Flows down Tenn-Tom," Atlanta (GA) *Journal and Constitution*, 9 August 1992.

10. Quoted in Greg Borzo, "Tenn-Tom Development Falters; Coal Down, Forest Products Up," *Traffic World*, 211 (September 14, 1987), 22.

11. See *ibid.*, p. 23; Semoon Chang and Philip R. Forbus, "Tenn-Tom Versus the Mississippi," *Transportation Journal*, 25 (Summer 1986), 47–54; and Semoon Chang and Philip R. Forbus, *Estimating Savings in Transportation Costs from the Use of the Tennessee-Tombigbee Waterway*, Research Report 3 (Mobile, AL: Center for Business and Economic Research, University of South Alabama, 1986).

12. See Fred Grimm, "South Rebels at Garbage from North," Miami (FL) *Herald*, 9 July 1989. For the Tenn-Tom's meager contribution to regional economic development, see Alice J. Hall, "The Hidden Tenn-Tom: Bypassed but Still Striving," *National Geographic*, 169 (March 1986), 384–387; and Yardley, "Poverty, Not Prosperity Flows down Tenn-Tom."

13. For a detailed account of this larger phenomena, see Robert D. Bullard, *Dumping in Dixie: Race, Class, and Environmental Quality* (Boulder, CO: Westview Press, 1990).

14. For discussion of the failure to pass an omnibus water bill, see Joseph A. Davis, "Prospects Uncertain: Water Projects Await Funding as Debate over Cost Sharing Stalls Action on Capitol Hill," *Congressional Quarterly Weekly Report*, 41 (July 30, 1983), 1551–1558. For the link between the Tenn-Tom controversy and the Corps's tarnished public image, see "Tainted Tenn-Tom Pork," Louisville (KY) *Courier-Journal*, 10 February 1987.

15. For the background to this legislation, see Martin Reuss, *Reshaping National Water Politics: The Emergence of the Water Resources Development Act of 1986*, IWR Policy Study 91–PS-1 (Washington: GPO, 1991).

Bibliography

Archival and Manuscript Collections

Thomas Abernethy collection. Archives and Special Collections, University of Mississippi. University, Mississippi.
Alabama Conservancy. Birmingham, Alabama.
Alabama Department of Archives and History. Governors' Official Files. Montgomery, Alabama.
Alabama Department of Conservation and Natural Resources. Game and Fish Division. Montgomery, Alabama.
Alabama Department of Economic and Community Affairs. Montgomery, Alabama.
Alabama Department of Environmental Management. Air Division and Water Division. Montgomery, Alabama.
Alabama Highway Department. Montgomery, Alabama.
Alabama State Docks Department. Mobile, Alabama.
American Folklife Center. Library of Congress. Washington, D.C.
Associated General Contractors of America. Washington, D.C.
Association of American Railroads. Washington, D.C.
Tom Bevill House of Representatives office papers. Washington, D.C.
Birmingham Audubon Society. Birmingham, Alabama.
Jimmy Carter Presidential Library. Atlanta, Georgia.
Charles B. Castner personal files. CSX Transportation. Louisville, Kentucky.
Arthur P. Cooley papers. Department of Special Collections, State University of New York at Stony Brook.
Council on Environmental Quality. Washington National Records Center. Suitland, Maryland.
Jack Edwards papers. University of South Alabama Archives. Mobile, Alabama.
Environmental and Energy Study Conference. U.S. Congress. Washington, D.C.
Environmental Defense Fund papers. Department of Special Collections, State University of New York at Stony Brook.
Environmental Policy Center. Washington, D.C.
Gerald R. Ford Presidential Library. Ann Arbor, Michigan.

Glover Wilkins Tennessee-Tombigbee Waterway Archives. Archives and Museums Department, Mississippi University for Women. Columbus, Mississippi.
Historic American Buildings Survey. National Park Service. Washington, D.C.
Historic American Engineering Record. National Park Service. Washington, D.C.
Lawrence Livermore National Laboratory. Office of History and Historical Records. Livermore, California.
Louisville and Nashville Railroad collection. University of Louisville Archives. Louisville, Kentucky.
Mississippi Department of Archives and History. Governors' Official Files. Jackson, Mississippi.
Mississippi Department of Economic Development. Jackson, Mississippi.
Mississippi Department of Wildlife Conservation. Bureau of Fisheries and Wildlife. Jackson, Mississippi.
Mississippi State Highway Department. Jackson, Mississippi.
Mobile Bay Audubon Society. Mobile, Alabama. (The Society subsequently donated its papers to the University Archives, University of South Alabama, Mobile.)
Mobile Municipal Archives. Mobile, Alabama.
Morrison-Knudsen Corporation. Boise, Idaho.
National Park Service. Southeast Regional Office, Archeological Services Branch. Atlanta, Georgia.
National Taxpayers Union. Washington, D.C.
National Waterways Conference. Washington, D.C.
National Wildlife Federation. Washington, D.C.
Gaylord Nelson papers. State Historical Society of Wisconsin. Madison, Wisconsin.
Gaylord Nelson personal papers. Washington, D.C.
Nixon Presidential Materials Project. National Archives and Records Administration. Washington, D.C.
Office of Management and Budget. Natural Resources Division. Record Group 51. National Archives. Suitland, Maryland.
Office of the Assistant Secretary of the Army for Civil Works. Department of the Army. Washington, D.C.
Office of the Secretary of the Army. Administrative Support Group. Department of the Army. Washington, D.C.
Paul E. Roberts personal papers. Washington, D.C.
Sierra Club records. Washington, D.C., office files. Bancroft Library. Berkeley, California.
Robert E. Smolker papers. Department of Special Collections, State University of New York at Stony Brook.
John J. Sparkman collection. Special Collections, University of Alabama Library. University, Alabama.
John C. Stennis collection. Public Series. Mitchell Memorial Library, Mississippi State University. Mississippi State, Mississippi.
John C. Stennis Senatorial office papers. Washington, D.C.
Tennessee-Tombigbee Waterway Development Authority. Columbus, Mississippi.
Tombigbee River Valley Water Management District. Tupelo, Mississippi.
U.S. Army Corps of Engineers. Board of Engineers for Rivers and Harbors. Fort Belvoir, Virginia.
U.S. Army Corps of Engineers. Mobile District. Mobile, Alabama.
U.S. Army Corps of Engineers. Nashville District. Nashville, Tennessee.
U.S. Army Corps of Engineers. Office of History, Headquarters. Fort Belvoir, Virginia.
U.S. Army Corps of Engineers. Office of the Chief of Engineers. Washington, D.C.
U.S. Army Corps of Engineers. South Atlantic Division. Atlanta, Georgia.
U.S. Army Corps of Engineers. South Atlantic Division Laboratory. Marietta, Georgia.
U.S. District Court for the Northern District of Mississippi. Oxford, Mississippi.
U.S. Environmental Protection Agency. Region IV. Atlanta, Georgia.

U.S. Fish and Wildlife Service. Division of Ecological Services. Daphne, Alabama.
U.S. Fish and Wildlife Service. Ecological Services Division. Washington, D.C.
U.S. Fish and Wildlife Service. Endangered Species Office. Jackson, Mississippi.
U.S. Fish and Wildlife Service. Endangered Species Section. Washington, D.C.
Water Resources Congress. Arlington, Virginia.
Wildlife Society. Bethesda, Maryland.
James D. Williams personal papers. Arlington, Virginia. (Williams subsequently donated his papers to the University of Alabama Archives, University, Alabama.)

Unpublished Materials

American Society of Civil Engineers, Mobile (AL) Chapter. "Tennessee-Tombigbee Waterway: Nomination for Outstanding Civil Engineering Award (OCEA)—1986," nomination report, 1986.
Association of American Railroads. "Tennessee-Tombigbee Waterway: Supplement to General Design Memorandum, Reevaluation of Project Economics," a preliminary review submitted to the Office of the Chief of Engineers on 15 December 1966.
Boschung and Associates, Inc. "A Study of the Fishes of the Upper Tombigbee River Drainage System South of Columbus Lock and Dam, Tennessee-Tombigbee Waterway," report prepared for U.S. Army Corps of Engineers, Mobile District, 1984.
Brandes, William F. "The Tennessee-Tombigbee Waterway Project v. The National Environmental Policy Act," term paper, EWRE 279, School of Engineering, Vanderbilt University, January 1975.
Caldwell, Richard Dale. "A Study of the Fishes of the Upper Tombigbee River and Yellow Creek Drainage Systems of Alabama and Mississippi," Ph.D. dissertation, University of Alabama, 1969.
Carroll, Joseph L. "A Critique of Waterway Benefit-Cost Analysis: The Tennessee-Tombigbee," Ph.D. dissertation, Indiana University, 1962.
Cooper, John E.; and Martha R. Cooper. "A Critique of Some Aspects of U.S. Army Corps of Engineers Environmental Statement, Tennessee-Tombigbee Waterway, March 1971," report submitted to the Environmental Defense Fund, 30 July 1971.
Dames and Moore, Consulting Engineers. "Engineering Study of Construction Methods for Canal Section," report prepared for U.S. Army Corps of Engineers, Mobile District, June 1975.
Doane Agricultural Service, Inc. "Review of Corps of Engineers Supplement to General Design Memorandum, Reevaluation of Project Economics, Tennessee-Tombigbee Waterway," report prepared for the Tennessee-Tombigbee Waterway Development Authority, 9 December 1966.
Grace, Randall. "The Fresh-Water Mussel Industry of the Lower Tennessee River: Ecology and Future," M.S. thesis, Western Kentucky University, 1974.
Hartley, Joseph R. "Analysis of the Tennessee-Tombigbee Waterway Reevaluation of Project Economics, Tennessee-Tombigbee Waterway, Corps of Engineers," consultant's report prepared for the Tennessee-Tombigbee Waterway Development Authority, December 1966.
Jones, James William. "An Analytical History of the Tennessee-Tombigbee Waterway," M.A. thesis, University of Mississippi, 1982.
Kemp, Emory L. "The Engineering Aspects of the History of the Tennessee-Tombigbee Waterway," two parts, consultant's report to History Associates Incorporated, Rockville, Maryland, March 1987 and April 1987.
Kitchens, James H., III. "An Outlet to the Gulf: The Tennessee-Tombigbee Waterway, 1541–1971," Office of History, U.S. Army Corps of Engineers, 1985.
Matthews, William M. "Land Use Planning in an Interstate Context: A Proposed Model for the Tennessee-Tombigbee Waterway Region," M.A. thesis, University of Tennessee, 1975.
McClure, N. D., IV. "Environmental Planning for Tennessee-Tombigbee Waterway: An Interdis-

ciplinary Approach," paper presented at the American Society of Civil Engineers Water Resources Planning and Management Division Specialty Conference, Houston, Texas, 25–28 February 1979.

McClure, N. D., IV. "A Winning Combination," report on the Tenn-Tom Litigation Management Unit, U.S. Army Corps of Engineers, Mobile District, September 1983.

Miles, Lee W. "Tenn Tom Constructors Equipment Department Final Report for Tennessee Tombigbee Waterway," in-house report, Morrison-Knudsen Company, Boise, Idaho, n.d. (probably 1984).

Miller, Frank. "Wildlife Habitat Mitigation Project, the Tennessee-Tombigbee Waterway: Demopolis to Pickwick Lake," report submitted to the U.S. Fish and Wildlife Service, Decatur, Alabama, 27 April 1978.

National Campaign to Stop the Tenn-Tom. "Why Fight the Tenn-Tom Ditch?" political tract, n.d. (probably June 1981).

Roberts, Paul. "Economic Analysis of the Navigation Benefits of the Tennessee-Tombigbee Waterway," report (dated 10 December 1975) released to the press by the Environmental Policy Center on 17 December 1975.

Roberts, Paul E. "Tennessee-Tombigbee Waterway: An Economic Background Report to the Louisville & Nashville Railroad Company," June 1977.

Russell, Stephen Blaine. "Tombigbee River Valley Water Management District and the Tennessee-Tombigbee Waterway," M.A. thesis, University of Mississippi, 1983.

Simpson, David. "Energy and Labor Requirements for the Construction and Annual Operation of the Tennessee-Tombigbee Waterway Project," Technical Memo No. 21, Energy Research Group, Center for Advanced Computation, University of Illinois at Urbana-Champaign, 7 May 1974.

Stewart, William Histaspas, Jr. "The Tennessee-Tombigbee Waterway: A Case Study in the Politics of Water Transportation," Ph.D. dissertation, University of Alabama, 1968.

Stine, Jeffrey K. "Environmental Politics and Water Resources Development: The Case of the Army Corps of Engineers during the 1970s," Ph.D. dissertation, University of California at Santa Barbara, 1984.

Tangerose, James G. "A History of the Corps of Engineers' Economic Evaluations of the Tennessee-Tombigbee Waterway and an Analysis of the Corps' 1976 Reevaluation," report prepared for the Association of American Railroads, August 1978.

Tennessee-Tombigbee Waterway Development Authority. "Environmental Policy Statement," Columbus, Mississippi, August 1971.

Tennessee-Tombigbee Waterway Development Authority. "Facts and Information on the Tennessee-Tombigbee Waterway and Its Opponents' Allegations," Columbus, Mississippi, 5 April 1979.

Tennessee-Tombigbee Waterway Development Authority. "Minority Participation on Tennessee-Tombigbee Waterway," Columbus, Mississippi, 25 March 1981.

Toutant, William Thomas. "Development of Mathematical Performance Models for River Tows: With Specific Application to the Tennessee-Tombigbee Waterway," Ph.D. dissertation, University of Louisville, 1985.

University of Alabama, Tuscaloosa. "Chemical and Biological Effects of the Mixing of the Tennessee and Tombigbee Water Systems," a project proposal submitted to the U.S. Army Corps of Engineers, Mobile District, March 1971.

Wilkins, Glover. "My Association with the Tennessee-Tombigbee Waterway Project," manuscript prepared for the U.S. Army Corps of Engineers, Mobile District, 1985.

Williams, James David. "Distribution and Habitat Observations of Selected Tombigbee River Unionid Mollusks," background report, U.S. Fish and Wildlife Service, Office of Endangered Species, Washington, D.C., n.d. (probably 1984 or 1985).

Williams, Ned. "Economic Justification of the Tennessee-Tombigbee Waterway," M.A. thesis, Columbia University, 1949.

Zippert, John. "The Minority Peoples Council on the Tennessee-Tombigbee Waterway: A Citizen's Response to Rural Development and Industrialization," paper presented at the Southern Association of Agricultural Scientists, Atlanta, Georgia, 8 February 1977.

Zippert, John; and Robert Valder. "The Tennessee Tombigbee Water Project—A 'White' Paper," report prepared under the auspices of the Minority Peoples Council on the Tennessee-Tombigbee Waterway and the NAACP Legal Defense and Educational Fund, August 1977.

Government Documents

Alabama Development Office. *The Tennessee-Tombigbee Waterway: A Chronological Bibliography in Report Form.* Alabama Planning Resource Checklist, State Series No. 2. Montgomery, AL: Alabama Development Office, April 1977.

Alabama Geological Survey. *A Preliminary Consideration of the Environmental Impact of the Tennessee-Tombigbee Waterway in Alabama.* University, AL: Alabama Geological Survey, 1970.

American Technical Services, Inc. *An Examination of Economic Development Potential in the Tennessee-Tombigbee Waterway Impact Area of Greene, Hale, Sumter, Pickens Counties, Alabama.* Montgomery, AL: Alabama Development Office, 1977.

American Technical Services, Inc. *Financing Tennessee-Tombigbee Waterway Related Growth.* Prepared for the Alabama Development Office. Huntsville, AL: American Technical Services, Inc., April 1977.

Baker, John A.; Carolyn L. Bond; and C. H. Pennington. *Tennessee-Tombigbee Waterway Bendway Study Biological Impact Assessment: A Letter Report to the U.S. Army Engineer District, Mobile.* Vicksburg, MS: U.S. Army Engineer Waterways Experiment Station, Aquatic Habitat Group, January 1983.

Brown, Howard A. *Brief History of Tennessee-Tombigbee Waterway Project.* Washington: Library of Congress, Congressional Research Service, Environmental Policy Division, 16 October 1974.

Church, Robert F.; and Charles J. Mundo. *Study Plan for Estimating the Impact on Railroads of the Tennessee-Tombigbee Waterway.* Cambridge, MA: U.S. Department of Transportation, Research and Special Programs Administration, Transportation Systems Center, 15 May 1980.

Congressional Record. 1970–1986.

Federal Register. 1970–1986.

Horne, Thad. *An Economic Base Study of the Tombigbee River Basin.* Little Rock, AK: U.S. Department of Agriculture, Natural Resources Economics Division, 1973.

Hughes, H. Steve. *Water Resources Projects: Tennessee-Tombigbee Waterway, Alabama, Mississippi.* Mini Brief No. MB82245. Washington: Library of Congress, Congressional Research Service, Environment and Natural Resources Policy Division, 29 October 1982.

Johnson, Leland R. *Engineers on the Twin Rivers: A History of the U.S. Army Engineers, Nashville District, 1769–1978.* Nashville, TN: U.S. Army Corps of Engineers, Nashville District, 1978.

Lloyd, Nelson A.; et al. *Mercury Concentrations in Sediment Samples from the Tennessee, Mobile, Warrior, and Tombigbee Rivers, Alabama.* Circular No. 79. University, AL: Geological Survey of Alabama, 1972.

Melvin, Bill. *Policy Position: Tennessee-Tombigbee Waterway.* Montgomery, AL: Alabama Development Office, June 1972.

Miller, W. F.; J. L. Tingle; and E. L. Blake, Jr. *Atlas of Selected Resources: Tennessee-Tombigbee Waterway.* Mobile, AL: U.S. Army Corps of Engineers, Mobile District, n.d. (probably 1974).

Mississippi Agricultural and Forestry Experiment Station, Mississippi State University. *Erosion Control Experimentation, Tennessee-Tombigbee Waterway Project, Divide Cut Section: Final Report.* Nashville, TN: U.S. Army Corps of Engineers, Nashville District, 1975.

Mississippi Economic Council, Resource Development Committee. *The Tennessee-Tombigbee Waterway: Problems and Potentials.* Jackson, MS: Mississippi Economic Council, Resource Development Committee, November 1984.

Northwest Alabama Council of Local Governments. *Economic Impact Study, Tennessee-Tombigbee Waterway*. Final Report. Muscle Shoals, AL: Northwest Alabama Council of Local Governments, October 1976.

Nungesser, M. K.; C. T. Hunsaker; and R. J. Olson. *Environmental Resources of the Tennessee-Tombigbee Corridor: Summary Report*. Mobile, AL, and Nashville, TN: U.S. Army Corps of Engineers, Mobile and Nashville Districts, December 1982.

Reuss, Martin. *Shaping Environmental Awareness: The United States Army Corps of Engineers Environmental Advisory Board, 1970–1980*. Environmental History Series. U.S. Army Corps of Engineers, Office of the Chief of Engineers, Historical Division. Washington: GPO, 1983.

Stine, Jeffrey K.; and Michael C. Robinson. *The U.S. Army Corps of Engineers and Environmental Issues in the Twentieth Century: A Bibliography*. Environmental History Series. U.S. Army Corps of Engineers, Office of the Chief of Engineers, Historical Division. Washington: GPO, 1984.

Strack, O. D. L.; and H. M. Haitjema. *Study of the Environmental Impact of the Divide-Cut Section of the Tennessee-Tombigbee Waterway*. Nashville, TN: U.S. Army Corps of Engineers, Nashville District, 1979.

Thompson, Duane A. *Factors Affecting Coal Traffic on the Tennessee-Tombigbee Waterway*. Washington: Library of Congress, Congressional Research Service, Environment and Natural Resources Division, 17 July 1981.

Till, Thomas; and Michael Welch. *A White House Rural Initiative in Transition: The Impact of Large-Scale Construction Projects on Rural Economic Development*. Final Report prepared for the White House and the U.S. Community Services Administration. Washington: National Rural Center, June 1981.

U.S. Army Audit Agency. *Report of Audit: Aliceville Lake Regional Visitor Center, Tennessee-Tombigbee Waterway Project*. Audit Report No. SO 84–707. Washington: U.S. Army Audit Agency, 5 June 1984.

U.S. Army Audit Agency. *Report of Audit: Contract Administration, Tennessee-Tombigbee Waterway Project*. Audit Report No. SO 81–1. Atlanta, GA: U.S. Army Audit Agency, Southern District, 24 December 1980.

U.S. Army Audit Agency. *Report of Audit: Education Facilities, Tennessee-Tombigbee Waterway Project*. Audit Report No. SO 84–207. Washington: U.S. Army Audit Agency, 14 May 1984.

U.S. Army Audit Agency. *Report of Audit: Tennessee-Tombigbee Waterway Project, U.S. Army Corps of Engineers*. Audit Report No. SO 76–408. Atlanta, GA: U.S. Army Audit Agency, Southern District, 17 September 1976.

U.S. Army Corps of Engineers, Mobile District. *Environmental Statement: Tennessee-Tombigbee Waterway, Alabama and Mississippi Navigation*. Mobile, AL: U.S. Army Corps of Engineers, Mobile District, March 1971.

U.S. Army Corps of Engineers, Mobile District. *An Evaluation of the Transportation Economics of the Tennessee-Tombigbee Waterway, Final Report*. Volume I, *Executive Summary*. Report prepared by A. T. Kearney Management Consultants, Inc. Mobile, AL: U.S. Army Corps of Engineers, Mobile District, February 1976.

U.S. Army Corps of Engineers, Mobile District. *An Evaluation of the Transportation Economics of the Tennessee-Tombigbee Waterway, Final Report*. Volume II, *Project Report*. Report prepared by A. T. Kearney Management Consultants, Inc. Mobile, AL: U.S. Army Corps of Engineers, Mobile District, April 1976.

U.S. Army Corps of Engineers, Mobile District. *First Supplemental Environmental Report: Continuing Environmental Studies, Tennessee-Tombigbee Waterway*. 9 volumes. Mobile, AL: U.S. Army Corps of Engineers, Mobile District, August 1975.

U.S. Army Corps of Engineers, Mobile District. *Operational Forecast for Initial Traffic on the Tennessee-Tombigbee Waterway*. Prepared by Henry K. Jackson, Sr., of BHS Economic Research. Mobile, AL: U.S. Army Corps of Engineers, Mobile District, August 1985.

U.S. Army Corps of Engineers, Mobile District. *Public Meeting on the Tennessee-Tombigbee*

Waterway, Alabama and Mississippi, Held in Columbus, Mississippi on 23 January 1982. Mobile, AL: U.S. Army Corps of Engineers, Mobile District, 1982.

U.S. Army Corps of Engineers, Mobile District. *Public Meeting on the Tennessee-Tombigbee Waterway, Alabama and Mississippi, Held at Columbus, Mississippi, on 29 March 1977.* 5 volumes. Mobile, AL: U.S. Army Corps of Engineers, Mobile District, 1977.

U.S. Army Corps of Engineers, Mobile District. *Second Supplemental Environmental Report: Continuing Environmental Studies, Tennessee-Tombigbee Waterway.* 9 volumes. Mobile, AL: U.S. Army Corps of Engineers, Mobile District, October 1977.

U.S. Army Corps of Engineers, Mobile District. *Studies of Project Costs and Benefits: Tennessee-Tombigbee Waterway.* Mobile, AL: U.S. Army Corps of Engineers, Mobile District, October 1975.

U.S. Army Corps of Engineers, Mobile District. *Tennessee-Tombigbee Corridor Study: Main Report.* Mobile, AL: U.S. Army Corps of Engineers, Mobile District, September 1985.

U.S. Army Corps of Engineers, Mobile District. *Tennessee-Tombigbee Waterway, Alabama and Mississippi: Design Memorandum No. 1—General Design.* Mobile, AL: U.S. Army Corps of Engineers, Mobile District, 30 June 1960. [Approved by the Chief of Engineers, 12 April 1962.]

U.S. Army Corps of Engineers, Mobile District. *Tennessee-Tombigbee Waterway, Alabama and Mississippi: Supplement to Design Memorandum No. 1—General Design. Reevaluation of Project Economics.* Mobile, AL: U.S. Army Corps of Engineers, Mobile District, 30 June 1966.

U.S. Army Corps of Engineers, Mobile District. *Tennessee-Tombigbee Waterway, Alabama and Mississippi: Tombigbee River Multi-Resource District Proposed Mitigation Plan.* 3 volumes. Mobile, AL: U.S. Army Corps of Engineers, Mobile District, November 1977.

U.S. Army Corps of Engineers, Mobile District. *Tennessee-Tombigbee Waterway—Divide Cut—Nuclear Excavation Feasibility Study.* Mobile, AL: U.S. Army Corps of Engineers, Mobile District, 1966.

U.S. Army Corps of Engineers, Mobile District. *Third Supplemental Environmental Report: Continuing Environmental Studies, Tennessee-Tombigbee Waterway.* 13 volumes. Mobile, AL: U.S. Army Corps of Engineers, Mobile District, 1984.

U.S. Army Corps of Engineers, Mobile District. *Wildlife Mitigation Feasibility Study and Environmental Impact Statement for the Tennessee-Tombigbee Waterway, Alabama-Mississippi.* 3 volumes. Mobile, AL: U.S. Army Corps of Engineers, Mobile District, July 1983.

U.S. Army Corps of Engineers, Mobile and Nashville Districts. *Comments and Responses on the Final Supplement to the Environmental Impact Statement: Tennessee-Tombigbee Waterway, Alabama and Mississippi Navigation.* Mobile, AL and Nashville, TN: U.S. Army Corps of Engineers, Mobile and Nashville Districts, July 1982.

U.S. Army Corps of Engineers, Mobile and Nashville Districts. *Final Supplement to the Environmental Impact Statement: Tennessee-Tombigbee Waterway, Alabama and Mississippi Navigation.* 2 volumes. Mobile, AL and Nashville, TN: U.S. Army Corps of Engineers, Mobile and Nashville Districts, April 1982.

U.S. Army Corps of Engineers, Office of the Chief of Engineers. *Annual Report of the Chief of Engineers.* Washington: GPO, 1970–1985.

U.S. Congress. Congressional Budget Office. *Financing Waterway Development: The User Charge Debate.* CBO Staff Working Paper, July 1977. Washington: GPO, 1977.

U.S. Congress. House of Representatives. Committee on Appropriations. Subcommittee on Deficiencies and Army Civil Functions. *Investigation of Corps of Engineers Civil Works Program. Part 2, Proposed Tennessee-Tombigbee Waterway.* 82nd Congress, 1st session. Washington: GPO, 1951.

U.S. Congress. House of Representatives. Committee on Appropriations. Subcommittee on Public Works. Hearings, 1967–1985.

U.S. Congress. House of Representatives. Committee on Rivers and Harbors. *Waterway Connecting the Tombigbee and Tennessee Rivers.* House Document No. 486. 79th Congress, 2nd session. Washington: GPO, 1946.

U.S. Congress. House of Representatives. Subcommittee on Water Resources. *Corps of Engineers Oversight Hearings.* 94th Congress, 1st session, 9 May 1975. Washington: GPO, 1978.

U.S. Congress. Senate. Committee on Appropriations. Hearings on Public Works appropriations, 1967–1985.

U.S. Congress. Senate. Committee on Appropriations. Subcommittee on Energy and Water Development. *Tennessee-Tombigbee Waterway: Allegations, Responses, Discussions.* 96th Congress, 2nd session. Washington: GPO, 1980.

U.S. Congress. Senate. Committee on Environment and Public Works. Subcommittee on Water Resources. *Transportation Needs of Increased Coal Production and Completion of the Tennessee-Tombigbee Waterway.* 96th Congress, 2nd session. Washington: GPO, 1980.

U.S. Department of Commerce, Bureau of Economic Analysis. *Update of the Tennessee-Tombigbee Waterway Base Year Traffic to 1971 and Projections for Selected Years 1980–2020.* Final Report. Prepared for the Mobile District, U.S. Army Corps of Engineers. Washington: U.S. Department of Commerce, Bureau of Economic Analysis, April 1974.

U.S. Fish and Wildlife Service, Division of Ecological Services. *Bendway Management Study, Tennessee-Tombigbee Waterway, Alabama and Mississippi: A Fish and Wildlife Coordination Act Report.* Submitted to the Mobile District, U.S. Army Corps of Engineers. Daphne, AL: U.S. Fish and Wildlife Service, Division of Ecological Services, March 1984.

U.S. Fish and Wildlife Service, Division of Ecological Services. *Fish and Wildlife Coordination Act Report: Tennessee-Tombigbee Waterway, Alabama and Mississippi.* Submitted to the Mobile District, U.S. Army Corps of Engineers. Jackson, MS: U.S. Fish and Wildlife Service, Jackson Area Office, February 1981.

U.S. Fish and Wildlife Service, Division of Ecological Services. *A Resource Inventory of the Tennessee-Tombigbee Corridor.* Submitted to the Mobile District, U.S. Army Corps of Engineers. Decatur, AL: U.S. Fish and Wildlife Service, Division of Ecological Services, December 1981.

U.S. General Accounting Office. *An Overview of Benefit-Cost Analysis for Water Resources Projects—Improvements Still Needed.* Washington: General Accounting Office, 7 August 1978.

U.S. General Accounting Office. *To Continue or Halt the Tenn-Tom Waterway? Information to Help the Congress Resolve the Controversy.* Washington: General Accounting Office, 15 May 1981.

The White House Rural Development Initiatives: Area Development from Large-Scale Construction. Planning and Implementation Guidelines. Washington: n.p., January 1980.

Wilson, Stephen R.; John Zippert; and Wendell Paris. *History and Accomplishments of Affirmative Action during the Construction of the Tennessee-Tombigbee Waterway.* Mobile, AL: U.S. Army Corps of Engineers, Mobile District, May 1986.

Books, Pamphlets, and Nongovernment Reports

Allen, Barry; and Mina Hamilton Haefele (eds.). *In Defense of Rivers: A Citizens' Workbook on Impacts of Dam and Canal Projects.* Stillwater, NJ: Delaware Valley Conservation Association, 1976.

American Rivers Conservation Council, et al. *Disasters in Water Development: Some of the Most Economically Wasteful and Environmentally Destructive Projects of the Corps of Engineers, the Bureau of Reclamation, and the Tennessee Valley Authority.* Washington: n.p., April 1973.

American Rivers Conservation Council, et al. *Disasters in Water Development II: A Description of Army Corps of Engineers and Bureau of Reclamation Projects which Will Destroy Irreplaceable Natural and Cultural Resources along Some of America's Finest Rivers and Valleys.* Washington: n.p., February 1977.

Anderson, Frederick R. *NEPA in the Courts: A Legal Analysis of the National Environmental Policy Act.* Baltimore: Johns Hopkins University Press for Resources for the Future, 1973.

Andrews, Richard N. L. *Environmental Policy and Administrative Change: Implementation of the National Environmental Policy Act.* Lexington, MA: D. C. Heath, 1976.

Bachtel, Douglas C.; and Joseph J. Molnar. *Industrialization along the Tennessee-Tombigbee*

Waterway: Perceptions and Preferences of West Alabama Leaders. Rural Sociology Series No. 4. Auburn, AL: Agricultural Experiment Station, Auburn University, 1980.

Brose, David S. *Yesterday's River: The Archaeology of 10,000 Years along the Tennessee-Tombigbee Waterway.* Cleveland, OH: Cleveland Museum of Natural History, 1991.

Caldwell, Lynton K. *Science and the National Environmental Policy Act: Redirecting Policy through Procedural Reform.* University, AL: University of Alabama Press, 1982.

Carroll, Joseph L.; and Chester L. Meade. *Review of Coal Forecasts on the Tennessee-Tombigbee Waterway.* PTI 7721, Final Report to the Louisville and Nashville Railroad Co. University Park, PA: Pennsylvania State University, Pennsylvania Transportation Institute, December 1977.

Carroll, Joseph L.; and Srikanth Rao. *A Sensitivity Analysis of the Economics of the Tennessee-Tombigbee Water Project.* PTI 7720, Final Report to the Louisville and Nashville Railroad Co. University Park, PA: Pennsylvania State University, Pennsylvania Transportation Institute, December 1977.

Chang, Semoon; and Philip R. Forbus. *Estimating Savings in Transportation Costs from Use of the Tennessee-Tombigbee Waterway.* Research Report 3. Mobile, AL: University of South Alabama, Center for Business and Economic Research, February 1986.

Clement, Thomas M., Jr.; and Glen Lopez. *Engineering a Victory for Our Environment: A Citizen's Guide to the U.S. Army Corps of Engineers.* Washington: Institute for the Study of Health and Society, 1972.

Coalition for Water Project Review. *Tennessee-Tombigbee Waterway: Background Report.* Washington: Coalition for Water Project Review, March 1979.

Cobb, James C. *The Selling of the South: The Southern Crusade for Industrial Development, 1936–1980.* Baton Rouge: Louisiana State University Press, 1982.

Cowdrey, Albert E. *This Land, This South: An Environmental History.* Lexington: University Press of Kentucky, 1983.

Doster, James F.; and David C. Weaver. *Tenn-Tom Country: The Upper Tombigbee Valley in History and Geography.* University, AL: University of Alabama Press, 1987.

Federation of Southern Cooperatives. *Peoples Guide to Tennessee Tombigbee Waterway.* Epes, AL: Cooperator Press, 1975 (revised, 1977).

Ferejohn, John A. *Pork Barrel Politics: Rivers and Harbors Legislation, 1947–1968.* Stanford, CA: Stanford University Press, 1974.

Germane, Gayton E. *Transportation Policy Issues for the 1980's.* Reading, MA: Addison-Wesley Publishing Co., 1983.

Goldsmith, Edward; and Nicholas Hildyard. *The Social and Environmental Effects of Large Dams.* San Francisco: Sierra Club Books, 1984.

Gulf Research Associates, Inc. *Socio-Economic Assessment of Impacts Resulting from Alternative Plans, Tennessee-Tombigbee Waterway Wildlife Mitigation Study.* Mobile, AL: Gulf Research Associates, Inc., June 1981.

Gullatt, Doswell; and Associates. *An Evaluation of the Economic Justification for the Tennessee-Tombigbee Waterway Claimed by the Corps of Engineers in Its Restudy Report Dated June 30, 1960.* Washington: Doswell Gullatt and Associates, n.d. (probably 1961).

Hays, Samuel P. *Beauty, Health, and Permanence: Environmental Politics in the United States, 1955–1985.* Cambridge: Cambridge University Press, 1987.

Jones, Arthur R.; and Howard M. McLeskey. *Water Politics in Mississippi: A Comparative Analysis of Two Water Resource Development Organizations.* State College, MS: Mississippi State University, Water Resources Research Institute, 1969.

Kelly, Brian. *Adventures in Porkland: How Washington Wastes Your Money and Why They Won't Stop.* New York: Villard Books, 1992.

Laycock, George. *The Diligent Destroyers.* New York: Doubleday, 1970.

Liroff, Richard A. *A National Policy for the Environment: NEPA and Its Aftermath.* Bloomington: Indiana University Press, 1976.

Maass, Arthur. *Muddy Waters: The Army Engineers and the Nation's Rivers.* Cambridge: Harvard University Press, 1951.

Mazmanian, Daniel A.; and Jeanne Nienaber. *Can Organizations Change? Environmental Protection, Citizen Participation, and the Corps of Engineers.* Washington: Brookings Institution, 1979.

Molnar, Joseph J.; Leisle A. Ewing; Brenda Clark; and Macon Tidwell. *Developing the Tennessee-Tombigbee Corridor: Perceptions and Preferences of West Alabama Residents.* Rural Sociology Series No. 5. Auburn, AL: Agricultural Experiment Station, Auburn University, 1981.

Palmer, Tim. *Endangered Rivers and the Conservation Movement.* Berkeley: University of California Press, 1986.

Perhac, Ralph M. *Heavy Metal Distribution in Bottom Sediment and Water in the Tennessee River-Loudon Lake Reservoir System.* Research Report No. 40. Knoxville, TN: University of Tennessee, Water Resources Research Center, December 1974.

Petersen, Margaret S. *River Engineering.* Englewood Cliffs, NJ: Prentice-Hall, 1986.

Powledge, Fred. *Water: The Nature, Uses, and Future of Our Most Precious and Abused Resource.* New York: Farrar, Straus, and Giroux, 1982.

Proceedings of the 10th Annual Environmental and Water Resources Engineering Conference, June 3–4, 1971. Technical Report No. 25. Nashville, TN: Department of Environmental and Water Resources Engineering, Vanderbilt University, 1971.

Rathlesberger, James (ed.). *Nixon and the Environment: The Politics of Devastation.* New York: Village Voice, 1972.

Reisner, Marc. *Cadillac Desert: The American West and Its Disappearing Water.* New York: Viking, 1986.

Shank, William H. *Towpaths to Tugboats: A History of American Canal Engineering.* York, PA: American Canal and Transportation Center, 1985.

Sherman, Harry L. (ed.). *Environmental Quality II: An MSCW Leadership Forum.* Columbus, MS: Mississippi State College for Women, 1972.

Stewart, William H., Jr. *The Tennessee-Tombigbee Waterway: A Case Study in the Politics of Water Transportation.* University, AL: Bureau of Public Administration, University of Alabama, 1971.

Taylor, Serge. *Making Bureaucracies Think: The Environmental Impact Statement Strategy of Administrative Reform.* Stanford, CA: Stanford University Press, 1984.

Tennessee-Tombigbee Waterway Development Authority. *The Tennessee-Tombigbee Waterway Story.* Columbus, MS: Tennessee-Tombigbee Waterway Development Authority, January 1969.

Tennessee-Tombigbee Waterway Development Authority. *The Truth about Tenn-Tom.* Columbus, MS: Tennessee-Tombigbee Waterway Development Authority, August 1980.

Weatherford, J. McIver. *Tribes on the Hill.* New York: Rawson, Wade Publishers, 1981.

Wenner, Lettie M. *The Environmental Decade in Court.* Bloomington: Indiana University Press, 1982.

Whitaker, John C. *Striking a Balance: Environmental and Natural Resources Policy in the Nixon-Ford Years.* Washington: American Enterprise Institute for Public Policy Research, 1976.

Wilkins, Glover. *Facts about the Tennessee-Tombigbee Waterway Development Authority.* Columbus, MS: Tennessee-Tombigbee Waterway Development Authority, 1968.

Worsham, John P., Jr. *The Tennessee-Tombigbee Waterway: A Chronological Bibliography in Report Form.* Exchange Bibliography No. 1189. Monticello, IL: Council of Planning Librarians, 1976.

Articles

"Action Now." *Sierra Club Bulletin,* 56 (August 1971), 27.

"AGC Challenges Government on EEO and Apprenticeship." *Engineering News-Record,* 197 (October 28, 1976), 41.

Ammons, Linda. "Waterway: A Way to More Jobs." *Black Enterprise,* 9 (October 1978), 26.

"Army Auditors Fault Corps on Tenn-Tom Estimating." *Engineering News-Record,* 200 (April 27, 1978), 11.

"Army Corps of Engineers Given Go-Ahead to Start Tennessee-Tombigbee Barge Canal." *Malacological Review,* 6 (1973), 74.

"An Atomic Blast to Help Build a U.S. Canal?" *U.S. News and World Report,* 54 (May 20, 1963), 14.

Auerbach, S. I.; P. E. LaMoreaux; and G. J. McLindon. "The Tennessee-Tombigbee Waterway: An Example of Synergistic Science and Engineering." *Environmental Geology and Water Sciences,* 7 (nos. 1/2, 1985), 3–4.

Barfield, Claude E.; and Richard Corrigan. "Environment Report: White House Seeks to Restrict Scope of Environmental Law." *National Journal,* 4 (February 26, 1972), 336–349.

"Behind All the Furor over 'Tenn-Tom.'" *U.S. News and World Report,* 86 (February 19, 1979), 65.

"Benefits or Baloney?" *Newsweek,* 78 (September 13, 1971), 65.

Bierman, Don E.; and W. Rydzkowski. "Regional Politics in Public Works Projects: The Tennessee-Tombigbee Waterway." *Transportation Quarterly,* 45 (April 1991), 169–180.

"Big Digging on 'Tenn-Tom' Waterway." *The eM-Kayan,* 38 (November 1979), 10–13.

"Big Digging on the Tenn-Tom Waterway." *The eM-Kayan,* 40 (January 1982), 3–5.

"The Big Tenn-Tom: Benefit or Boondoggle?" *Business Week,* no. 2520 (February 6, 1978), 94 and 98.

"Biggest Waterway in the South." *Port Construction International,* 2 (September–November 1985), 50–51.

"Boondoggle." *Forbes,* 117 (March 1, 1976), 54–55.

Borzo, Greg. "Tenn-Tom Development Falters; Coal Down, Forest Products Up." *Traffic World* (September 14, 1987), 22–24.

Bragonier, Reg. "A New Shortcut from the Great Lakes to the Gulf." *Motor Boating and Sailing,* 155 (April 1985), 48–53, 140, 142.

Brumfield, Carroll. "The Tenn-Tom." *Environmental Action,* 4 (July 22, 1972), 11.

Bryan, Jack H. "Hydrogeological and Geotechnical Aspects of the Tennessee-Tombigbee Waterway." *Environmental Geology and Water Sciences,* 7 (nos. 1/2, 1985), 25–50.

"The C&G: Linking the Tenn-Tom and the Mississippi." *Mainstream Mississippi,* 3 (Winter 1983–84), 6–8.

"A Canal Plan Makes Waves in Washington." *Electrical World,* 191 (April 1, 1979), 31–32.

Carroll, Joseph L. "Tennessee-Tombigbee Waterway Revisited." *Transportation Journal,* 22 (Winter 1982), 5–20.

Carroll, Joseph L.; and Srikanth Rao. "Economics of Public Investment in Inland Navigation: Unanswered Questions." *Transportation Journal,* 17 (Spring 1978), 27–54.

Carroll, Joseph L.; and Srikanth Rao. "The Role of Sensitivity in Inland Water Transport Planning: The Case of the Tennessee-Tombigbee." *Transportation Research Forum,* 19 (Proceedings—19th Annual Meeting, 1978), 46–54.

Chang, Semoon; and Philip R. Forbus. "Tenn-Tom Versus the Mississippi River." *Transportation Journal,* 25 (Summer 1986), 47–54.

Chaze, William L.; Robert Barr; and Robert S. Dudney. "The $75 Billion Pork-Barrel Ripoff." *U.S. News and World Report,* 94 (May 2, 1983), 18–21.

Clemmer, Glenn H. "The Tennessee-Tombigbee Waterway: Are We Sure of the Costs?" In Harry L. Sherman (ed.), *Environmental Quality II: An MSCW Leadership Forum* (Columbus, MS: Mississippi State College for Women, 1972), pp. 105–110.

Clemmer, Sherrill M. "It's Time to Stop the 'Tenn-Tom' Waterway." *Sierra Club Bulletin,* 62 (February 1977), 23.

Coan, Gene. "Congress and the Environment." *Sierra Club Bulletin,* 67 (January–February 1982), 32–33.

"Conservationists Ask Court to Block Construction of Tennessee-Tombigbee Link." *Traffic World,* 147 (July 19, 1971), 14.

"Corps of Engineers: Digging Their Own Grave." *The Economist* (February 17, 1979), 47–48.

"Corps Says Environmental Effects of Tombigbee Project Not Detrimental." *Environmental Reporter* (June 11, 1972), 149.

"Corps' Tenn-Tom Job Called Illegal." *Engineering News-Record*, 200 (May 11, 1978), 9.

"Cost-Benefit Trips up the Corps." *Business Week*, no. 2573 (February 19, 1979), 96–97.

Davis, Joseph A. "Water Projects Await Funding as Debate over Cost Sharing Stalls Action on Capitol Hill." *Congressional Quarterly Weekly Report*, 41 (July 30, 1983), 1551–1558.

"Debut of the Giant 'Tenn-Tom'." *U.S. News and World Report*, 98 (January 28, 1985), 14.

Denning, Frank C. "Tennessee-Tombigbee Waterway." *Bulletin of the Permanent International Association of Navigation Congresses*, no. 23 (1976), 16–22.

"Dreaming of the Golden Gulf." *Time*, 113 (January 15, 1979), 15.

Drew, Cynthia A. "The Tennessee-Tombigbee Waterway: Divide Section." *Military Engineer*, 77 (May–June 1985), 170–174.

Dunn, Herbert D. "Tenn-Tom Waterway Criticized." *Alabama Conservation*, 42 (February–April 1972), 12.

"Excavation of 95-Million Yards Nears End on Tenn-Tom Project." *The eM-Kayan*, 41 (November 1982), 6–9.

"Five Clams Believed in Danger of Extinction." *Endangered Species Technical Bulletin*, 11 (May 1986), 3–4.

"Flooding Marks Finale of Tenn-Tom Project." *The eM-Kayan*, 42 (November 1983), 3.

"Folklore of the Tenn-Tom." *Earthwatch* (Spring 1981), 48.

Garnett, Bernard. "Black Judge Raps Foes of Waterway." *Race Relations Reporter*, 3 (January 17, 1972), 5–6.

Gay, Forest T., III. "Tenn-Tom Waterway Opens." *Constructor*, 67 (May 1985), 58–61.

Godfrey, K. A., Jr. "Waterway Is Public Works Landmark." *Civil Engineering*, 56 (July 1986), 42–45.

Grace, Randall. "For the Betterment of the Nation: The Selling of the Tennessee-Tombigbee Waterway." In Barry Allen and Mina Hamilton Haefele (eds.), *In Defense of Rivers: A Citizens' Workbook on Impacts of Dam and Canal Projects* (Stillwater, NJ: Delaware Valley Conservation Association, 1976), pp. 146–153.

Green, Samuel R. "An Overview of the Tennessee-Tombigbee Waterway." *Environmental Geology and Water Sciences*, 7 (nos. 1/2, 1985), 9–13.

Greene, Johnny. "The Corps of Engineers in Fantasy Land." *Inquiry*, 2 (May 14, 1979), 13–17.

Greene, Johnny. "Selling Off the Old South." *Harper's*, 254 (April 1977), 39–46, 51–58.

Greer, Julian A. "The Tennessee-Tombigbee Waterway." *Military Engineer*, 65 (September–October 1973), 336–339.

Griffith, Harry A. "The Tennessee-Tombigbee Waterway Project and Its Effects upon the Environment." In Harry L. Sherman (ed.), *Environmental Quality II: An MSCU Leadership Forum* (Columbus, MS: Mississippi State College for Women, 1972), pp. 91–104.

"The Growing Clamor to Halt the Tenn-Tom." *Business Week*, no. 2711 (October 26, 1981), 58–59.

Hall, Alice J. "The Hidden Tenn-Tom: Bypassed but Still Striving." *National Geographic*, 169 (March 1986), 384–387.

Hamilton, Katie. "The Tenn-Tom: New Route through the Old South." *Lakeland Boating*, 37 (September 1982), 44–45.

Hazard, John L. "Comments on 'Unfulfilled Promises' and 'Unanswered Questions.'" *Transportation Journal*, 17 (Spring 1978), 55–64.

Herndon, G. Melvin. "A 1796 Proposal for a Tennessee-Tombigbee Waterway." *Alabama Historical Quarterly*, 37 (Fall 1975), 176–182.

Josephy, Alvin M., Jr. "The South's Unstoppable Waterway." *Fortune*, 100 (August 27, 1979), 80–82.

"'Justifications' for Waterway Projects." *Traffic World*, 181 (March 31, 1980), 5.

Kennedy, William T. "Tennessee-Tombigbee Waterway Links Heartland to World Markets." *Business America*, 8 (June 24, 1985), 16.

Kirschten, Dick. "Congress Makes Waves over Carter's Water Policy." *National Journal*, 10 (August 1, 1978), 1052–1056.

Kirschten, Dick. "The Less Painful Strategy." *National Journal*, 13 (May 16, 1981), 891.

Kirschten, Dick. "Playing Water Politics." *National Journal*, 11 (April 7, 1979), 570.

Kirschten, Dick. "A Trial Run for a New Water Policy." *National Journal*, 11 (January 20, 1979), 88.

Kirschten, J. Dicken. "Draining the Water Projects Out of the Pork Barrel." *National Journal*, 9 (April 9, 1977), 540–548.

Kizzia, Tom. "Will Tenn-Tom's Approval Spawn a 'Tenn-Tom North'?" *Railway Age*, 181 (July 4, 1980), 12.

"Lawsuit Challenges Proposed Tennessee-Tombigbee Waterway." *Conservation News*, 36 (August 15, 1971), 6.

LoDico, John. "Historic Tenn-Tom Waterway Opens Amid Cautious Forecasts for Growth." *Traffic World*, 202 (June 10, 1985), 23–26.

Love, Sam. "A Native Returns to the Tombigbee." *Living Wilderness*, 37 (Summer 1973), 24–26, 28, 30–31.

"Major Shipping Route Unfolds: Earthmoving on Grand Scale along 'Tenn-Tom' Waterway." *The eM-Kayan*, 39 (August 1980), 3–7.

McClure, Nathaniel D., IV. "A Major Project in the Age of the Environment: Out of Controversy, Complexity, and Challenge." *Environmental Geology and Water Sciences*, 7 (nos. 1/2, 1985), 15–24.

McClure, Nathaniel D., IV. "A Summary of Environmental Issues and Findings: Tennessee-Tombigbee Waterway." *Environmental Geology and Water Sciences*, 7 (nos. 1/2, 1985), 109–124.

McClure, Nathaniel Dehass, IV. "Tennessee-Tombigbee Waterway and the Environment." In Steven R. Abt and Johannes Gessler (eds.), *Hydraulic Engineering* (New York: American Society of Civil Engineers, 1988), pp. 328–333.

McIntyre, Kenneth E. "Managing Macroproject Externals: The Tennessee-Tombigbee Waterway." In Frank P. Davidson and C. Lawrence Meador (eds.), *Macro-Engineering: Global Infrastructure Solutions* (New York: Ellis Horwood, 1992), pp. 43–50.

McLindon, Gerald J. "Creative Spoil: Design Concepts, Construction Techniques, and Disposal of Excavated Materials." *Environmental Geology and Water Sciences*, 7 (nos. 1/2, 1985), 91–108.

Miller, James Nathan. "Bitter Battle of the Waterways." *Reader's Digest*, 111 (September 1977), 83–87.

Miller, James Nathan. "Half a Billion Dollars Down the Drain." *Reader's Digest*, 109 (November 1976), 143–148.

Miller, James Nathan. "Trickery on the Tenn-Tom." *Reader's Digest*, 113 (September 1978), 138–143.

Moser, Don. "Dig They Must, the Army Engineers, Securing Allies and Acquiring Enemies." *Smithsonian*, 7 (December 1976), 40–51.

Mosher, Lawrence. "Water Politics as Usual May Be Losing Ground in Congress." *National Journal*, 12 (July 19, 1980), 1187–1190.

Nixon, Richard. "Remarks at the Dedication of the Tennessee-Tombigbee Waterway in Mobile, Alabama. May 25, 1971," in *Public Papers of the Presidents of the United States: Richard Nixon, 1971* (Washington: GPO, 1972), pp. 658–662.

"NTU to Fight Tenn-Tom in Senate." *Dollars & Sense*, 12 (September 1981), 5.

O'Brien, Kathleen M. "Tenn-Tom Plans 87% Coal Traffic, 1986 Completion." *Coal Age*, 83 (July 1978), 126–128.

"Outlays for Environmental Items Stabilize: Stage Is Set for Congressional Battle." *National Journal*, 4 (January 29, 1972), 202.

Ovechka, Greg. "Tenn-Tom: America's New Waterway." *Marine Engineering/Log*, 186 (July 1981), 46–50, 52, 54, 57–58.

Parfit, Michael. "The Army Corps of Engineers: Flooding America in Order to Save It." *New Times*, 7 (November 12, 1976), 25–26, 30, 34–37.

Patterson, Carolyn Bennett. "The Tennessee-Tombigbee Waterway: Bounty or Boondoggle?" *National Geographic*, 169 (March 1986), 364–383.

Pelham, Ann. "Water Policy: Battle over Benefits." *Congressional Quarterly Weekly Report*, 36 (March 4, 1978), 565–574.

Perry, James M. "The Great Washington Pork Feast." *Audubon*, 81 (July 1979), 102–107.

Price, Steve. "Exploring the Tenn-Tom Waterway." *Sports Afield*, 189 (March 1983), 104–106, 120.

"Project Checklist: Tennessee-Tombigbee Waterway." *World Dredging and Marine Construction*, 12 (October 1976), 39–40.

"Rail Spokesman Labels 'Tenn-Tom' Waterway 'Waste of Tax Money'." *Traffic World*, 161 (March 17, 1975), 31.

Reiger, George. "Let's Terminate the Tenn-Tom!" *Georgia Sportsman*, 2 (February 1977), 22–23.

"Riprap Machine to 'Armor' Tenn-Tom Slopes." *The eM-Kayan*, 40 (May 1981), 3.

Roberts, Paul E. "Benefit-Cost Analysis: Its Use (Misuse) in Evaluating Water Resource Projects." *American Business Law Journal*, 14 (Spring 1976), 73–84.

Roberts, Paul E. "A Primer of Environmental Economics." *Naturalist*, 46 (February 1973), 15–19.

Roberts, Paul E. "The Role of Economics in the Environmental Law Suit." *Proceedings of the 1971 DuPont Environmental Engineering Seminar*, 26 (November 1972), 37–42.

Rodeffer, Stephanie Holschlag; and Lloyd N. Chapman. "Managing to Manage: A Case Study of the Tennessee-Tombigbee Waterway." *Contract Abstracts and CRM Archeology*, 1 (May 1980), 19–23.

Rohan, Thomas M. "A River-Barge Shortcut: Will 'Tenn-Tom' Waterway Start a Boom?" *Industry Week*, 225 (May 27, 1985), 28.

Sagner, James S. "Benefit/Cost Analysis: Efficiency-Equity Issues in Transportation." *Logistics and Transportation Review*, 16 (no. 4, 1980), 339–388.

Simmons, Marvin D. "Unwatering the Divide Cut of the Tennessee-Tombigbee Waterway: A Major Challenge to Construction." *Environmental Geology and Water Sciences*, 7 (nos. 1/2, 1985), 51–67.

Smith, R. Jeffrey. "The Waterway That Cannot Be Stopped." *Science*, 213 (August 14, 1981), 741–742, 744.

Stine, Jeffrey K. "Environmental Politics in the American South: The Fight over the Tennessee-Tombigbee Waterway." *Environmental History Review*, 15 (Spring 1991), 1–24.

Stine, Jeffrey K. "Regulating Wetlands in the 1970s: U.S. Army Corps of Engineers and the Environmental Organizations." *Journal of Forest History*, 27 (April 1983), 60–75.

Stine, Jeffrey K. "The Tennessee-Tombigbee Waterway and the Evolution of Cultural Resources Management." *The Public Historian*, 14 (Spring 1992), 7–30.

"Sue the Bastards" *Time*, 98 (October 18, 1971), 54 and 57.

"The 'Tenn-Tom': As Waterways Go, Benefits Grow." *World Dredging and Marine Construction*, 12 (October 1976), 26–29.

"Tenn-Tom Challenge: Keep Huge Earthmoving Fleet Rolling." *The eM-Kayan*, 39 (August 1980), 7.

"Tenn-Tom Construction Is Officially Under Way." *Waterways Journal*, 86 (December 23, 1972), 5.

"Tenn-Tom: Lose a Billion, or Keep Digging?" *U.S. News & World Report*, 90 (June 15, 1981), 12.

"Tenn-Tom Taken to Court." *Environmental Action*, 3 (July 24, 1971), 6.

"Tenn-Tom Traps Corps in Political Web." *Engineering News-Record*, 202 (February 8, 1979), 8–9.

"Tenn-Tom Waterway: Was Congress Misled?" *Railway Age*, 179 (December 11, 1978), 10.

"Tenn-Tom Will Be a Sportsman's Paradise." *Waterways Journal*, 86 (December 2, 1972), 30.

"Tenn-Tom's Soil Headaches Get Heavy Doses of Hardhat Know-how." *Engineering News-Record*, 204 (January 17, 1980), 56–58.

"Tenn-Tom's Trials." *Time,* 109 (April 4, 1977), 19.
"The Tennessee-Tombigbee: The Largest Earth-Moving Project in U.S. History Will Link Mid-America with the Gulf of Mexico." *Work Boat,* 33 (February 1976), 29, 59–60.
"Tennessee-Tombigbee Waterway: Disaster in Water Development." *National Parks and Conservation Magazine,* 47 (July 1973), 30.
"Tennessee-Tombigbee Waterway: A Start." *Environmental Science and Technology,* 5 (July 1971), 581–583.
"Time for Tom-Tenn?" *Chemical Week,* 106 (May 27, 1970), 23.
"Tombigbee's Goal for Coal Movements: A 'Pipe Dream'." *Coal Week* (February 12, 1979), 3.
Underwood, Kenneth D.; and F. Dewayne Imsand. "Hydrology, Hydraulic, and Sediment Considerations of the Tennessee-Tombigbee Waterway." *Environmental Geology and Water Sciences,* 7 (nos. 1/2, 1985), 69–90.
Van Brahana, John. "Beneath the Tenn-Tom." *Water Spectrum,* 7 (Winter 1975–76), 17–24.
Van Brahana, John. "Tenn-Tom Investigation Goes Underground," *World Dredging and Marine Construction,* 12 (October 1976), 32–34.
Waldon, Donald G. "The Tennessee-Tombigbee Waterway: America's Next Transportation Artery." *Constructor,* 64 (May 1982), 38–42.
"Water: A Billion Dollar Battleground." *Time,* 109 (April 4, 1977), 17–20.
"A Water Tour along Proposed Route of the Tennessee-Tombigbee Waterway." *Waterways Journal,* 86 (December 2, 1972), 20–21, 32.
"Waterway Work Goes on Despite Environmental Protest." *Engineering News-Record,* 186 (June 10, 1971), 14.
Werneth, George. "Trouble Along the Tenn-Tom." *Atlanta Magazine,* 22 (October 1982), 76–80, 107–108, 110.
"West Alabama's Short Cut to Progress: Special Report, the Tenn-Tom Impact." *Alabama News Magazine,* 42 (June 1976), 1–20.
White, David Fairbank. "The Ditch that Could Change the Country," *Parade Magazine* (August 12, 1984), 10–11.
"Wildlife on the Tenn-Tom." *Southern Living,* 20 (December 1985), 34.
"Women and Minorities Receive Training for Operation of Earthmoving Equipment." *The eM-Kayan,* 39 (August 1980), 4.

Interviews

Amonett, D. Ray, 17 September 1986.
Antle, Lloyd George, 7 January 1987.
Arnold, Richard S., 1 June 1987.
Atkins, George H., 18 November 1986.
Belcher, Owen Doug, 10 October 1986.
Benforado, David J., 22 May 1987.
Bevill, Tom, 20 May 1987.
Birindelli, Joseph R., Jr., 20 May 1986.
Blackwelder, Brent, 12 April 1988.
Blanding, Roland E., Jr., 7 October 1986.
Boatman, Howard, 16 September 1986.
Brandes, William, 18 September 1986.
Bratton, Joseph K., 22 May 1987.
Bryan, Jack H., 17 November 1986.
Burke, Roger A., 17 November 1986.
Burns, Daniel R., 22 May 1986.
Carroll, Joseph L., 19 and 20 November 1987.
Cathey, H. Joe, 15 September 1986.
Clarke, Frederick J., 15 December 1987.

Clemmer, Glenn H., 15 April 1987.
Connell, Norman., 28 May 1986.
Connell, Richard J., Jr., 6 October 1986.
Couch, Frank B., Jr., 17 September 1986.
Crisp, Robert L., Jr., 8 October 1986.
Currie, F. L. (Les), 15 May 1986.
Deming, Frank C., 21 November 1986.
Dola, Steven, 9 January 1987.
Domenici, Pete V., 3 August 1987.
Eaton, Harold C., 30 May 1986.
Edgar, C. Ernest, III, 8 October 1986.
Edwards, Jack, 20 November 1986.
Epps, Tom P., 27 May 1986.
Etheridge, Nick J., III, 14 May 1986.
Galdis, Alan V., 17 November 1986.
Gay, Forrest T., III, 21 May 1987.
Gholson, Hunter M., 28 May 1986.
Gillespie, Norman L., 7 May 1986.
Goad, James G., 18 and 19 September 1986.

Grantham, Billy, 16 September 1986.
Gray, Herman, 16 September 1986.
Green, Lawrence R., 21 November 1986.
Hall, Daniel F., 15 September 1986.
Hatch, Henry J., 16 December 1987.
Hearrean, William F., 21 March 1986.
Hewell, Joe E., 28 May 1986.
Hill, William, 19 September 1986.
Holland, Robert G., 9 October 1986.
Holmes, Vernon S., 30 May 1986.
Hooper, Charles B., 19 September 1986.
Huneke, William F., 22 May 1987.
Hunt, Joycelyn W., 19 November 1986.
Jabbour, Alan, 6 January 1987.
Jackson, Henry K., 18 November 1986.
Jones, Freddy R., 21 November 1986.
Jordan, James C., 11 June 1986.
Karwedsky, Robert, 18 September 1986.
Keady, William C., 2 May 1986.
Keel, Bennie C., 8 January 1987.
Kelly, Patrick J., 11 June 1986.
Kimberl, Richard T., 21 November 1986.
Kuehnert, Cecil W., 29 May 1987.
Lancelot, Jill, 6 August 1987.
Lanier, Philip M., 23 November 1987.
LeTellier, Carroll N., 14 April 1988.
Leverty, Richard, 10 June 1986.
Mallory, Jack C., 16 May 1986 and 30 June 1986.
McClure, Nathaniel D. (Skeeter), IV, 16 May 1986.
McIntyre, Kenneth E., 14 November 1986.
Miller, Gerald, 10 April 1986.
Miller, John U., III, 17 September 1986.
Miller, Robert C., 16 September 1986.
Mims, Lambert C., 25 March 1986.
Misso, Robert, 8 January 1987.
Morris, John W., 15 December 1987.
Nelson, Gaylord, 15 April 1988.

Nielsen, Jerry J., 20 March 1986 and 14 May 1986.
Nottingham, Jonathan D., 6 October 1986.
Odom, Benton Wayne, Jr., 20 May 1986.
Osborne, William H., 9 October 1986.
Ottinger, Gene A., 19 September 1986.
Patton, Walter S., 18 November 1986.
Pierce, Tommie E., Jr., 30 May 1986.
Roach, Randy, 29 May 1986.
Roady, Stephen E., 20 May 1987.
Roberts, Paul E., 5 August 1987.
Robertson, William L., 9 June 1986.
Ross, Pat, 27 May 1986.
Rucker, Marc D., 8 October 1986.
Ruland, Willis E., 19 November 1986.
Rushing, John Wayne, 6 October 1986.
Seckinger, Ernest W., Jr., 16 May 1986.
Smith, Joe M., 29 May 1986.
Smyth, Robert V., 16 September 1986.
Steinberg, Bory, 9 January 1987.
Stennis, John C., 3 August 1987.
Stephenson, Robert J., 7 October 1986.
Stoudenmire, George D., 19 November 1986.
Thompson, Frederick G., 21 May 1986.
Tidmore, John Jeffrey, 20 May 1986.
Underwood, Kenneth D., 21 May 1986.
Veysey, Victor V., 30 January 1986.
Waldon, Donald G., 27 and 28 May 1986.
Wall, James, 7 October 1986.
Wall, John F., 6 January 1987.
Wallace, George C., 19 May 1986.
White, Christopher G., 9 October 1986.
Whitten, Jamie, 11 December 1986.
Wilkins, Glover, 15 and 16 November 1986.
Williams, Dennis R., 18 September 1986.
Williams, James D., 7 January 1987 and 11 February 1987.
Wodder, Rebecca R., 15 April 1988.
Wood, Lance D., 19 May 1987.

Index

A. T. Kearney Management Consultants, Inc., 158. *See also* Kearney Report
AAA. *See* Army Audit Agency
AAR. *See* Association of American Railroads
Aberdeen, Mississippi, 122
Aberdeen Lock, 37, 249
Abernethy, Thomas G., 17, 50, 94
AEC. *See* Atomic Energy Commission
Affirmative action. *See* employment goals for minorities and local workers
African-Americans. *See* minorities
Alabama Conservancy, 94–5, 104, 132–33, 141; and second Tenn-Tom litigation, 147, 206, 212, 297 n.19
Alabama Department of Conservation and Natural Resources, 81
Alabama League of Women Voters, 104
Alabama News Magazine, 176
Alexander, Clifford L., Jr., 174–75, 189–91
Aliceville Lock and Dam, 37, 41, 46–7; cost increase, 154–55; photo, 251
Allen, Jim, 106
American Commercial Barge Line, 249
American Rivers Conservation Council, 166, 220
Americans for Democratic Action, 241
An Overview of Benefit-Cost Analysis for Water Resources Projects—Improvements Still Needed, 163
Appalachian mountain chain, 2, 34
Appleton, Richard, 120
Appropriations. *See* construction funds

Arab oil embargo, 154
Area Development from Large-Scale Construction. *See* Carter Administration
Area redevelopment: as economic justification, 5, 18, 20, 25
Areawide Affirmative Action Plan, 188
Arkansas River project, 101
Army Audit Agency (AAA), 9, 207, 223, 231; evaluation of Tenn-Tom economics, 161–162
Army Corps of Engineers, 6, 10, 18, 23, 84, 161; pre-World War II activities, 2, 3, 13, 14–15; Board of Engineers for Rivers and Harbors, 21–22; Facilities Engineer Support Activity, 22; organization, 34–35. *See also* Cross-Florida Barge Canal; Environmental Advisory Board; Gillham Dam; Locks and Dam 26; Mobile District; Nashville District; Nuclear Cratering Group; Ohio River Division; quarterly meetings; Red River Waterway; Secretary of the Army; South Atlantic Division; St. Lawrence Seaway; "308" studies; Trinity River navigation project; Waterways Experiment Station
Arnold, Richard S., 110, 135, 278 n.4; and first Tenn-Tom litigation, 113–114, 118, 120–122, 127
Ash, Roy, 140
Askew, Reubin, 106
Association of American Railroads (AAR), 15, 17, 130–134, 220, 237; Economics and

325

Finance Department, 21; Tennessee-Tombigbee Waterway Project Committee, 21–22, 133–134
Association of Southeastern Biologists, 136, 137
Atlanta Journal and Constitution, 249
Atomic Energy Commission (AEC), 54
Audubon Society. *See* National Audubon Society
Austin, J. A., 201
Authorization. *See* Congress

Baker, Howard H., Jr., 186–187, 228, 234
Bankhead Lock, 63
Bankhead Wilderness, 95
Barge industry: competition with railroads, 8. *See also* lobbying
Bay Springs Lock and Dam, 4, 35, 47, 54; design, 41, 60–64; lockages, 48; photo, 63; navigational limits, 160
Beal, Thaddeus R., 180
Beaver Lake Recreation Area, 52
Bedell, Berkley, 232–233
Bell, Griffin, 205
Bendways, 37–39
Benefit/cost ratio, 3, 24, 62, 103, 114; criticism, 16–18, 19, 20, 22; exaggeration, 112, 162, 163; and first Tenn–Tom litigation, 123, 126; challenged by railroads, 132, 139; challenged by Environmental Defense Fund, 135–136; criteria, 138, 151, 202; Congressional oversight, 162–164, 223; erosion, 155, 157, 200; recalculation, 156–161; and second Tenn-Tom litigation, 208, 210, 212, 213, 214; updated benefit/cost ratios, 229–230. *See also* economic evaluations; interest rate; Kearney Report; social benefits
Bennett, James B., Jr., 131
Bevill, Tom, 68, 158, 210, 227, 234–235; chairmanship of House Subcommittee on Public Works, 174; and minority participation, 193, 194, 195; and 1980 supplemental appropriation, 226; and multiple-year appropriations, 243–245
Big Bear Creek, 13,
Birmingham, Alabama, 106, 186
Birmingham Audubon Society, 147, 168, 206, 212, 297 n.19
Bisha, R. E., 138
Black Warrior River, 36, 63, 250
Black Warrior-Tombigbee Waterway, 5, 6, 89, 146, 159; opposition to expansion, 145, 206–208; plans to expand, 160–161, 200–201, 208, 253
Blackwelder, Brent, 99, 173, 245; and Office of Management and Budget, 139–140, 166; and L&N Railroad, 140–142; and minority participation, 184, 195

Board of Engineers for Rivers and Harbors. *See* Army Corps of Engineers
BoB. *See* Bureau of the Budget
Boise, Idaho, 76
Bonneville Dam, 72
Bottomland hardwoods, 36, 39, 86. *See also* wetlands
Branch, William McKinley, 180–181, 185
Brandes, William F., 122, 124–125, 128
Brennan, William J., Jr., 110
Bridge replacements, 66, 142, 162, 164–167
Briggs, Richard, 148
Brooke, Edward W., 184
Brotherhood of Locomotive Engineers, 141, 143
Brotherhood of Railway and Airline Clerks, 143, 242
Browder, Joe, 104
Brown, Harold, 54
Brown, Jon T. (Rick), 111, 142, 198, 224
Brown, Robert J., 180
Brown and Root Inc., 55, 71, 78
Burch, Jack, 124
Burdin, Robert P., 124
Bureau of Reclamation (BuRec), 15, 168, 248
Bureau of the Budget (BoB), 19, 20, 21; as target of lobbying, 22, 24; and Tennessee-Tombigbee Waterway Development Authority, 26–27, 30; controversy surrounding, 162–163. *See also* Office of Management and Budget
BuRec. *See* Bureau of Reclamation
Burgin, William G., Jr., 115–116, 119
Bush, Fred M., Jr., 119

Caldwell, R. Dale, 87–88
Cameron, Roderick A., 135
Canal section, 33–34, 47–54, 213; length, 4, 265 n.37; illustrations, 51. *See also* chain-of-lakes; Locks A, B, C, D, and E; perched canal
Carlson, Peter, 173
Carroll, Joseph L., 143–144
Carter, Jimmy: as target of lobbying, 146, 168; reevaluation of Tenn-Tom, 167–175, 210; establishment of Rural Initiative Program, 178, 191; concern for civil rights, 189–191, 192. *See also* water projects review list
Carter Administration, 9, 203, 254; Small Community and Rural Development Policy, 192; Area Development from Large-Scale Construction, 192. *See also* Rural Initiative Program
Cassidy, William, 57
Cathey, Joe, 80
CBC. *See* Congressional Black Caucus
CEQ. *See* Council on Environmental Quality
Chafee, John H., 228, 235

326 Index

Chain-of-lakes: design, 48–50, 51, 52, 201, 213; illustration, 51; injunction, 214
Chambers, Samuel, 76–77
Change of venue: in first Tenn-Tom litigation, 119–122; in second Tenn-Tom litigation, 205
Channel dimensions, 5; expansion, 9, 18–19, 21, 206–207; of 1875 plan, 13; of 1913 plan, 14; as considered in 308 studies, 14
Chicago Sun-Times, 153–154
Christenbury, Edward S., 205
Christian Science Monitor, 248
Circeo, Louis J., Jr., 55
Civil rights movement, 9, 253; and arguments for building the Tenn-Tom, 7; as target of pro-Tenn-Tom lobbying, 177; legacy, 177–178, 183, 184
Clark, Charles, 126–127
Clarke, Frederick J., 96, 209
CLEAN. *See* Committee for Leaving the Environment of America Natural
Clean Water Act (1977), 212
Clemmer, Glenn H., 84–85, 93, 136–137, 144–145, 199
Clemmer, Sherrie, 209
Cleveland *Plain Dealer,* 194
Cleveland harbor improvement, 195
Clinch River Breeder Reactor, 243
Coalition for Water Project Review, 173, 237; constituent members, 292 n.83
Coalition of American Rivers, 143
Cochran, Thad, 165, 166, 222
Cohen, William S., 225
Columbus, Mississippi, 23, 43, 168–172, 247
Columbus Chamber of Commerce, 23, 171
Columbus Lock and Dam, 37, 80–81; threat to Plymouth Bluff, 43–44; relocation, 44, 45, 50, 64
Committee for Leaving the Environment of America Natural (CLEAN), 8, 98, 99, 169, 209; establishment, 84–85; early opposition to Tenn-Tom, 85, 93–95, 107–108; and President Nixon, 104; joining Environmental Defense Fund, 109–111, 135, 198, 220; and the L&N Railroad, 132–137, 141, 144–145; and minority issues, 181, 254; and second Tenn-Tom litigation, 199; committee structure, 273 n.6
Common Cause, 237, 242
Congress, 8, 13, 22; influence of Southern members, 3, 6, 28, 153, 230; advocacy of Southern members, 15, 17; authorizing Tenn-Tom, 16, 105; as focus of Tenn-Tom opponents, 99, 142, 218, 219; and oversight of benefit/cost ratio, 123, 148, 150, 213, 223; and budget deficit, 167; ability to kill water projects, 209; 1980 supplemental appropriation, 225–229. *See also* Congressional authorization; Congressional Black

Caucus; cost increases; various House and Senate committees and subcommittees
Congressional appropriations. *See* construction funds
Congressional authorization, 2–3, 7, 149–151; need for, 53–54, 206, 207, 211; abuse, 112, 212; of Locks and Dam 26, 201; litigation, 208–212, 213; lack of, 220
Congressional Black Caucus (CBC), 7; as target of lobbying, 178, 238; push for minority participation on Tenn-Tom, 187, 193–195, 196
Congressional Record, 174, 236, 238
Congressional Research Service (CRS), 9, 239–240
Conrail, 228
Conservation Caucus, 242
Construction Cost Index. *See Engineering News-Record* Construction Cost Index
Construction funds: first appropriation, 7, 16, 31, 130; annual appropriations, 149, 225, 232
Construction management, 68–70, 82
Construction sequencing, 66–68
Cook, Harry N., 233–234
Cooper, John E., 91–92, 113, 133
Cooper, Martha R., 113
Corps of Engineers. *See* Army Corps of Engineers
Cossatot River, 110. *See also* Gillham Dam
Cost increases: during 1970s, 8–9, 44, 154–158, 219–220; concealment from Congress, 9, 148, 155–157, 162, 206; during 1940s and 1950s, 16; criticism, 44, 140–141, 157–158, 161
Cost-sharing. *See* local cost sharing
Council on Environmental Quality (CEQ), 94, 96, 120, 134; ruling on Tenn-Tom environmental impact statement, 97–98, 100, 107, 109; and Nixon Administration, 102, 104, 105, 115; and Environmental Policy Center, 141; review of water projects, 168
Court of Appeals for the Fifth Circuit. *See* Fifth Circuit Court of Appeals
Craig, Sara, 191
Crisp, Robert L., Jr., 69–71
Cronenberg, Arthur M. (Bud), 48
Cross-Florida Barge Canal, 60, 98, 143, 158, 229; stopped by President Nixon, 30, 102–103, 104; its stoppage as precedent, 91, 94–95, 99–100; and Florida Defenders of the Environment, 92, 133; authorization, 105; injunction, 107, 123, 133; litigation, 111; economics, 113; minority involvement, 181
CRS. *See* Congressional Research Service
Cutoffs, 36, 37, 87. *See also* bendways; Rattlesnake Bend cutoff

Index 327

Dawson, Robert K., 230
DDT, 109
Dedication: of Tenn-Tom, 246–247
Deferred maintenance, 228, 230–232
Demopolis, Alabama, 36, 66
Demopolis Lock and Dam, 4, 20, 199, 200
Dent, Harry, 29
Department of Agriculture (USDA), 93, 109–110
Department of Conservation and Natural Resources. *See* Alabama Department of Conservation and Natural Resources
Department of Justice, 114–116, 128; and first Tenn-Tom litigation, 117, 119–122; and second Tenn-Tom litigation, 203–205, 210, 215; turnover rate, 204–205
Department of Labor, 188
Department of Transportation, 165, 167
Design memoranda. *See* feature design memoranda; general design memorandum
Dewatering: in river section, 44–46; in divide cut, 58–59, 71, 73–74; altering techniques, 230
Diacou, Nicholas, 204
Dimensions. *See* channel dimensions
Dingell, John D., 237–238
Discretionary authority: abuse by Corps, 9; and the canal section, 49, 52; questioning, 201–202, 208, 211. *See also* Locks and Dam, 26
Divide cut, 188, 207; length, 1, 4, 34; width, 5, 71; depth, 14, 54, 71; design, 57–60, 61, 273 n.10; construction, 67, 68, 70–79; photos, 73, 74, 78; and groundwater, 88–89; environmental impact, 139–140, 201; protection of banks, 231. *See also* divide section; nuclear explosives
Divide section, 33–34, 54–64, 213; length, 4; construction sequencing, 67. *See also* Bay Springs Lock and Dam; divide cut; Mackeys Creek; Yellow Creek
Domenici, Pete, 237
Doswell Gullatt and Associates, 17–18
Draft EIS, 102. *See also* environmental impact statement
Dredged material: disposal of in river section, 39–41, 86
Dredging, 37, 39–41; photo, 38
Drinkard, William H., 23
Duncan, John J., 25–26

EAB. *See* Environmental Advisory Board
Eagan, Mike, 205
Earth Day, 7, 150, 151
Eastland, James O., 28, 174; and Department of Justice, 115, 205; and Judge John Lewis Smith, 121; and Mississippi bridge replacements, 165

Economic benefits. *See* benefit/cost ratio; economic evaluations
Economic controversy: over Tenn-Tom, 8–9, 16–17. *See also* benefit/cost ratio; economic evaluations
Economic costs. *See* benefit/cost ratio; cost increases; economic evaluations
Economic evaluations: prior to 1970, 13–14, 16, 17, 18; criticisms by railroads, 21–22; in 1970s, 161–164. *See also* benefit/cost ratio; Doswell Gullatt and Associates; Kearney Report; litigation
Economic justification: of Tenn-Tom, 5
Ecotactics, 133
Eddie Waxler, 1, 4, 9, 247, 257 n.1; photo, 247
Edey, Marion, 100
EDF. *See* Environmental Defense Fund
Edgar, Robert, 173, 226–228, 232–233, 242–243
Edwards, Jack, 29, 152, 179–180, 227; and President Nixon, 29, 101, 102, 115; and the Alabama Conservancy, 95
Ehrlichman, John D., 103, 106
EIS. *See* environmental impact statement
Eisenhower, Dwight D., 17
Eizenstat, Stuart E., 175
Emelle, Alabama, 251
Employment goals for minorities and local workers, 190, 193–194, 195; absence of, 177. *See also* minimum employment goal; Minority Peoples Council on the Tennessee-Tombigbee Waterway
Endangered Species Act (1973), 140
Energy and Water Development Appropriations Bill of 1981, 235
Engineering: attitudes toward during 1930s and 1940s, 7; culture of, 35; approach to channelization, 37; and environmental politics, 255. *See also* public works
Engineering News-Record Construction Cost Index, 154–155
Environmental Advisory Board (EAB), 96
Environmental criticism: of Tenn-Tom, 42, 111–112
Environmental Defense Fund (EDF), 60, 116, 229; and first Tenn-Tom litigation, 49, 110–114, 118, 120–127; and second Tenn-Tom litigation, 52, 199–218 *passim*; and early considerations of Tenn-Tom, 92, 107, 108; lawsuit to halt Mirex poisoning, 93, 109–110; joining CLEAN, 109–111, 135, 198, 220; and Gillham Dam litigation, 110, 123; and scientific community, 113; and Cross-Florida Barge Canal, 123; and environmental litigation, 128, 143; and L&N Railroad, 135–136, 144, 198; and Gaylord Nelson, 153, 220, 224; and Congress, 235, 241
Environmental impact: controversy, 1, 153; of Lock D, 53; of divide cut, 59–60, 139–140;

328   Index

attempts to minimize, 79–81; of Tenn-Tom, 86–90, 113
Environmental impact statement (EIS), 115, 117, 120, 151; and dredging, 39; and Plymouth Bluff, 43; and divide cut, 59, 60; and expansion of public involvement, 85–86; demands for its completion, 91–95, 140; reception, 96–100; completion, 103, 105; approval, 107, 108, 109; inadequacy, 111, 118, 212–213; critique, 113, 123; defense of, 114, 126, 128; failure to address design changes, 199, 201. *See also* litigation; supplemental EIS; three-phase approach
Environmental movement, 9, 10, 15; emergence, 3, 32, 83, 84; values, 3, 30, 33, 89–90; celebration of Earth Day, 7; in the South, 7–8, 30; and minorities, 180–181, 253
Environmental Policy Center (EPC), 139, 173, 237; and Council on Environmental Quality, 141; opposition to bridge replacement legislation, 166; and affirmative action, 184, 195; and National Campaign to Stop the Tenn-Tom, 238
Environmental Protection Agency (EPA), 41, 98; and dredging, 40; evaluation of Tenn-Tom environmental impact statement, 99, 103; and Nixon Administration, 115
Environmental protest: to Tenn-Tom, 8, 83–108 *passim*
Environmental Quality Improvement Act (1970), 112
EPA. *See* Environmental Protection Agency
EPC. *See* Environmental Policy Center
Epes, Alabama, 183
Equitable doctrine of laches, 211–212
Euclid Creek flood control projects, 195
Evans, Rowland, 106
Evins, Joseph, 152, 158, 174, 184
Excavation: earth-moving machinery, 14; in canal section, 48, 53; in divide section, 54–60; Tenn-Tom compared to Panama Canal, 72, 87. *See also* divide cut

Facilities Engineer Support Activity. *See* Army Corps of Engineers
Fairbanks, Richard M., 115
Feature design memoranda, 36
Federal Aid Highway Act (1976), 167
Federal Highway Act (1975), 165
Federal Water Pollution Control Act, 212
Federal Water Pollution Control Administration (FWPCA), 41–42
Federation of Southern Cooperatives, 183
Ferguson, Denzel E., 84
Field and Stream, 100
Fifth Circuit Court of Appeals, 224; and first Tenn-Tom litigation, 125–127, 137; and second Tenn-Tom litigation, 212, 214, 215–216

Finch, Cliff, 171–172; photo, 171
Fire ant eradication program. *See* Mirex poisoning
First Peoples Conference on the Tennessee-Tombigbee Waterway, 183. *See also* Minority Peoples Council on the Tennessee-Tombigbee Waterway
Fiscal conservatives. *See* National Taxpayers Union
Fish and Wildlife Coordination Act (1934), 112, 212
Fish and wildlife enhancement, 5, 41, 60
Fish and wildlife habitat: associated with bendways, 37, 39; loss, 39, 102, 112
Fish and wildlife mitigation, 89, 214
Fish and Wildlife Service, 39, 40, 86, 93
Flood control, 14, 15, 25
Florida: environmentalists' strength within, 30. *See also* Cross-Florida Barge Canal
Florida Conservation Foundation, 100
Florida Defenders of the Environment, 92, 133
Ford, Gerald, 140, 143, 167
Foster, Charles H. W., 96
Foster, Luther, 185
Frazier, Tom, 249
Friends of the Earth, 99, 104, 166, 220
FWPCA. *See* Federal Water Pollution Control Administration

G.O.P.'s Southern Strategy, 29–31, 101
Gainesville, Alabama, 41
Gainesville Lock and Dam, 37, 46, 62, 81; impact on water quality, 41–43; photo, 42; cost increase, 155; employment practices, 180–181
GAO. *See* General Accounting Office
Garbage dumps, 250–251
Garrison Diversion project, 243
Gatlin, Boyd, 94
General Accounting Office (GAO), 9, 54, 144, 210, 223; Tenn-Tom audit, 162–164, 239–240
General design memorandum, 18, 36, 53, 58, 59, 159
Geological Survey, 78
Georgia Sportsman, 167–168
Gholson, Hunter M., 119, 123–124, 233; and second Tenn-Tom litigation, 203, 204, 205, 215
Giattina, A. Theodore, 120
Gillespie, Norman, 301 n.80
Gillham Dam: injunction, 107, 110, 123; minority involvement, 181
Godwin, Lamond, 185
Goldberg, Irving L., 126
Goldbloom, Irwin, 115, 119–120
Goldwater, Barry, 29
Gordon, Brooks H., 131, 133, 134

Index 329

Grace, G. Randall, 141, 142, 285 n.51; and Tombigbee River Conservation Council, 145–146; and water projects review, 169, 172; and second Tenn-Tom litigation, 199
Grand Coulee Dam, 72
Gravel, Mike, 222–224
Graves, Ernest, Jr., 56, 157, 160, 186
Gray, L. Patrick, 119–120
Great Lakes, 88
Great Lakes Sport Fishermen Club, 221
Green, Lawrence R., 81
Green, Samuel R., 246–247
Greene County, 180, 185, 250
Grenn, Bill, 227
Griffith, Harry A., 90
Grimm, Fred, 247–248
Ground breaking ceremony, 100–106, 181; photo, 105
Groundwater: in river section, 44–46, 52; in divide section, 77–79; endangerment, 88–89, 102, 112
Gulf, Mobile and Ohio Railroad Company, 21, 131
Gullatt, Doswell. *See* Doswell Gullatt and Associates
Gullett, John H., 115–116, 119–120, 203–204
Guy H. James Construction Company, 181

Haldeman, H. R., 101
Hall, Katie, 195
Harrington, Joseph, 124
Harrison, Esther M., 183
Hartley, Joseph R., 24, 204
Hawkins, Augustus F., 187
Hawkins, Paula, 248
Hazardous waste dump, 251
Heflin, Howell, 228, 240; photo, 171
Heiberg, E. R., 217
Henry, Aaron, 189
Hertzler, Richard A., 209
Historic Sites Act (1960), 112
Hit List. See water projects review list
Holcut, Mississippi, 189
Holmes, Alfred, 120, 128
Holmes, Vernon S., 69, 70
Hooper, Virginia, 29–30
Hoover Dam, 72
House Committee on Appropriations, 16, 156, 174, 194, 242
House Committee on Education and Labor, 187
House Committee on Public Works and Transportation, 165–166
House Committee on Rivers and Harbors, 14, 15, 16
House Committee on Rules, 243
House Subcommittee on Deficiencies and Army Civil Functions, 16–17
House Subcommittee on Equal Opportunities, 187
House Subcommittee on Public Works, 24, 25, 99, 131–132; and nuclear excavation, 57; and construction sequencing, 67–68; 1973 testimony, 137; 1975 testimony, 141–142; and 1976 cost increases, 157–158; and Tom Bevill, 174, 195; and minority participation, 184, 195
Hydroelectric power generation, 14, 15

Illinois Central Railroad Company, 21, 131, 132
Industrial and Technological Research Commission. *See* Mississippi
Industrial development: promise of, 6, 24; of the South, 7; and pollution, 89
Industrialization. *See* industrial development; public works
Injunction: against Tenn-Tom, 118, 152–153, 215–217. *See also* Cross-Florida Barge Canal; Gillham Dam; Locks and Dam, 26
Interagency Coordinating Council, 191–192
Interest rate: recommended in 1960s, 22; in 1971, 112; challenge of Corps's 3.25 percent rate, 113–114; court order to increase, 214, 216

Jacobs Associates, 270 n.5
Jamie L. Whitten Natchez Trace Parkway Span, 248
Johnson, Gerald W., 55
Johnson, Lyndon B., 28, 101
Johnson, Paul, 55
Jones, Bob, 28
Jones, Robert E., 165
Jordan, Robert E., III, 31

Karaganis, Joseph V., 198–199, 202
Keady, William C.: and first Tenn-Tom litigation, 122–126; and second Tenn-Tom litigation, 205–218 *passim*; retention of trial records, 301 n.80
Kearney Report, 158–161, 200
Kelly, James L., 68
Knudsen, Morris, 72
Koester, E. Leo, 131–148 *passim*
Koisch, Francis P., 96
Kuehne, Robert A., 91–93, 99–100, 133, 136–137

L&N. *See* Louisville and Nashville Railroad
Labor unions. *See* Brotherhood of Locomotive Engineers; Brotherhood of Railway and Airline Clerks
Laches. *See* equitable doctrine of laches
Lanier, Philip M., 134, 135–136, 140, 142–143
Lawrence Radiation Laboratory (LRL), 54. *See also* Project Plowshare

Lawsuit. *See* litigation
League of Conservation Voters, 100, 238, 241
League of Women Voters, 237, 242
Length of waterway, 1, 4, 257 n.3. *See also* canal section; divide section; river section
LeTellier, Carroll N., 69, 160–161; concern with cost increases, 155–157; and minority participation, 186
Liming, F. Glenn, 199
Limited Nuclear Test Ban Treaty, 56
Litigation, 49, 52, 60, 144, 175, 222; to stop Mirex poisoning, 93; first Tenn-Tom lawsuit, 109–129, 198, 199, 218; second Tenn-Tom lawsuit, 146, 147, 198–218
Lobbying, 2, 32, 255; by Tenn-Tom opponents, 8, 142, 145; by railroads, 15, 22; by local supporters, 22–28; by barge industry, 104, 233; James Eastland and Department of Justice, 115; by Tombigbee River Conservation Council, 147; by Tenn-Tom proponents, 153–154, 234–235, 240; of Congressional Black Caucus, 238. *See also* Tennessee-Tombigbee Waterway Development Authority; Glover Wilkins
Local cost sharing, 142, 146, 154, 164–167, 252. *See also* bridge replacements
Local support. *See* regional support
Lock A, 47
Lock B, 47
Lock C, 47, 52
Lock D, 41, 47, 50, 52–54; photo, 250
Lock E, 47, 52, 53
Locks: size, 5, 16, 41; reducing number, 17
Locks and Dam, 26, 202, 223; injunction, 143, 201–202
Los Angeles Times, 236–237
Louisville and Nashville Railroad (L&N), 21, 147, 227; route, 6, 21; joining with environmentalists, 131–148; and Environmental Defense Fund, 135–136, 144, 198; and second Tenn-Tom litigation, 52, 199–218 *passim*; and Gaylord Nelson, 220, 224
Louisville *Courier-Journal*, 104, 106, 229
Louisville *Times*, 111
LRL. *See* Lawrence Radiation Laboratory
Lynn, James, 166

Mackeys Creek, 4, 47, 71
Maps, xiii, xiv
Marlin, John, 143
Martin K. Eby Construction Company, 71, 78
Mayo, Robert P., 30
McDonald, Sid, 202–203
McIntyre, Kenneth E., 186, 190
MEDC. *See* Mississippi Economic Development Corporation
Media. *See* news media

Mercury pollution, 60, 88
Meta Systems, 124
Metzenbaum, Howard, 221
Meyer, M. Barry, 165
Miami Herald, 247
Miles, Lee W., 75
Miller, James Nathan, 174, 210
Minimum employment goal, 184, 186, 187, 188–189
Minorities: and the Tenn-Tom, 176–197. *See also* civil rights movement; Congressional Black Caucus
Minority Peoples Council on the Tennessee-Tombigbee Waterway: establishment, 178, 183–184; objectives, 184, 185, 186; activities, 187–197
Minority Resource and Oversight Center, 191
Mirex poisoning, 93, 109–110
Mississippi: Industrial and Technological Research Commission, 55; economic development, 182
Mississippi Clone, 225, 228
Mississippi Economic Development Corporation (MEDC), 182–183
Mississippi River, 2, 5, 90; competition with Tenn-Tom, 32, 239, 250; Upper Mississippi River Navigation Project, 143
Mississippi State Board of Water Commissioners, 31
Mississippi State College for Women, 43, 98. *See also* Mississippi University for Women
Mississippi State Legislature, 164–165
Mississippi State Park Commission, 60
Mississippi State University, 84–85, 92
Mississippi University for Women, 169. *See also* Mississippi State College for Women; Tenn-Tom public hearing
Missouri Pacific Railroad Company, 201–202
Mitchell, John N., 106
Mitigation. *See* fish and wildlife mitigation
Mittendorf, George H., 46
Mobile, Alabama, 2, 4, 249; and ground breaking ceremony, 104, 105, 106
Mobile Area Chamber of Commerce, 171
Mobile Bay, 89
Mobile Bay Audubon Society, 141
Mobile District, 17, 146, 156; jurisdiction, 34–36; and riprap, 59, 231–232; and Tenn-Tom environmental impact statement, 95–98; and Department of Justice, 117, 120, 128, 203; and Tenn-Tom public hearing, 172; and second Tenn-Tom litigation, 217; employment discrimination, 189–190, 254; Tuscaloosa Area Office, 189
Mobile *Register*, 202
Monnet, Pat, 76–77
Morris, John W., 155, 156, 204
Morrison, Harry, 72
Morrison-Knudsen Corporation, 71–78

Index 331

Moynihan, Daniel Patrick, 226, 228, 235–236, 239
Multi-purpose river development, 15

NAACP. *See* National Association for the Advancement of Colored People
NAACP Legal Defense and Education Fund, 191; Southeast Regional Office, 190
Nader's Raiders, 133
Nashville District: jurisdiction, 34–36; and riprap, 59, 231; domestic well inventory, 77–79
National Academy of Sciences: Highway Research Board, 55
National Association for the Advancement of Colored People (NAACP), 7, 189. *See also* NAACP Legal Defense and Education Fund
National Audubon Society, 93, 100, 131, 132; and second Tenn-Tom litigation, 147, 206, 212, 220, 297 n.19
National Campaign to Stop the Tenn-Tom, 238–239
National Environmental Policy Act (NEPA), 82, 99, 152, 218; enactment, 3, 30, 36, 85; empowerment of environmentalists, 8, 107, 251; impact, 42, 49, 106, 108; requirements, 90–91, 103, 118, 123, 125, 128, 151, 156; and environmental litigation, 109, 201, 209; and Gillham Dam litigation, 110; violation, 112, 123, 126, 208, 212, 214
National Gray Panthers, 242
National Resources Defense Council, 220
National Speleological Society, 91
National Taxpayers Union (NTU), 224, 227, 237, 240, 241, 242; joins Tenn-Tom opposition, 173
National Water Commission, 136, 138, 284 n.27
National Waterways Conference (NWC), 233
National Wildlife Federation, 173, 220, 237
Navigation channel. *See* channel dimensions
Nedelman, Jeff, 224
Nelson, Gaylord, 118; early opposition of Tenn-Tom, 138, 150–151, 152–153, 220; and Earth Day, 150, 151; and Environmental Defense Fund, 153; opposition of Tenn-Tom in 1979–1980, 220–225, 226; photo, 221
NEPA. *See* National Environmental Policy Act
New York Times, 210, 228, 241
News media, 69; use by Tenn-Tom opponents, 8, 99, 209–210; critical coverage of Tenn-Tom, 9, 84, 148, 177; and the Tennessee-Tombigbee Waterway Development Authority, 24, 116–117; investigation of Tenn-Tom, 147; and second Tenn-Tom litigation, 210–215

Nixon, Richard M., 32, 139; pressure to fund Tenn-Tom, 29–31; enactment of NEPA, 30, 85; stoppage of Cross-Florida Barge Canal, 30; appropriation of Tenn-Tom construction funds, 31; as target of environmental lobbying, 91, 94–95, 100; and ground breaking ceremony, 100–106, 107; photo, 105
Nixon Administration: funding of Tenn-Tom, 30–31; as target of environmental lobbying, 100–106; as target of pro-Tenn-Tom lobbying, 115, 120; and promotion of Tenn-Tom employment opportunities, 180
Northeast Mississippi Daily Journal, 78
Novak, Robert, 106
NTU. *See* National Taxpayers Union
Nuclear Cratering Group, 55–56
Nuclear explosives for excavation, 54–57
Nunn, Louie B., 101, 105
NWC. *See* National Waterways Conference

O'Neal Dam, 243
O'Neill, Thomas P., 244
Office of Civil Functions. *See* Secretary of the Army
Office of Federal Contracts Compliance, 186
Office of Management and Budget (OMB), 24, 31, 69, 103, 152; and Tennessee-Tombigbee Waterway Development Authority, 25; standards of, 112; as target of environmental lobbying, 139–140, 166; review of water projects, 168; under David Stockman, 236. *See also* Bureau of the Budget
Ohio River, 2, 5, 249
Ohio River Division, 35
Oklahoma, 181
Operation and maintenance costs, 53
Osann, Edward, 173
Osborn, Prime F., 202, 235

PAC. *See* Tenn-Tom Project Area Council
Panama Canal. *See* excavation
Paris, Wendell H., 184, 185, 189, 192–193
Partington, William M., 100, 133
Penn Central Railroad, 134
Perched canal, 47–50, 51, 201, 213; illustration, 51. *See also* canal section
Percy, Charles, 239
Peterson, Russell, 140
Pickwick Dam, 14
Pickwick Lake, 4, 54, 61, 201, 206; and mercury pollution, 60, 88
Pickwick Lock, 75
Planning. *See* preconstruction planning
Plans and specifications, 36
Plowshare. *See* Project Plowshare
Plug, the: photo, 244

Plymouth Bluff, 43–44, 50, 64; photo, 45
Pork barrel politics, 210, 220–221
Potter, W. C., 54
Preconstruction planning, 16; resumption, 18, 28
Press, Robert, 248
Prichard, Joel, 173, 226–228, 232–233, 242–243
Prigmore, Charles S., 95
Project Plowshare, 54–56
Proxmire, William, 174, 221
Public works: politics, 1; competition for, 2–3, 32, 179; value of, 3, 15–16, 178; as element of Southern industrialization, 6; impact of societal values, 10; 1930s attitudes, 14; and need to balance priorities, 33, 149, 150; management, 65–66; and Congressional process, 149, 150. *See also* engineering; transportation revolution
Public Works Appropriations Act (1956), 17
Puleston, Dennis, 113, 118, 127, 143

Quarterly meetings, 69–70

Railroads, 8, 69; as critics of navigation projects, 15, 130; opposition to Tenn-Tom; 21–22, 130–148 *passim*. *See also* Association of American Railroads; Conrail; Gulf, Mobile and Ohio Railroad Company; Illinois Central Railroad Company; Louisville and Nashville Railroad; Missouri Pacific Railroad Company; Penn Central Railroad; Seaboard Coast Line Industries; Southern Railroad System; St. Louis-San Francisco Railway Company
Railway Labor Executives Association, 237
Ramsey, John S., 93, 94
Randolph, Jennings, 222–223, 224–225, 235–236
Rattlesnake Bend cutoff, 230; photo, 38
Raymond, Daniel A., 67–68, 155
Reader's Digest, 174, 210
Reagan, Ronald, 193, 236
Reagan Administration, 192–193, 236–237
Record of Decision, 217
Recreation: as an economic justification, 5, 15, 18, 25; overstatement of benefits, 21; and canal section, 49
Red River Waterway, 192
Regional support: of Tenn-Tom, 6–7; organization, 22–28. *See also* lobbying; Tennessee-Tombigbee Waterway Development Authority
Reiger, George, 167–168
Reigle, Donald W., 221
Republican party. *See* G.O.P.'s Southern Strategy
Resor, Stanley R., 20–21, 209
Rippa, Len, 240, 241, 242

Riprap, 89; and divide cut, 59, 71, 75–77, 231; reduction, 230, 231, 232
River section, 33–34, 36–47, 213; length, 4, 37, 263 n.8; photo, 249. *See also* Aberdeen Lock; Aliceville Lock and Dam; bendways; Columbus Lock and Dam; cutoffs; Demopolis Lock and Dam; Gainesville Lock and Dam; Plymouth Bluff; Rattlesnake Bend cutoff; slurry trench construction
Rivers and Harbors Act (1936), 112
Rivers and Harbors Act (1946), 16
Roady, Stephen E., 198
Roberts, Paul, 113
Rock Monster, 76–77; photo, 76
Rogers, David, 248
Roney, Paul H., 126
Ruckelshaus, William D., 98–99
Rural Initiative Program, 178, 191–193
Rushing, John W., 117
Ryan, Claude, 133

San Francisco *Examiner*, 249
San Francisco-Oakland Bay Bridge, 72
Sasaki, Wesley K., 103
Schardt, Arlie, 143–144
SCS. *See* Soil Conservation Service
Seaboard Coast Line Industries, 235
Seaborg, Glenn T., 56
Secretary of the Army, 20–21, 59, 208, 211; Office of Civil Functions, 19
Secretary of War, 13
Seltzer, Manning, 203
Senate Committee on Appropriations, 156
Senate Committee on Environment and Public Works, 222, 228, 235, 237
Senate Committee on Public Works, 142, 165–166
Senate Committee on Small Business, 223–225
Senate Committee on the Judiciary, 28
Senate Subcommittee on Public Works: 1971 hearings, 99; 1970 hearings, 131–132; 1973 testimony, 137; 1975 testimony, 141–142; and cost increases, 160–161; 1977 Tenn-Tom hearing, 172; and minority participation, 184
Senate Subcommittee on Water Resources, 222, 223, 224, 228, 235
Shaffer, Denny, 242
Shultz, George, 31
Sierra Club, 131, 166, 237, 241, 242
Simmons, Marvin, 78–79
Simpson, Alan, 225, 237
Simpson, Bill, 205
Sinclair, Ward, 153, 163
Six Companies, Inc., 72
Size of waterway. *See* channel dimensions
Slurry trench construction, 45–47

Index 333

Small Community and Rural Development Policy. *See* Carter Administration
Smith, John F., Jr., 131
Smith, John Lewis, Jr.: and Tenn-Tom litigation, 112–113, 118, 119, 124; injunction of Tenn-Tom, 118, 127, 152–153; and James Eastland, 121; grant of venue change, 121–122
Social benefits: failure to deliver, 9, 196; promise of the Tenn-Tom, 176–182, 183
Soil Conservation Service (SCS), 152, 162–163
Sontag, Paul D., 90
South Atlantic Division, 35, 46, 59, 69–70
Southeastern Federal Regional Council, 191
Southern Congressional members. *See* Congress
Southern environmentalism. *See* environmental movement
Southern Railway System, 136
Southern Strategy. *See* G.O.P.'s Southern Strategy
Sparkman, John J., 28–29, 90, 106, 121
Spencer, Charles, 76–77
SST. *See* supersonic transport
St. Lawrence Seaway, 5, 222
St. Louis-San Francisco Railway Company, 21, 131
St. Petersburg Times, 237, 248
Staats, Elmer B., 163
Stahr, Elvis J., 132
Stansbery, David, 124
Starkville, Mississippi, 85, 92
Steak luncheons, 234
Steinberg, Bory, 155
Stennis, John C., 151–152, 153, 172; chairmanships, 28, 222; and construction sequencing, 68; as friend of Judge William Keady, 122; and Tennessee-Tombigbee Waterway Development Authority, 153, 234; and GAO investigation, 163; photo, 171; and minority participation, 186; and Gaylord Nelson, 222; and final Tenn-Tom appropriations, 228, 245
Stockman, David A., 236–237
Stokes, Louis, 194–195
Suez Canal, 54
Sumter County, 251
Sunk costs, 223, 228–230, 232
Supersonic transport (SST), 229
Supplemental Appropriations and Rescission Bill of 1980, 226–229, 232
Supplemental EIS, 213–217. *See also* environmental impact statement
Supreme Court, 127, 212

Tabb, R. P., 95–98
Tangerose, James G., 21–22, 130–134
Technology: and the environment, 3, 11, 255; reliance upon, 113; and politics, 255

Teller, Edward, 55
Tellico Dam, 143, 223
Tenn-Tom Constructors, 71–78
Tenn-Tom Project Area Council (PAC), 192–193
Tenn-Tom public hearing (1976), 168–172; photos, 158–159
Tenn-Tom Review, 146
Tenn-Tom-Topics, 146
Tennessee River, 4, 5, 54; early plans to connect with Tombigbee River, 1–2, 4, 13, 14; as a water source, 48, 58; environmental impact of mixing with Tombigbee River, 87–88, 112. *See also* Pickwick Lake
Tennessee-Tombigbee Construction Assistance Center, 183
Tennessee-Tombigbee Waterway Deauthorization Act of 1979, 221–222
Tennessee-Tombigbee Waterway Development Authority (TTWDA), 5, 50, 84, 145, 224, 245; organization, 22–23; strategy, 23–28; and nuclear excavation, 43; environmental optimism, 83; and President Nixon, 102, and first Tenn-Tom litigation, 114–117, 119–122; expenditure of state funds, 116, 133; packing courtroom, 124–125; newsletter, 146; and John Stennis, 153; and water projects review, 169–173; and social issues, 179; and minority participation, 184–186, 193; and second Tenn-Tom litigation, 203–205, 210, 211, 215; public relations campaign, 227–228; and 1981 appropriations, 234–236; steak luncheons, 234; and Tenn-Tom dedication, 246–247
Tennessee-Tombigbee Waterway Project Committee. *See* Association of American Railroads
Tennessee Valley Authority (TVA), 15, 90, 143, 162, 177; and Pickwick Dam, 14; Yellow Creek nuclear power plant, 75; response to mercury pollution, 88; and Tenn-Tom Project Area Council, 192
The Truth about Tenn-Tom, 227–228
Thoman, John R., 41, 99
Thomas, Harold, 124
Three-billion-dollar bill, 238; photo, 239
308 studies, 14
Three-phase approach: of Tenn-Tom environmental impact study, 96–98, 117–118, 125, 128
Tidmore, John Jeffrey, 69
Time/Life, 50
Timmons, William E., 101–102
Tombigbee River, 113, 114, 125; early plans to connect with Tennessee River, 1–2, 4, 13, 14; physical characteristics, 34, 36–37, 44, 47, 119; free-flowing nature, 85, 86–87, 89, 111–112, 136; species diversity, 87; environ-

mental impact of mixing with Tennessee River, 87–88, 112
Tombigbee River Conservation Council (TRCC), 144–148, 169, 172
Tombigbee River Valley Water Management District, 119, 297 n.19
Tombigbee Valley Development Authority, 297 n.19
Tozzi, James J., 19, 59–60
Train, Russell E., 102, 103, 106, 115, 120, 134; and Tenn-Tom environmental impact statement, 97–98
Transfer of venue. *See* change of venue
Transportation revolution: of the 19th century, 13
TRCC. *See* Tombigbee River Conservation Council
Trinity River navigation project, 58
Tripp, James T. B., 143, 199
Trout Unlimited, 100
TTWDA. *See* Tennessee-Tombigbee Waterway Development Authority
Tupelo, Mississippi, 189
Tuscaloosa Area Office. *See* Mobile District
Tuskegee Institute, 185
TVA. *See* Tennessee Valley Authority

U.S. District Court for the District of Columbia: and fire ant eradication complaint, 110; and first Tenn-Tom complaint, 111; preferred venue for environmentalists, 119; and second Tenn-Tom complaint, 144, 199. *See also* Smith, John Lewis, Jr.
U.S. District Court for the Northern District of Mississippi: and first Tenn-Tom litigation, 120, 121, 122, 127; and second Tenn-Tom litigation, 205, 212, 214, 215–217. *See also* Keady, William C.
U.S. District Court for the Southern District of Alabama, 120, 121
U.S. News & World Report, 56
University of California, 54
University of Kentucky in Lexington, 91
USDA. *See* Department of Agriculture

Valder, Robert, 190
Venue. *See* change of venue
Veysey, Victor V.: and Tenn-Tom cost increase, 161; and minority participation, 184–185, 187, 189

Waddy, Joseph C., 205
Wagner, William C., 131
Waldon, Donald G., 19–20, 26–27, 193
Walker, George H., 57
Wall, James D., 44, 59
Wall, John F., 194, 196
Wall Street Journal, 238, 248
Wallace, George C., 102, 105, 185; presidential candidacy, 29, 106
Walter F. George Lock, 268 n.87
Washington Post, 163, 233, 249
Water projects hit list. *See* water projects review list
Water projects review list, 9, 168–169, 172–175. *See also* Tenn-Tom public hearing
Water quality, 25, 80; impact of dredging, 39, 102; in river section, 40, 41–43; and Bay Springs Lake, 64; reduction, 89
Water resources appropriations bill (1971), 151
Water Resources Council, 22, 112, 114, 138
Water resources development. *See* public works
Waterway. *See* length of waterway
Waterway user fee, 15, 138–139, 148, 252
Waterways Experiment Station (WES), 61–63
Watson, Jack H., Jr., 191–192
Waxler Towing Company, 257 n.1. *See also* Eddie Waxler
Webb, David H., 52, 203
Weinberger, Caspar, 27, 152
Welland Canal, 88
Werner, Robert R., 96
WES. *See* Waterways Experiment Station
Western water projects: post-World War II battles over, 3
Wetlands: damage from dredging, 39, 201; protection, 40, 50, 64, 89; destruction, 83, 86, 217; tidal, 89. *See also* bottomland hardwoods
Whitaker, John C., 101–106, 120
White, C. E., 81
White, N. E., 132
White Paper, 190, 193. *See also* Minority Peoples Council on the Tennessee-Tombigbee Waterway
Whitsett, W. Gavin, 132
Whitten, Jamie L., 27, 152, 227; chairmanships, 28, 157, 174; and construction sequencing, 67–68; and domestic water wells, 79; photo, 157; reaction to cost increases, 158; and Mississippi bridge replacements, 165, 167; and minority participation, 194; and 1981 appropriations, 233; and Tenn-Tom dedication, 248. *See also* Jamie L. Whitten Natchez Trace Parkway Span
Wilderness Society, 100
Wilkins, Glover, 23, 27, 50, 146–147; and Congressional lobbying, 24–28, 227; and nuclear excavation, 56; and ground breaking ceremony, 100–101; and first Tenn-Tom litigation, 114–117, 119–122; and John Stennis, 153; lobbying Jimmy Carter, 168; and water projects review, 169–173;

photo, 171; and minority participation, 185–186; and 1981 appropriations, 233–234
Williams, James D., 98–99, 111; and Environmental Defense Fund, 135, 136; and African-American community, 181, 254
Williams, John Bell, 105
Wilson Lock, 268 n.87
Wodder, Rebecca, 220

Yellow Creek, 4, 14
Yellow Creek nuclear power plant. *See* Tennessee Valley Authority

Zippert, John, 183, 190

ABOUT THE AUTHOR

Jeffrey K. Stine is Curator of Engineering at the National Museum of American History, Smithsonian Institution. He received a Ph.D. in the History of Technology and American History in 1984 from the University of California at Santa Barbara. He has been awarded several prizes for his scholarship, including the 1992 James Madison Prize from the Society for History in the Federal Government, and the 1993 G. Wesley Johnson Prize from the National Council on Public History.